U. Holzgrabe, I. Wawer, B. Diehl

NMR Spectroscopy in Drug Development and Analysis

U. Holzgrabe, I. Wawer, B. Diehl

NMR Spectroscopy in Drug Development and Analysis

WILEY-VCH

Weinheim · New York · Chichester · Brisbane · Singapore · Toronto

Prof. Dr. Ulrike Holzgrabe
University of Würzburg
Department of Pharmaceutical
Chemistry
Pharmaceutical Institute
Am Hubland
D-97074 Würzburg
Germany

Prof. Dr. Iwona Wawer
Medical University of Warsaw
Department of Physical
Chemistry
Faculty of Pharmacy
Banacha 1
PL-02097 Warsaw
Poland

Dr. Bernd W. K. Diehl
Spectral Service
Vogelsanger Str. 250
D-50825 Köln
Germany

This book was carefully produced. Nevertheless, authors and publisher do not warrant the information contained therein to be free of errors. Readers are advised to keep in mind that statements, data, illustrations, procedural details or other items may inadvertently be inaccurate.

Parts of this book are reprinted from a review in the Journal of Pharmaceutical & Biomedical Analysis, Vol. 17, U. Holzgrabe, I. Wawer, B. W. K. Diehl, *"NMR spectroscopy in pharmacy"*, pp 557–616, (1998) with permission from Elsevier Science.

Library of Congress Card No.: applied for

British Library Cataloguing-in-Publication Data:
A catalogue record for this book
is available from the British Library

Die Deutsche Bibliothek – CIP-Einheitsaufnahme
A catalogue record for this book is available from the Deutsche Bibliothek

© WILEY-VCH Verlag GmbH, D-69469 Weinheim (Federal Republic of Germany), 1999

Printed on acid-free and chlorine-free paper

All rights reserved (including those of translation into other languages). No part of this book may be reproduced in any form – by photoprinting, microfilm, or any other means – nor transmitted or translated into a machine language without written permission from the publishers. Registered names, trademarks, etc. used in this book, even when not specifically marked as such are not to be considered unprotected by law.
Composition: Spectral Service GmbH, D-50825 Köln
Printing: Strauss Offsetdruck GmbH, D-69509 Mörlenbach
Bookbindung: Großbuchbinderei J. Schäffer, D-67269 Grünstadt
Printed in the Federal Republic of Germany

Preface

Only few fields of scientific techniques had such an astonishing and wide spread development into different areas as nuclear magnetic resonance. Starting more than fifty years ago as a purely physical technique it soon entered the field of organic and inorganic chemistry, passed over to structural biology and is now probably the most important imaging technique in medicine, since it can create images from the inner of the human body without any invasive means.

The possibilities of NMR even in a more limited field such as in pharmaceutical and drug research are countless, as will be demonstrated in this volume which was collected and edited by U. Holzgrabe, I. Wawer and one of my first Ph.D. students B. K. Diehl, which made a great effort to present here a number of timely and well researched articles which demonstrate the state of the art of nuclear magnetic resonance applications in this area.

The volume starts with a report on the contributions offered by NMR in the field of drug admission and control (S. K. Branch) and describes how NMR has already invaded the international regulations as documented in the Pharmacopoeias (Diehl and Holzgrabe). We see a review on pH dependent NMR measurements (Haegele and Holzgrabe) which demonstrates the fundus of information which can be gained from NMR titration curves. Similar, articles how NMR is used to observe complexation and stereochemical information are provided (Diehl and Holzgrabe). The rapid development of on line coupling between chromatographic techniques and NMR as the possibly best chromatographic detector is covered by one of the major contributors in this field (Albert). It is this area where we may expect further significant technical breakthroughs in the next decade. The important results of these techniques are covered in a review of NMR of Body Fluids (Holzgrabe) which shows the exciting possibilities NMR can offer to medical research.

In every volume of collected reviews one finds one article which best meets ones own personal interest. In this volume for me it is the outstanding article by Gemmecker which outlines the current possibilities of SAR by NMR and the use of ^{15}N labelled proteins to demonstrate their interaction with drugs.

There can be no volume in modern NMR without considering the tremendous advantages obtained by the solid state methods and this is covered by Waver shedding light on galenic questions which can be solved by solid state NMR. Finally the volume ends with an review of the current state of imaging and localized spectroscopy with respect to pharmaceutical questions.

In summary, the collection of articles presented in this volume seems to be well chosen in topics covering the most pertinent questions of current pharmaceutical applications. It is a volume from which much can be learned and many stimulating ideas may emerge. It demonstrates the liveliness of NMR and its continuing importance in all fields of structural and analytical problems.

S. Berger, Leipzig

Contributors

Prof. Dr. Klaus Albert   Universität Tübingen,
Institut für Organische Chemie,
Auf der Morgenstelle 18,
72076 Tübingen, Germany

Prof. Dr. Gottfried Blaschke   University of Münster
Institut für Pharmazeutische Chemie,
Hittorfstr. 56,
48149 Münster, Germany

Dr. Sarah K. Branch   Medicines Control Agency
Market Towers
1 Nine Elms Lane
SW8 5NQ London, England

Prof. Dr. Bezhan Chankvetadze   Tbilisi State University,
Department of Chemistry,
Chavchavadze Ave 1,
380028 Tbilisi, Georgia

Dr. Bernd W.K. Diehl   Spectral Service,
Vogelsanger Str. 250,
50825 Köln, Germany

PD Dr. Gerd Gemmecker   TU München,
Insti. für Org. Chemie und Biochemie II,
Lichtenbergstr. 4,
85747 Garching, Germany

Prof. Dr. Ulrike Holzgrabe   University of Würzburg, Lehrstuhl f.
Pharmazeutische Chemie,
Pharmazeutisches Institut, Am Hubland,
97074 Würzburg, Germany

Prof. Dr. Giorgio Pintore   University of Sassari
                             Dipartimento Farmaco Chimico
                             Tossicoligio
                             Faculty of Pharmacy
                             Sassari, Italy

Prof. Dr. J. K. Seydel   Forschungszentrum Borstel,
                             Zentrum für Medizin und
                             Biowissenschaften, Parkallee 1-40,
                             23845 Borstel, Germany

Prof. Dr. Iwona Wawer   Medical University of Warsaw,
                             Department of Physical Chemistry,
                             Faculty of Pharmacy, Banacha 1,
                             02097 Warsaw, Poland

Contents

Preface

Contributors

Chapter 1 Introduction 1
B.W.K. Diehl

1.1	The Instrument	2
1.2	Principles	2
1.2.1	Spectra	2
1.2.2	Response	3
1.2.3	Reproducibility	3
1.2.4	Calibration	3
1.3	Experimental	4

Chapter 2 NMR Spectroscopy in the European Regulatory Dossier 5
S. Branch

2.1	Directives, Rules and Guidelines	5
2.2	Information Required to Establish Quality	6
2.3	Control of Starting Materials	8
2.3.1	Specifications and Routine Tests	8
2.3.2	Scientific Data	10
2.4	Control of the Finished Product	13
2.5	Stability Studies	14
2.6	Concluding Remarks	15

Chapter 3 Analysis of Drugs 16
B.W.K. Diehl and U. Holzgrabe

3.1	NMR Spectroscopy in International Pharmacopoeias and Related Applications	16
3..1.1	Polymers	22
3.1.2	Biomolecules	26
3.2	Identification and Quantification of Impurities in Drugs	32
3.2.1	Impurities Resulting from the Synthesis Pathway	32
3.2.2	Decomposition Reactions	36
3.3	Quantification of Drugs in Dosage Forms	43
3.3.1	Analysis of Single Drugs in Dosage Forms	43

3.3.2	Analysis of Drug Mixtures in Dosage Forms	45
3.3.3	Analysis of Drugs and Decomposition or Isomer Traces in Dosage Forms	48
3.4	Analysis of Complex Mixtures, e. g. Excipients	49
3.4.1	Phospholipids	50
3.4.2	Silicones	52
3.4.3	Fluorine Containing Substances	55
3.5	Concluding Remarks	57

Chapter 4 pH-Dependent NMR Measurements   61
G. Hägele and U. Holzgrabe

4.1	Introduction	61
4.2	Determination of the Site of Protonation	62
4.3	Determination of Dissociation Constants and Stability Constants by NMR-Controlled Titrations	65
4.4	Applications in Pharmacy	68
4.4.1	Macroscopic Dissociation Equilibria	68
4.4.2	Microscopic Dissociation Equilibria	69

Chapter 5 Complexation Behaviour of Drugs Studied by NMR   77
B.W.K. Diehl and U. Holzgrabe

5.1	Selfassociation of Drugs and Association with Other Components of a Formulation	77
5.2	Complexation with Cations	78

Chapter 6 Determination of the Isomeric Composition of Drugs   82
B.W.K. Diehl and U. Holzgrabe

6.1	Determination of the Enantiomeric Excess	85
6.2	Comparison of Different CSAs	87

Chapter 7 On-line Coupling of HPLC or SFC   101
K. Albert

7.1	Introduction	101
7.2	HPLC-NMR Coupling	104
7.2.1	Continuous-flow Measurements	104
7.2.2	Stopped-flow Measurements	107
7.3	SPE-HPLC-NMR Coupling	109
7.4	SFE-NMR and SFC-NMR Coupling	111

| 7.5 | Capillary Separations | 113 |

Chapter 8 NMR of Body Fluids   118
U. Holzgrabe

8.1	NMR of Urine - Studies of Metabolism	118
8.1.1	^{1}H NMR Spectroscopy	118
8.1.2	^{19}F NMR Spectroscopy.	122
8.1.3	^{15}N NMR Spectroscopy	124
8.1.4	^{31}P NMR Spectroscopy	124
8.2	NMR of Bile, Blood Plasma, Cerebrospinal and Seminal Fluids	125
8.3	LC NMR Hyphenation	128

Chapter 9 NMR as a Tool in Drug Research   134
G. Gemmecker

9.1	Introduction	134
9.2	Protein Structures from NMR	135
9.3	Protein-Ligand Interactions	136
9.3.1	NMR of Molecular Complexes	136
9.3.2	Aspects of Binding Affinity	137
9.3.3	Exchange Time Scales for NMR Parameters	139
9.3.4	Transfer NOE	140
9.3.5	Isotope Filters	142
9.3.6	Measuring Binding Affinities by NMR	144
9.3.7	Binding Site Localization	147
9.3.8	Solvent-Accessible Surfaces	149
9.3.9	Screening Protein-Ligand Interactions	149
9.3.10	Conclusion	151

Chapter 10 Ligand-Cyclodextrin complexes   154
B. Chankvetadze, G. Blaschke and G. Pintore

10.1	Introduction	154
10.2	Cyclodextrins and their Properties	155
10.3	Application of Cyclodextrins in Drug Development and Drug Analysis	157
10.4	NMR Spectroscopic Studies of Ligand-CD Complexes	159
10.4.1	Stoichiometry of Ligand-CD Complexes	159
10.4.2	Binding Constants of Ligand-CD Complexes	164
10.4.3	Structure and Dynamics of Ligand-CD Complexes	166

Chapter 11 Ligand-Membrane Interaction 174
J.K. Seydel

11.1	Introduction	174
11.2	Functioning, Composition and Organization of Membranes	174
11.2.1	The Physiology of Cells and the Importance of Membranes for their Functioning	174
11.2.2	Composition and Organization of Membranes	175
11.2.2.1	Mammal Membranes	175
11.2.2.2	Artificial Membranes and Liposome Preparation	183
11.2.3	Dynamic Molecular Organisation of Membranes	184
11.2.3.1	Thermotropic and Lysotropic Mesomorphism of Phospholipids	185
11.2.3.2	Phase separation and domain formation	186
11.2.4	Possible Effects of Drugs on Membranes and Effects of Membranes on Drugs	187
11.3	NMR Spectrometry; an Analytical Tool for the Study and Quantification of Ligand-Membrane Interactions	190
11.3.1	Tools for the Analysis and Quantification of Drug-Membrane Interactions	190
11.3.2	Study of Membrane Polymorphism by ^{31}P-NMR	192
11.3.3	Effect of Cholesterol and Diacylglycerols	194
11.3.4	Effect of Drugs	195
11.3.4.1	^{31}P-NMR to Study Changes in Orientation of Phospholipid Head Groups.	196
11.3.4.2	^{31}P-NMR to Study Drug Transmembrane Transport	199
11.3.4.3	^{1}H-NMR in Combination with Pr3+ fot the Studying of Drug-Location	203
11.3.4.4	The Use of ^{2}H-NMR and ^{13}C-NMR to Determine the Degree of Order and the Molecular Dynamics of Membranes	206
11.3.4.5	Change in Relaxation Rate 1/T2; a Method of Quantifying Drug-Membrane Interaction	208
11.3.4.6	NOE-NMR in the Study of Membrane Induced Changes in Drug Conformation	217
11.3.4.7	Combination of Methods and Techniques for the Studying of Drug-Membrane Interactions	220
11.4	Concluding remarks	226

Chapter 12 Solid-State NMR in Drug Analysis 231
I. Wawer

12.1	Fundamentals and Techniques of Solid State NMR	231
12.2	Solid State Conformations of Drugs and Biologically Active Molecules	241
12.3	Polymorphism/Pseudopolymorphism of Drugs	245
12.4	Studies of Tablets and Excipients	252

Chapter 13 MR Imaging and MR Spectroscopy 257
I. Wawer

13.1	In vitro and in vivo Measurement Methods	257
13.2	^{1}H MR Imaging of Tablets	261
13.3	^{1}H Imaging of Plants	265
13.4	Studies on Biopsy Samples and Tissues	269
13.5	Multinuclear Preclinical Measurements in Animals	273
13.6	MRI and MRS in Humans	277
13.6.1	Studies of brain.	277
13.6.2	MR Imaging of Breast	283
13.6.3	Musculoskeletal Tumors	284
13.6.4	Other Studies	286

Concluding remarks 294
U. Holzgrabe

Chapter 1

B.W.K. Diehl

1   Introduction

Since the development of the high-resolution NMR spectrometer in the 1950s, NMR spectra have been a major tool for the study of both newly synthesized and natural products isolated from plants, bacteria etc. In the 1980s a second revolution occurred. The introduction of reliable superconducting magnets combined with newly developed, highly sophisticated pulse techniques and the associated Fourier transformation provided the chemist with a method suitable for determining the 3-dimensional structure of very large molecules, e.g. biomacromolecules such as oligopeptides, in solution. An interesting development of NMR spectroscopy is demonstrated in Figure 1-1. It shows a nearly linear increase of the used magnetic field strength of superconducting magnets with time, this relation seems to be valid at least until the end of this millennium.

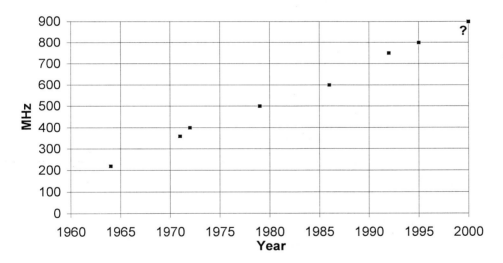

Figure 1-1:   Evolution of the magnetic-field strength of superconducting magnets.

Since drugs in clinical use are mostly synthetic or natural products, NMR spectroscopy has been mainly used for the elucidation and confirmation of structures. For the last decade, NMR methods have been introduced to quantitative analysis in order to determine the impurity profile of a drug, to characterize the composition of drug products, and to investigate metabolites of drugs in body fluids. For pharmaceutical technologists, solid-state measurements can provide information about polymorphism of drug powders, conformation of drugs in tablets etc. Micro-imaging can be used to study the dissolution of tablets, and whole-body imaging is a powerful tool in clinical diagnostics. Taken together, this review will cover applications of NMR spectroscopy in drug analysis, in particular methods of international pharmacopeias, pharmaceutics and pharmacokinetics. The authors have repeated many of the methods described here in their own laboratories.

1.1 The Instrument

Organic compounds are composed basically of the elements hydrogen, carbon, phosphorus, nitrogen and oxygen. Additionally, there are the halogens fluorine, chlorine, bromine and iodine and sometimes metal atoms. Each of these elements has an isotopic nucleus which can be detected by the NMR experiment. The low natural abundance of ^{15}N and ^{17}O in nature prevents NMR being routinely applied to these elements without the use of labeled substances, but ^{1}H, ^{13}C, ^{19}F and ^{31}P NMR spectroscopy are daily routine work. Many instruments are equipped with a so-called QNP (quattro nuclei probe) for sequential NMR analysis of ^{1}H, ^{13}C, ^{31}P and ^{19}F, without the hardware having to be changed. Modern NMR spectrometers are available up to field strengths of 18.8 Tesla or a proton resonance frequency of 800 MHz. Routine analysis is made at proton frequencies between 300 and 500 MHz.

Depending on the kind of experiments, the high-field instruments allow analysis of concentrations down to some µg/ml, but the "normal" case is a concentration of 1–100 mg/ml. The data are recorded using 32-bit ADCs (analog-digital converters). This results in a high spectral dynamic range.

1.2 Principles

The NMR experiment makes the direct observation of atoms possible. The integral of an NMR signal is strictly linearly proportional to the amount of atoms in the probe volume. The signals are a measure of molar ratios of molecules, independently of the molecular weight. There are no response factors such as those in UV-detection caused by varying extinctions dependent on molecular structures; non-linear calibration curves such as those found with light-scattering detectors are unknown to NMR spectroscopy.

1.2.1 Spectra

The frequency at which an NMR signal appears depends mainly on the magnetic field strength. For example, protons have a resonance frequency of 300 MHz at 7.05 Tesla. The chemical environment of an active nucleus leads to a small shift in the resonance frequency, the "chemical shift". Functional groups find their expression in the chemical shift. The result is an intensity/frequency diagram, the NMR spectrum. This collection of more or less separated NMR signals is analogous to intensity/time diagrams in chromatography, in which one component is represented by one signal. In ^{1}H NMR spectroscopy, each H atom leads to at least one signal. Since most molecules of analytical interest contain more than one H atom, spectra are more complex than chromatograms. What is crucial to the information taken from NMR spectra is the spectral dispersion, which is a linear function of the magnetic field strength. The homonuclear spin coupling of protons leads to a low dispersion of ^{1}H NMR spectroscopy. In ^{1}H NMR spectra of complex mixtures, it is often not possible to detect single components, but the sum of functional groups in the mixture can be determined. In ^{13}C NMR spectra, the dispersion is much higher. The low natural abundance of the NMR-active ^{13}C isotope has a detrimental effect on the sensitivity, but signals are singlets after heteronuclear decoupling. The high spectral dispersion makes parts of the ^{13}C NMR spectra directly comparable to chromatograms. An example is the carbonyl region in a ^{13}C NMR spectrum of a lipid mixture, where each fatty acid is represented by a specific signal.

^{31}P NMR spectroscopy is the method of choice for phospholipids or any other phosphorus-containing compound. Most phospholipids contain only one phosphorus atom, so the ^{31}P NMR spectrum of lecithin reads like an HPLC chromatogram. There are some advantages in comparison with HPLC: specific detection of the phosphorus nucleus, high dispersion and high dynamics. The role of ^{31}P NMR spectroscopy will be discussed later in detail.

1.2.2 Response

The area of an NMR signal is directly proportional to the molar amount of the detected isotope. The ratio between two different signals of one molecule should be 1:1, the number of represented atoms being taken into account. In practice, there are differences caused by different relaxation times. This is the time an excited atom needs to fall to the ground state. In case of heteronuclear decoupling, the nuclear Overhauser effect can cause response factors as well. These response problems are influenced by the measuring parameters, they disappear or minimize with the correct (problem-oriented) choice. Within a family of atoms in similar chemical surroundings, e.g. all end-positioned methyl groups in the ^{13}C NMR spectra of fatty acid-containing material, these effects may be neglected. The same applies to corresponding carbonyl groups, but it is incorrect to compare areas of carbonyl and methyl signals without an appropriate experimental design or experimentally determined response factors. The response factors change from ±10% in ^{1}H NMR spectra up to ±50% in ^{13}C NMR spectra, and this is in fact advantageous in comparison to HPLC/UV detection.

1.2.3 Reproducibility

During an NMR experiment, there is no contamination of sample and probe head. The electronic stability of NMR spectrometers is very good. The spectra of a stable sample stored in a sealed tube are reproducible over many years with a variation of less than 1%. These facts allow minima expenditure on validation measurements when using NMR methods.

1.2.4 Calibration

Like all classical quantitative analysis methods, NMR spectroscopy needs calibration, calibration standards and a validation procedure. The standard techniques are used for calibration: external calibration, the standard addition method and the internal standard method. A fourth is a special NMR calibration method, the tube-in-tube technique. A small glass tube (capillary) containing a defined amount of standard is put into the normal, larger NMR tube filled with the sample for analysis. In most cases, there are slight differences in the chemical shift of corresponding signals of the same molecule in the inner and outer tube. The spectrum shows two signals at different frequencies, and evaluation of the signal ratio allows quantification.

1.3 Experimental

Methods marked with "SSL" were developed by Spectral Service. To illustrate the published NMR methods, some spectra shown in the Figures were recorded by Spectral Service and replace those of the original studies. The 300-MHz NMR spectra were measured at Spectral Service GmbH, Cologne (Germany) on an NMR Spectrometer ACP 300, at 7.05 Tesla (BRUKER, Karlsruhe, Germany) equipped with automated sample changer and QNP head for nuclei ^{1}H, ^{13}C, ^{19}F and ^{31}P. The data processing was performed using BRUKER WIN NMR 5.0 software under Microsoft Windows 95.

1.4 Acknowledgment

Thanks to Dr. Werner Ockels and the Spectral Service team, for their support during the preparation of this book.

Chapter 2

S. K. Branch

2 NMR Spectroscopy in the European Regulatory Dossier

The development of NMR spectroscopy since its inception in the 1950s has been remarkable in demonstrating its power and versatility in the fields of chemistry, biochemistry, biology and medicine, as exemplified in other chapters of this book. The pharmaceutical industry has employed various aspects of the technique during development of drug substances and new medicinal products and thus NMR spectroscopy has an established role in the regulation of medicinal products which is described in this chapter.

2.1 Directives, Rules and Guidelines

There are various routes by which the sale of a medicinal product can be authorised in the European Union (EU). Some applications are made to the European Medicines Evaluation Agency (EMEA) and are dealt with by the Committee for Proprietary Medicinal Products (CPMP) through the 'centralised procedure', whilst others are made directly to the national or 'competent' authorities of the EU member states. The original legal basis of applications for marketing authorisations in the EU was set out in Council Directive 65/65/EEC together with a brief description of the documents and particulars which should accompany such applications in order to establish the quality, safety and efficacy of the product. Subsequent directives have extended the legal system for authorising medicines, introducing new procedures and expanding on existing legislation. The basic requirements for the contents of the dossier of information accompanying the application are the same whatever the route and are laid out in detail in Directive 75/318/EEC and its subsequent amendments. Provisions are made in Directive 65/65/EEC for the omission of data in certain circumstances where information is available to the regulatory authorities from other sources, for example, in the case of applications for line extensions to existing products or for generic drugs. Applications which do not include a full dossier of information are referred to as 'abridged applications'. The legal basis for the different types of applications for product licences will not be considered further since this chapter will concentrate on the technical aspects of the dossier. Further information may be found in *The rules governing medicinal products in the European Union*, published by the European Commission [1] The relevant legislation is presented in Vol. 1 while guidance on submission of applications is given in Vol. 2, the *Notice to applicants for marketing authorisations for medicinal products for human use in the European Union.*

Directive 75/318/EEC indicates that the dossier should be presented in four parts, Part I taking the form of a summary of the information presented. Part II relates to the quality of the product and gives details of its chemical, pharmaceutical and biological testing. In cases where the active ingredient is made by a manufacturer other than the applicant or product manufacturer, some of the information required in Part II may be presented in a separate file, a Drug Master File (DMF), to maintain the confidential nature of the synthetic process. Part III describes the toxicological and pharmacological tests conducted with the drug (pre-clinical tests) and the clinical documentation is presented in Part IV. Vol. 2B of the *Rules governing medicinal products* gives a detailed breakdown of the structure of a European regulatory dossier.

Vol. 3 of the *Rules* comprises *Guidelines on the Quality, Safety and Efficacy of Medicinal Products for Human Use* and is a compilation of the notes for guidance produced by the CPMP through its Working Parties or its membership of the International Conference on Harmonisation (ICH). The ICH is a tripartite body which is committed to harmonising the technical requirements for registration of pharmaceuticals in the EU, USA and Japan. The aim is to avoid unnecessary experimental duplication and to streamline the process of drug development world-wide. None of these guidelines are legally binding but are intended to be sufficiently flexible so as not to impede scientific progress in drug development. However, where an applicant chooses not to apply a guideline, the decision must be explained and justified in the dossier. The notes for guidance are continually being updated and added to, and applicants need to be aware of the current versions when preparing their dossiers. In the EU, the current guidelines and draft versions which have been released for consultation are available from EuroDirect [2] or the EMEA website [3].

The following discussion will focus on the information required in Part II to establish the quality of a medicinal product containing a chemical active substance. Additional regulations apply to radiochemical and biological medicinal products. Sections of the dossier other than Part II may also feature NMR spectroscopy, for example, reports of clinical studies using *in vivo* NMR imaging as a diagnostic tool or where the product is an NMR imaging agent itself. NMR spectroscopy is also used to identify drug metabolites isolated during animal and human pharmacokinetic studies, an area where hyphenated techniques are increasingly being used (see Chapter 8 on the NMR of biofluids).

2.2 Information Required to Establish Quality

Table 2-1 lists the headings under which information should be provided in Part II, according to the *Notice for Applicants*. NMR methods are most likely to appear in Parts IIC, IIE and IIF which describe respectively the control of starting materials, control of finished product and stability of active ingredient and finished product. As a general rule, any analytical methods should be described in sufficient detail to enable the procedures to be repeated if necessary, for example, by an official laboratory.

It should be noted that one of the ICH topics (M4) currently under discussion is a common technical document suitable for registration of medicinal products in the EC, USA and Japan. The draft requirements are at an early stage, and any harmonisation of requirements is likely to be an involved procedure given the currently differing regulatory practices in the three participating regions.

A selection of guidelines issued by the CPMP relevant to Part II of the dossier are listed in Table 2-2. The following sections will describe in more detail some of the information expected in Part II of the dossier, highlighting areas where NMR spectroscopy makes a

A selection of guidelines issued by the CPMP relevant to Part II of the dossier are listed in Table 2-2. The following sections will describe in more detail some of the information expected in Part II of the dossier, highlighting areas where NMR spectroscopy makes a contribution. The requirements of the different Directives and relevant guidelines will be drawn together under headings selected from those presented in the *Notice to Applicants*.

Table 2-1:   Requirements for the product quality aspects of the European regulatory dossier

CHEMICAL, PHARMACEUTICAL AND BIOLOGICAL TESTING OF MEDICINAL PRODUCTS

A.   Qualitative and quantitative particulars of the constituents

   1. Composition of the medicinal product

   2. Brief description of container

   3. Clinical trial formulae

   4. Development pharmaceutics

B.   Description of the method of preparation

   1. Manufacturing formula

   2. Manufacturing process

   3. Validation of the process

C.   Control of starting materials

   1. Active substance

   2. Excipients

   3. Immediate packaging material

D.   Control tests on intermediate products

E.   Control tests on the finished product

   1. Specifications and routine tests

   2. Scientific data

F.   Stability tests

   1. Stability tests on the active substance

   2. Stability tests on the finished product

2.3 Control of Starting Materials

2.3.1 Specifications and Routine Tests

Starting materials, whether the active ingredient itself or an excipient used in the manufacture of the finished product, are controlled by a specification comprising a list of tests and associated limits. The substance must comply with these limits before a batch can be deemed of suitable quality for use in the manufacture of the proposed medicinal product. The information required differs according to whether or not a substance appears in a pharmacopoeia.

2.3.1.1 Starting Materials Listed in a Pharmacopoeia

Compliance with monographs of the European Pharmacopoeia (Ph Eur) applies to all substances appearing in it and, for other substances, national pharmacopoeias may be enforced. Thus, where a pharmacopoeial monograph for an active substance or pharmaceutical excipient employs NMR spectroscopy in a test method, the substance must comply with this test. However, a test other than the pharmacopoeial test may be used if proof is supplied that the starting material meets the quality requirements of the relevant pharmacopoeia. If a pharmacopoeial monograph is applicable to a substance, then there is no need for the applicant to provide full details of the analytical tests or their validation, as reference to the pharmacopoeia in question is deemed sufficient.

However, where a starting material has been prepared by a method liable to leave impurities not controlled by the monograph, these impurities and their maximum tolerance limits must be declared and a suitable test procedure described. For example, changing a route of synthesis might lead to different solvent impurities, catalysts or related substances. NMR techniques are, of course, particularly useful for identifying new organic impurities if they can be isolated in sufficient quantities.

A competent authority is also at liberty to request more appropriate specifications if it considers that the monograph is insufficient to assure adequate quality of the substance. Further to the examples given above which result from differences in the synthesis route, additional tests may be required for particle size, polymorphic form, microbial contamination and sterility as necessary to ensure the correct performance of the starting material in the finished medicinal product. Limits which are tighter than the pharmacopoeial specification may be imposed if appropriate for the particular product in question. In addition, the competent authority will require stability data for active substances on which to base the storage conditions for the drug substance and its re-test period (the period of time for which it is expected to remain within specification and after which it must be re-tested for compliance and used immediately).

These requirements also apply to drug substances for which a Certificate of Suitability has been issued by the European Pharmacopoeia. This scheme was introduced in 1994 and certifies that the Ph Eur monograph for a substance is suitable for the control of that substance using *a particular method of manufacture*. Presentation of Certificates of Suitability are the preferred way for applicants to satisfy the guideline for such active substances (Table 2-2). A Certificate assures that the pharmacopoeial tests are adequate to control the drug substance manufactured by a particular synthetic route, even though that route may be different to the one used when the monograph was originally devised.

Table 2-2: Some guidelines relevant to Part II of the regulatory dossier.

Title	Reference number	Relevant section
Development pharmaceutics	CPMP/QWP/155/96	Part IIA
Manufacture of the finished dosage form	CPMP/QWP/486/95	Part IIB
Chemistry of new active substances	CPMP/QWP/130/96*	Part IIC
Summary of requirements for active substances	CPMP/QWP/297/97	Part IIC
Impurities in new drug substances	CPMP/ICH/142/95	Part IIC
Investigation of chiral active substances	CPMP/III/3501/91	Part IIC and Part IIF (also Parts III and IV)
Validation of analytical procedures: Definition and terminology	CPMP/III/5626/93	Part II: all sections
Validation of analytical procedures: Methodology	CPMP/ICH/281/95	Part II: all sections
Excipients in the dossier for application for marketing authorisation of a medicinal product	CPMP/III/3196/91	Part IIC
Specifications and control tests on the finished product	CPMP/III/3324/89	Part IIE
Impurities in new medicinal products	CPMP/ICH/282/95	Part IIE
Stability testing of new active substances and medicinal products	CPMP/ICH/380/95	Part IIF
Stability testing of existing active substances and related finished products	CPMP/QWP/556/96	Part IIF

*Draft revised guideline

NMR is most frequently used as an identity test in pharmacopoeial monographs, the spectrum of the sample being compared to that of a reference standard. Examples where active ingredients are subject to such pharmacopoeial NMR tests are licensed products containing tobramycin or one of the low-molecular-weight heparins (classified as biological medicinal products). Monographs for tobramycin use ^{1}H NMR spectroscopy, while ^{13}C NMR spectroscopy is used for heparins. Poloxamer is an example of an excipient with a monograph containing an NMR test: in this case ^{1}H NMR spectroscopy is used to determine the percentage of oxyethylene in the starting material. Chapter 3 gives further details of these particular tests and other examples of pharmacopoeial NMR methods, including some under discussion for inclusion in future editions.

2.3.1.2 Starting Materials not Listed in a Pharmacopoeia

Constituents not listed in a pharmacopoeia should be described in the same form as a monograph. The note for guidance on *Chemistry of new active substances* (Table 2-2) provides current recommendations for developing drug substance specifications, but the draft ICH guideline on *Specifications: test procedures and acceptance criteria for new drug substances and new drug products: chemical substances* should also be noted [4]. The latter seeks to provide guidance on the setting and justification of acceptance criteria and the selection of test procedures for new drug substances of synthetic chemical origin with the intention of establishing a single set of global specifications.

The most frequent use of NMR spectroscopy in drug substance specifications is as an identity test: ^{1}H, ^{13}C or multinuclear spectroscopy may be used as appropriate. A possible advantage of multinuclear NMR is the increased specificity afforded by the wider spectral width and the ability to distinguish the active ingredient from compounds not containing the observed heteroatom. NMR may offer greater structural specificity than other spectroscopic techniques, and it has therefore been used in identity tests for more complex molecules such as peptides and proteins as well as heparins. Another use of NMR in specifications has been to confirm that the drug substance is present in the correct polymorphic form (see below).

2.3.2 Scientific Data

A variety of chemical data are required under this heading, including information on the nomenclature of the drug substance, its description, its manufacture and quality control during manufacture, development chemistry, details of impurities and results of batch analyses. Some of this information may also be necessary for excipients which have not been previously used in a medicinal product.

Quality control of the drug substance manufacturing process requires the application of specifications to starting materials, intermediates, solvents and reagents. Any of these may include NMR tests, usually to provide identification of the substance in question. NMR spectroscopy comes into its own in the section on development chemistry, which should include evidence of chemical structure, discussion of potential isomerism, physicochemical characterisation, characterisation of reference materials and analytical validation data.

2.3.2.1 Evidence of Structure

Evidence of structure provided for new active substances should be related to the actual material to be used in the marketed product, particularly when complex molecules are involved. Where the data provided in this section relates to substance produced by a different route of synthesis, the structural identity of the different materials in question should be confirmed, and this principle should also be applied to existing active ingredients. It is expected that NMR (^{1}H and ^{13}C), together with other spectroscopic techniques and elemental analysis, should be used as a matter of routine to confirm the structure of the drug substance, where applicable. Care should be taken to ensure that reproductions of spectra are completely legible (a common problem in dossiers), and full assignments should be given where possible. The route of synthesis can also be used as supporting evidence for proof of structure, and NMR can help confirm that the reactions involved have led to the expected structures by establishing the identity of intermediate products.

A wide range of proton and carbon NMR techniques, almost exclusively using Fourier Transform (FT) spectroscopy, are presented in applications to support proposed chemical structures, with multinuclear NMR being used where appropriate. Polarisation transfer experiments (e.g. DEPT) are used to indicate carbon multiplicity and the use of two-dimensional techniques, which facilitate signal assignment, is seen more frequently in applications. Typical techniques include ^{1}H-^{1}H correlation (COSY and related phase-sensitive experiments) ^{1}H-^{13}C proton-carbon heteronuclear correlation (including inverse spectroscopy) to identify short- and long-range couplings, and experiments to identify intramolecular Nuclear Overhauser Enhancement (NOESY) and its rotating frame equivalent (ROESY).

2.3.2.2 Potential Isomerism

The potential isomerism of the active ingredient should be discussed by the applicant, and NMR spectroscopy is of value in establishing the stereochemistry of the molecule. An example here is the use of difference NOE spectroscopy to identify *cis-trans* isomerism. The absolute configuration of molecules containing chiral centres is best achieved by single-crystal X-ray diffraction studies, but NMR methods such as those employing chiral shift reagents may also be of value in establishing the presence or absence of optical isomers (see Chapter 6 for examples).

Conformational data may be required for macromolecules, e.g. proteins, particularly where the correct conformation is essential for activity. As reported in the literature, NMR has been of substantial value in this area through the analysis of coupling constants, use of NOE and relaxation measurements amongst other techniques, though this type of data has not yet been seen in regulatory dossiers (in any case, demonstration of appropriate biological activity would be necessary).

2.3.2.3 Physico-chemical Characterization

Physico-chemical data are needed (whether or not the active ingredient is listed in a pharmacopoeia) if the bio-availability of the product depends on them. The information to be presented includes the crystalline form and solubility of the drug substance, its particle size (after pulverisation if necessary), state of solvation, partition coefficient, pH and pK_a, even if these tests are not included in the final specification. The existence of different polymorphic forms of a drug substance can be established by examination of the spectroscopic and thermal characteristics of material re-crystallised in a variety of solvents and conditions. Control of polymorphs is particularly important for those active ingredients of low solubility where dissolution of the drug may have a significant effect on its bio-availability from the dosage form. Solid-state NMR, particularly the ^{13}C-MAS technique, is being used more frequently to characterise the polymorphic forms of drug substances, and in some cases has been the routine test method chosen for the specification where its specificity and sensitivity have proved superior to alternative methods such as IR spectroscopy or X-ray powder diffraction (see Chapter 12 for further discussion of solid state NMR in drug analysis).

2.3.2.4 Analytical Validation

Where an NMR method is used in the specification for a drug substance, it should be validated according to the ICH guidelines on analytical validation (Table 2-2), in the same way as other analytical methods. Revalidation may be necessary following changes in the synthesis of the drug substance or in the analytical method. The specificity of a method used for an identity test needs to be established to ensure lack of interference from related substances or other impurities. The power of NMR in distinguishing between even closely related structures is well recognised, and thus specificity is less likely to be a problem in an NMR test compared to other spectroscopic or chromatographic techniques. Indeed, the specificity of NMR has been used to advantage in the validation of test methods for other techniques, for example to ensure peak purity in chromatographic methods. Hyphenated methods such as HPLC-NMR (discussed in Chapters 7 and 8) may be used for this purpose.

With NMR methods, it is particularly important to ensure that the magnetic field is reproducible or that any fluctuations are compensated for by the use of appropriate standards. Full qualitative and, where necessary, quantitative characterisation of any reference standard is required. NMR spectroscopy has traditionally been limited in quantitative application by its relative lack of sensitivity compared to other methods, however, advances in technology, such as the introduction of high-field super-conducting magnets and FT spectroscopy, have overcome this problem to some extent. NMR has thus become a feasible option for quantitative methods in some cases (see Chapter 3), although the cost and availability of the instrumentation may limit its wider application to testing drug substances. The advantage of using NMR for determining related substances is the opportunity for simultaneous measurement and identification of the compounds present in the sample.

2.3.2.5 Impurities

The guideline on *Chemistry of new active substances* (Table 2-2) requires the applicant to discuss potential impurities and give details of any which have been synthesised and analytical methods used to detect them. Impurities may arise from the raw materials and solvents or reagents used in the manufacture of the active ingredient, from intermediates or by-products of the synthesis, and from degradation of the substance. According to the note for guidance on *Impurities in new drug substances* (Table 2-2), structural characterization is required for organic impurities at or above an apparent level of 0.1% (assuming the same analytical response factor as the drug substance), or lower if they are expected to be particularly toxic. Identification of related substances in these circumstances normally includes NMR methods in conjunction with other spectroscopic techniques and confirmation by independent synthesis.

The setting of impurity limits in the specification should take into account both the *quality* of the drug substance, i.e. the actual levels of related substances found in batch analysis, and also the *safety* of an individual impurity or given impurity profile at the specified levels. The acquisition and evaluation of data which establishes this safety is referred to as *qualification*. An impurity or impurity profile is considered qualified if it has been adequately tested in pre-clinical or clinical studies and such studies are needed for impurities present above the dose-dependent qualification thresholds prescribed in the note for guidance.

2.3.2.6 Batch Analysis

The information provided in this section should illustrate the actual results, including those from any NMR tests, which have been obtained from routine quality control of the active ingredient to establish compliance with the proposed specification. Explanations should be included in circumstances where earlier batches were tested against slightly different specifications, e.g. with wider assay or higher impurity limits. Data for batches used in toxicity tests and clinical trials should be reported, including the actual levels of impurities found, to facilitate assessment of the qualification process.

2.4 Control of the Finished Product

The dosage form administered to a patient is controlled by a finished product specification with which a batch of product must comply before it can be released for marketing. There may be justification for some tests not being applied on a routine basis, but the frequency of such skip or periodic testing must be stated. The general monographs for pharmaceutical forms (tablets, capsules, creams etc.) of the European or national pharmacopoeias are applicable to the finished product. In some member states, for example the UK and Germany, the national pharmacopoeias have monographs for specific preparations. Products are required to comply with these monographs, where applicable. Tests other than those in the pharmacopoeia may be used, but it must be shown that the product would comply with the monograph if tested. It should be noted that pharmacopoeial monographs apply to the preparation throughout its shelf-life, but competent authorities require release specifications to be stated as well, and these may be tighter than the compendial limits. As with the drug substance, the national authority may prescribe additional tests to those in the pharmacopoeia if they are considered necessary to assure the quality of the product. Full details of analytical tests are required if a pharmacopoeial method is not available or applicable.

In addition to the pharmacopoeial monographs, the guideline on *Specifications and control of finished products* (Table 2-2) provides details of requirements. Reference should also be made to the ICH draft note for guidance on *Specifications* [4]. The latter details the tests which might be expected for the control of different types of dosage forms.

Validation of analytical methods used in the finished product specification is required in the same way as for drug substance. Revalidation of these methods may also be necessary if the composition of the product has been changed during development, e.g. to reassess specificity.

As for drug substances, stressed stability studies (see below) are used to establish likely degradation products. NMR spectroscopy has been used to assist characterisation of any compounds which can be isolated. Decomposition may result from the usual mechanisms of chemical breakdown or may be due to specific interactions with excipients. The process may occur during manufacture of the formulation (e.g. induced by heating), thus requiring control in the release specification, or on storage of the product, in which case limits would be needed in the shelf-life specification. (Impurities which are limited in the drug substance specification and arise only from the synthetic process do not require further control in the finished product.) Degradation products should be identified and qualified as indicated in the guideline on *Impurities in new drug substances and new drug products* (Table 2-2), in the same way as discussed for the drug substance itself. The reporting, identification and qualification thresholds appropriate for different maximum daily doses of drug substances are presented in the note for guidance.

NMR, including multinuclear spectroscopy, has mainly been used as an identity test in the specifications for licensed products either for the active ingredient or an excipient. It has also been used to control the composition of polymeric excipients in the dosage form. Paramagnetic agents allow the use of NMR relaxation measurements in control of the quality of finished products.

2.5 Stability Studies

Stability studies fall into two categories. Firstly, 'stressed' studies referred to above are conducted in which the active substance or the finished product are subjected to extreme heat, light, acidic, basic and oxidative conditions with the intention of forcing degradation. Such experiments, when coupled with the isolation and identification of decomposition products (with the aid of NMR spectroscopy where appropriate) may allow the elucidation of degradation pathways in the drug substance and dosage form. In addition, these degraded samples may be used to test the specificity of the analytical methods applied to the drug substance and product.

The second type of stability studies are used to establish the re-test period for the active ingredient and the shelf-life of the finished product. The ICH guideline on stability testing (Table 2-2) stipulates standard conditions for storage (25°C/60%RH long-term and 40°C/75%RH accelerated) and gives recommendations on the number of batches, testing frequency and evaluation of results. The tests used should be the same as those in the drug substance and finished product specifications, including any based on NMR techniques, but additional stability-indicating methods may be included in the protocol. A relevant NMR example of the latter might be the inclusion of a solid-state method for identifying polymorphs in the stability trials for an active ingredient, even though the results eventually indicate it to be unnecessary in the final specification.

Full information on the batches tested and their packaging, the test methods (description and validation) and results and proposals for re-test period and shelf-life must be included in the dossier, together with details of any on-going studies.

2.6 Concluding Remarks

Regulatory dossiers, particularly for new active substances, tend to reflect state-of-the-art analytical techniques from a few years prior to the application because the drug development takes place over a relatively long time-scale. The unique power of NMR spectroscopy in structure elucidation is already well recognised, and its versatility is gradually becoming more widely utilised in other areas. Regulators can expect to see a wider range of techniques described in the future as advances are made in NMR technology and methodology. It is anticipated that the more sophisticated experiments for structure elucidation will be applied to a wider range of drug substances and will thus appear more frequently in descriptions of development chemistry. It is interesting to speculate on the use of imaging techniques, such as those described in Chapter 13, for examining tablets and excipients, and whether these techniques will be able to address problems encountered in pharmaceutical development. NMR techniques may become more common as routine analytical methods for control of drug substance or finished product as the sensitivity of the method increases and instruments become more accessible.

References

1. Eudralex: The rules governing medicinal products in the European Union, Vol.s 1-9, Office for Official Publications of the European Communities, Luxembourg, 1998.
2. Eurodirect Publication Service, Room 1207, Market Towers, 1 Nine Elms Lane, London, SW8 5NQ.
3. http://www.eudra.org/emea.html
4. ICH Topic Q6A *Specifications: test procedures and acceptance criteria for new drug substances and new drug products: chemical substances*, Step 3, draft, 16 July 1997 (CPMP/ICH/367/96).

Chapter 3

B. W. K. Diehl and U. Holzgrabe

3 Analysis of Drugs

3.1 NMR Spectroscopy in International Pharmacopoeias and Related Applications

Even though the reproducibility of NMR spectroscopic methods in terms of qualitative and quantitative analysis is proved to be very high, the European pharmacopoeias use NMR spectroscopy mostly for the identification of drugs and reagents, for instance, in the cases of tobramycin (Pharmacopoeia Europaea, Ph. Eur.) and hydrocortisone sodium phosphate (BP 93). Due to heavy signal overlapping, the spectra of these compounds are very complicated (see Figure 3-1 and Figure 3-2) and could be assigned only by means of 2D experiments. Thus, the ^{1}H NMR spectra are used in the same manner as IR spectra, which can be described as a sort of pattern recognition. In addition, an increasing number of reagents, e.g. adenine, adenosine, aesculin, butoxycaine, chamazulene, guaiazulene, nitrilotriacetic acid etc., are identified by ^{1}H and ^{13}C NMR spectra in European pharmacopoeias.

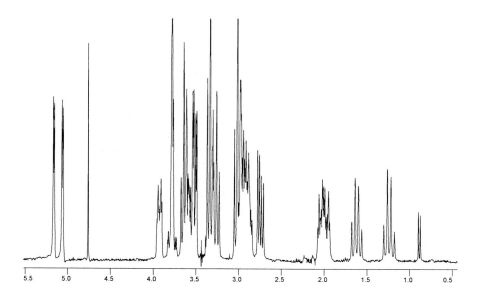

Figure 3-1: Expansion of the ^{1}H NMR spectrum of tobramycin, 300 MHz, solvent D$_2$O, "SSL".

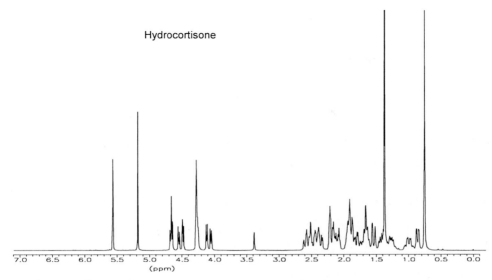

Figure 3-2: ^{1}H NMR spectrum of hydrocortisone sodium phosphate, 300 MHz, solvent DMSO-d$_6$, "SSL"

Whereas the European pharmacopoeias describe the method of NMR spectroscopy only in principle, the United States Pharmacopoeia 23 gives detailed information about the procedures of qualitative and quantitative applications. In the section on qualitative application, the correlation between chemical shifts and coupling constants on the one hand and the structure of a molecule on the other hand is stressed. For quantitative applications, an absolute method, utilizing an internal standard, and a relative method are given. Consequently, NMR spectroscopy is used in the USP for identification of drugs and their impurities (see test section) and for quantification (see assay section of a monograph).

For example, amyl nitrite, a mixture consisting chiefly of *iso*-amyl nitrite [(CH$_3$)$_2$CH–CH$_2$-CH$_2$–O–N=O] as well as other isomers, is identified by an ^{1}H NMR spectrum which is characterized among other peaks by a doublet centered at about 1 ppm and a multiplet centered at about 4.8 ppm representing the methyl hydrogens and methylene protons in α-position to the nitrite group. In the assay the substance is subjected to the absolute method, using benzyl benzoate as an internal standard. The quantity of amyl nitrite is calculated from the signal area of the α-methylene group of the drug (at 4.8 ppm) and the signal area of the methylene hydrogens of benzyl benzoate at 5.3 ppm.

The relative method of quantification is used in the orphenadrine citrate monograph in order to determine the content of *meta-* and *para-*methylphenyl isomer in the *ortho-*methylphenyl substituted drug (see Figure 3-3).

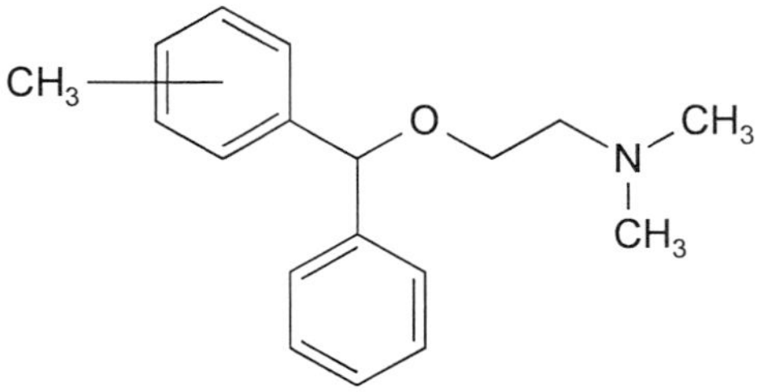

Figure 3-3: Orphenadrine.

The signals of the benzylic methine hydrogen atoms are well separated. Hence, from the areas of the signal of the *meta/para*-methyl substituted compound appearing at about 5.23 ppm and of the orphenadrine signal appearing at 5.47, the content of the impurities can be determined. After an alkaline extraction the sample is measured in CCl_4. The limit of quantitation using the CW method (continuous wave method) is specified to be 3%. Thus, the *meta*- and *para*-substituted isomers are limited to 3 percent.

The modern FT method allows a direct determination of orphenadrine citrate in the aqueous formulation. The method was tested with a sample of Norflex®. Two ^{1}H NMR spectra are shown: Figure 3-4a shows the ^{1}H NMR spectrum of a CCl_4 extract (free base) prepared according to the USP instruction. A capillary inside the NMR tube, which is filled with benzene-d_6, serves as the external deuterium lock. In Figure 3-4b the injection solution (0.5 ml) was treated with 0.2 ml D_2O and measured directly. Because of the fact that orphenadrine citrate was observed, considerable changes in the chemical shift of the signals are found. However, this has no influence on the analysis of the isomeric distribution. Additionally, the characteristic AB system of citric acid is found. The signals labeled with an asterisk are the corresponding ^{13}C NMR satellites, which have, because of the natural abundance of the ^{13}C isotope, an intensity of 0.56% of the main signal. This demonstrates that the detection limit is in fact lower than 0.5% with all three methods. With respect to an estimated quantitation limit of 0.2% no *meta*- or *para*-methyl compounds were found in the sample studied.

It is expected that further applications of NMR spectroscopy will be introduced to the Ph. Eur. and the USP in the future. For example, it is discussed to quantify dihydrolovastatin in lovastatin with a limit of quantitation at 0.1%. The advantage of higher magnetic field strength is demonstrated: 600-MHz spectra [1] are compared with 300-MHz spectra "SSL" [2] in Figure 3-5 and Figure 3-7.

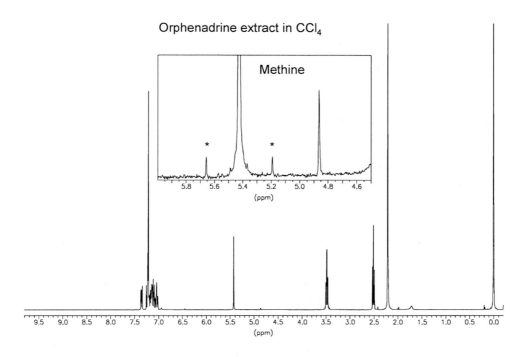
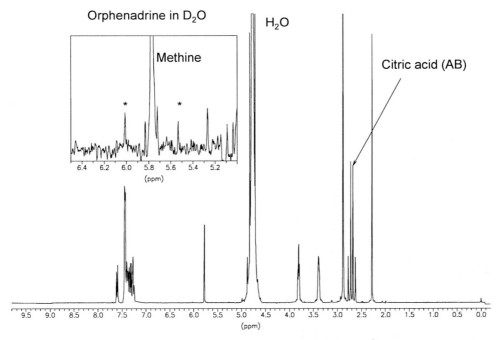

Figure 3-4a/b: ¹H NMR spectra of orphenadrine citrate (Norflex®), 300 MHz, solvent CDCl₃, "SSL".

Analysis of Drugs

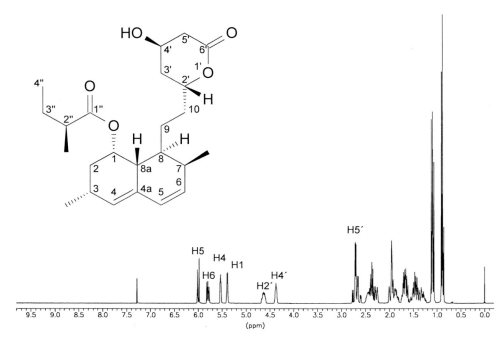

Figure 3-5: ¹H NMR spectrum of lovastatin, 300 MHz, solvent $CDCl_3$, "SSL".

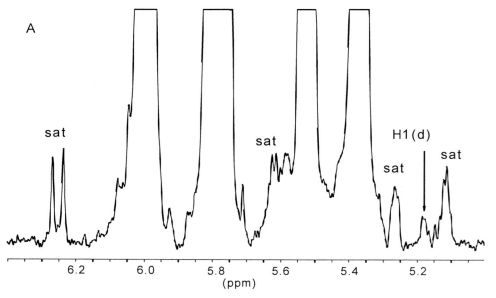

Figure 3-6 A: 300-MHz ¹H NMR spectrum of lovastatin, "SSL".

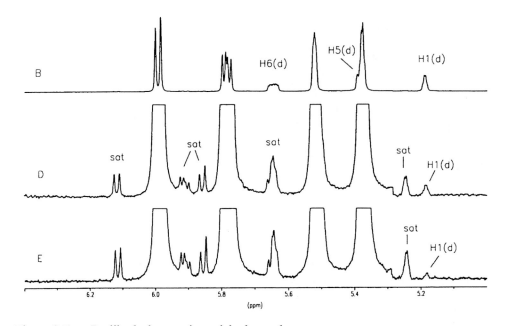

Figure 3-7: B: dihydrolovastatin enriched sample,
D: Gist-brocades pilot product,
E: 2 parts lovastatin USP and 1 part Gist-brocades pilot product.
B, D, E; 600-MHz ^{1}H NMR spectrum taken from [1], "Copied with permission from Pharmacopoeial Forum, 22 (3). All rights reserved © 1996. The United States Pharmacopoeial Convention, Inc.".

Some preliminary remarks are made to understand the comparison of a 300-MHz spectrum with a 600-MHz spectrum. All ^{1}H NMR signals of protons, which are bound directly to a C atom, show ^{13}C satellites of 0.56% of the intensity of the main signal. In practice, this theoretical ratio is found with good precision demonstrating the linearity of NMR signals over two orders of magnitude. Thus, the ^{13}C satellites can be used for NMR calibration.
The chemical shift of a proton is the same on a 300-MHz instrument as that on a 600-MHz spectrometer. But the distance of the ^{13}C NMR satellites to the parent signal is halved, changing from 300 MHz to 600 MHz. This becomes evident in the comparison of spectra in Figure 3-7. The H1 proton of dihydrolovastatin retains its chemical shift independently of the magnetic field, but whereas in the 300-MHz spectrum it appears between the main signal and the satellite, in the 600-MHz spectrum its position is high field (right) of the satellite. At 300 MHz the signals interfere, and an evaluation is practically impossible. Modern spectrometers are able to avoid this conflict using ^{13}C decoupling. Using this technique, the satellite signals are removed completely, and the main signal represents the protons bound to ^{12}C and ^{13}C.

3.2 Polymers

Besides the analysis of chemically defined drugs, the NMR spectroscopic method is extremely useful for the characterization of mixtures of synthetic polymers or partially synthetic biomolecules, such as heparins.

In poloxamer, a synthetic block copolymer of ethylene oxide (EO) and propylene oxide (PO) (USP XXIII, NF18), the oxypropylene/oxyethylene ratio is determined using a ^{1}H NMR spectrum measured in CDCl$_3$. The oxypropylene units are characterized by a narrow signal at about 1.1 ppm due to the CH$_3$ group. The CH$_2$O/CHO units cause a multiplet between 3.2 and 3.8 ppm. The percentage of oxyethylene can be calculated from the equation

$$\% \text{ oxyethylene} = \frac{3300 \cdot \alpha}{33\alpha + 58} \quad \text{with } \alpha = \frac{\text{area (CH}_2\text{O/CHO)}}{\text{area (CH}_3\text{)}} - 1$$

The content of oxypropylene in the sample of Pluronic F 68 examined here (Figure 3-8) amounts to 83% which is in accordance with the requirement of the pharmacopoeia.

In addition to a quantitative determination of EO and PO taken from ^{1}H NMR spectra, ^{13}C NMR spectra (Figure 3-9) enable us to see the type of polymerization, e.g. block- or mixed-polymerization, the type and ratio of endgroups (EO or PO) and the stereochemistry in polyethylene glycols (tactic or atactic) [3].

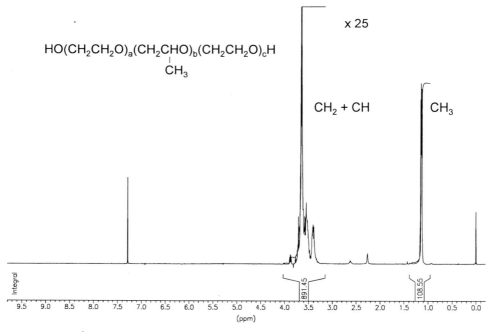

Figure 3-8: ^{1}H NMR spectrum of pluronic, 300 MHz, solvent CDCl$_3$, "SSL".

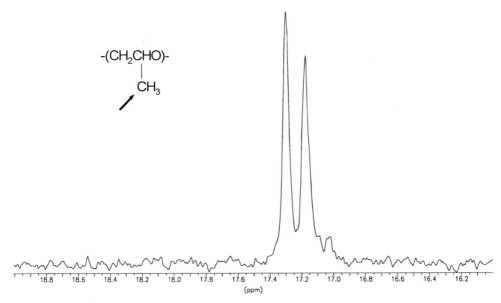

Figure 3-9: ^{13}C NMR spectrum of pluronic (16 to 19 ppm), methyl groups of PO, 75 MHz, solvent CDCl$_3$, "SSL".

The doublet structure of the methyl signal at 17 ppm is caused by the stereochemistry. Polylactides, not yet described in any pharmacopoeias, are rather new biodegradable polyesters derived from the chiral lactic acid and used, e. g., in drug delivery systems. The stereochemistry of the polymer is important to the physical and chemical behavior, especially the polymer properties. Pure tactic polymerization can be differentiated from atactic or mixed polymers by simple comparison of the ^{13}C NMR spectra (Figure 3-10) [4].
Homodecoupled ^{1}H NMR spectra can be used to quantitatively determine the composition of D-lactide and meso-lactide stereoisomer impurities in polylactide containing predominantly L-lactide [5].

Even mixed polylactide/polyglycollide copolymers can be characterized by NMR spectroscopy [2]. Figure 3-11 shows the ^{1}H NMR spectrum of a polylactide/glycollide copolymer. By comparing the integral areas it is possible to determine the ratio of both monomer units.

The ultra high resolved lactide carbonyl region of the ^{13}C-NMR spectrum demonstrates the complex structure, caused by stereochemistry and the polylactide/glycollide sequence Figure 3-12.

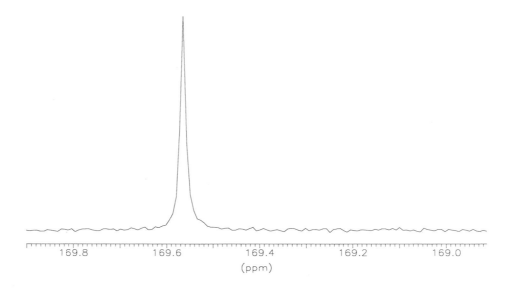
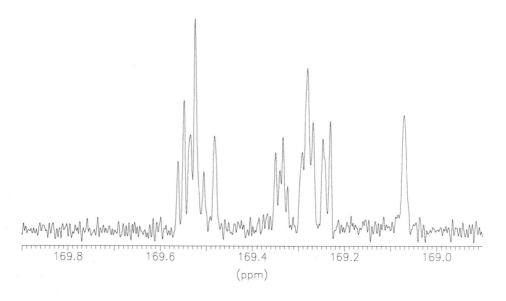

Figure 3-10: Tactic and atactic polymerization, ^{13}C NMR spectra at 75 MHz, solvent CDCl$_3$, "SSL". Carbonyl regions of tactic (*top*) and atactic polylactide (*botton*).

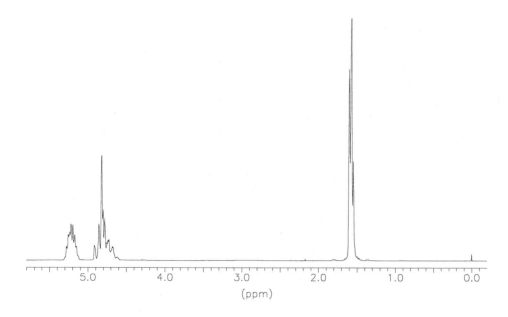

Figure 3-11: ¹H NMR spectrum of a polylactide/glycollide copolymer, 300 MHz, "SSL".

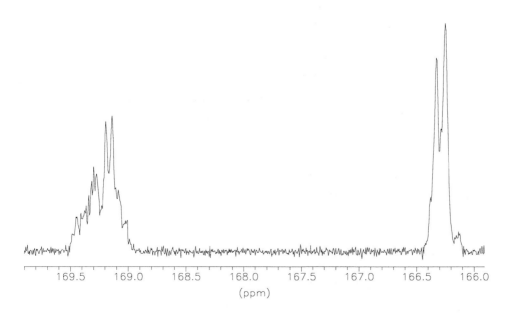

Figure 3-12: ¹³C NMR spectrum of a polylactide/glycollide copolymer, 75 MHz, ultra high resolved carbonyl region, "SSL".

3.2.1 Biomolecules

Low-molecular-weight (LMW) heparins (M_w < 8000), attracting interest in the management of thromboembolic diseases, are obtained from different depolymerization procedures of heparin, a polysulfated glycosaminoglycan. The group of Neville [6] was able to exactly characterize these compounds by means of ^{1}H and ^{13}C NMR spectroscopy. Hence, ^{13}C NMR spectroscopy (75 MHz) is used in international pharmacopoeias to identify LMW heparins: see general monograph, Ph. Eur. 1997. Four different forms of LMW heparins were introduced to the Ph. Eur. 1998: dalteparin, enoxaparin, parnaparin sodium and tinzaparin. The monographs of further LWM heparins are in progress. The LMW heparins differ in the procedure of depolymerization of the heparin: certoparin sodium is obtained by isoamyl nitrite depolymerization of heparin sodium from porcine intestinal mucosa, parnaparin by radical-catalyzed decomposition (H_2O_2 and cupric salts), dalteparin sodium and nadroparin calcium by nitrous acid depolymerization, tinzaparin by enzymatic degradation using heparinase, and enoxaparin by alkaline depolymerization of heparin benzyl ester. According to the different procedures, the LWM heparins differ in the structure of the non-reducing end (2-O-sulfo-α-L-idopyranosuronic acid or 4-enopyranose uronate) and the reducing end (6-sulfo-2,5-anhydro-D-mannose, the corresponding mannitol or 2-sulfamoyl-2-deoxy-D-glucose-6-sulfate) as well as in the molecular weight, see Figure 3-13.

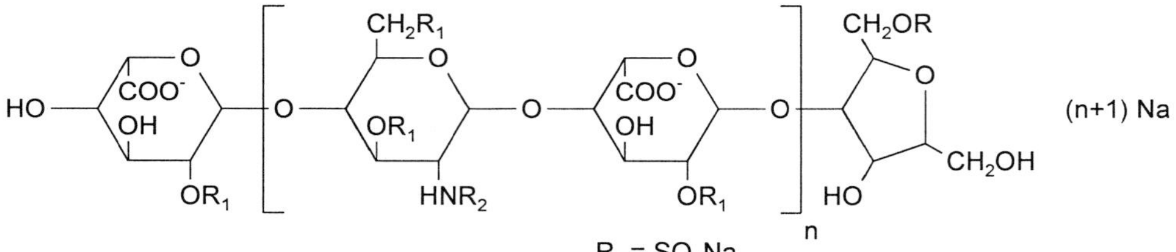

DALTEPARIN

$R = SO_3Na$
$R_1 = H, SO_3Na$
$R_2 = H, SO_3Na, COCH_3$

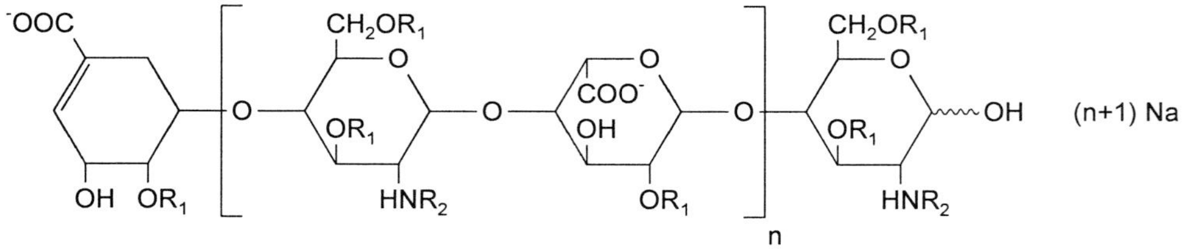

TINZAPARIN

$R_1 = H, SO_3Na$
$R_2 = H, SO_3Na, COCH_3$

Figure 3-13: Types of heparins.

Thus, signals of those rings appearing in the range between 80 and 92 ppm can be taken to identify the corresponding heparin. For example, the ^{13}C NMR spectrum of certoparin is characterized by four distinct signals corresponding to C3 (~81.6 ppm), C2 (~86.3 ppm), C4 (~86.6 ppm), and C1 (~90.4 ppm) for the anhydromannose skeleton, and a signal at 90.4 ppm for the hydrated aldehyde functional group. ^{1}H and ^{13}C NMR spectra of tinzaparin and dalteparin are displayed in Figure 3-14 and Figure 3-15. As expected, the signals of the anhydromannose moiety of dalteparin are similar to those of certoparin.

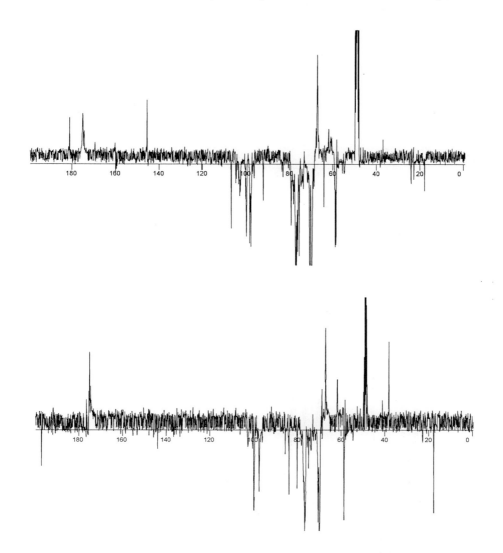

Figure 3-14: ^{13}C NMR DEPT spectra of tinzaparin (*top*) and dalteparin (*bottom*), 75 MHz, solvent D$_2$O.

28   *Analysis of Drugs*

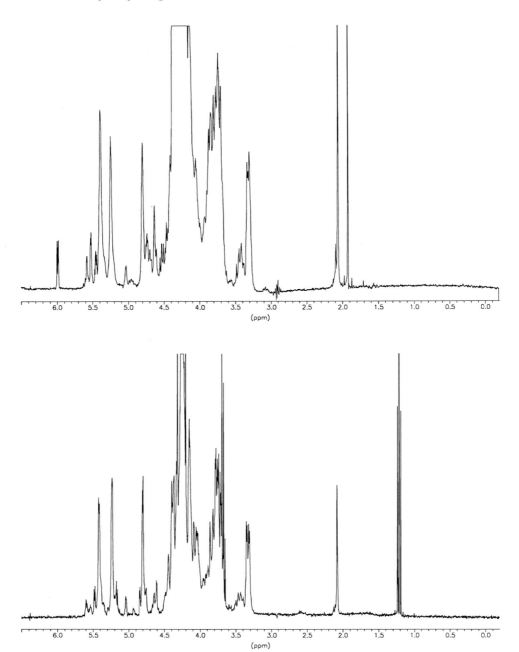

Figure 3-15: ^{1}H NMR spectra of tinzaparin (*top*) and dalteparin (*bottom*), 300 MHz, solvent D$_2$O at 353°K, "SSL".

Depending on the depolymerization methods, various amounts of the contaminant dermatan sulfate (chondroitin sulfate) were found by means of 300-MHz ^{1}H NMR spectra [7], [8]. The molecular weight can be determined using ^{13}C NMR [9]. Here, the signal intensities of the reducing end and internal anomeric carbons, having distinct chemical shifts, were compared in DEPT spectra. The method is as exact as size exclusion chromatography, which is often used for the purpose of weight determination of huge molecules.

It is noteworthy that bovine mucosal heparin and porcine mucosal heparin can be distinguished by ^{1}H and ^{13}C NMR spectra [10], [11], [12]. Since they differ in the extent of sulfation, they show a different "fingerprint". In addition, the chemical shift of H1 and H5 in iduronic acid as well as the magnitude of separation of the glucosamine/iduronic acid H1 pair are appropriate to discriminate between the counter-ion sodium or calcium. [10] In a similar manner, the extent and site of sulfation of dextran sulfate could be described by 300-MHz ^{1}H NMR spectra. Additionally, the manufacturer could be determined [13], [14] by comparison of the pattern.

To record ^{13}C NMR spectra of bovine or porcine mucosal heparin, the accumulation of a large number of transitions is necessary, which leads to very long measuring times up to 20 h. The higher the magnetic field strength of the NMR spectrometer the better are the results. Alternatively, ^{1}H NMR spectroscopy at higher temperature improves the spectral resolution and allows the differentiation between bovine and porcine mucosal heparin and other kinds of heparinoides. Figure 3-16 shows the high-temperature ^{1}H NMR spectra (353°K) of bovine and porcine mucosal heparin

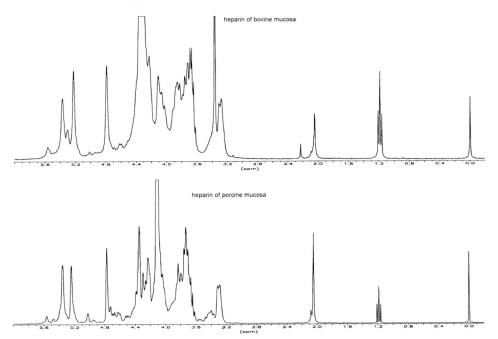

Figure 3-16: ^{1}H NMR spectrum of Heparin from bovine (*top*) and porcine (*bottom*) mucosa, 300 MHz, solvent D$_2$O at 353°K, "SSL".

The sucrose octasulfate anion, which is believed to be the active component of sulcrafate, an antiulcer drug, is hypothesized by means of FAB-MS to consist of large amounts of hepta- and hexasulfate derivatives. The technique of 2D and deuterium-induced, differential-isotope-shift (DIS) ^{13}C NMR spectroscopy (chemical shift induced by H/D exchange of the OH groups) provides a picture of sample identity and purity [15]. Without going into detail, it should be emphasized that undersulfation of sucrose did not occur before hydrolysis; all drug samples were found to be pure. Consequently, the hexa- and heptasulfated sucroses postulated by FAB-MS experiments were artifacts of this method!

High temperature ^{1}H NMR spectroscopy of complex polysaccharides is used to determine the identity of plant extracts. Aloe Vera gels contain glucose, malic acid and acetylated polymannose (Acemannan® Figure 3-17) as main components. Acemannan has a mean acetylation of 1.1 acetyl groups for each mannose unit. The distribution of these acetic groups (mono-, di and triacetylation) in positions 2,3 and 6 is characteristic of Aloe Vera. The signal of the acetate protons can be used as a significant fingerprint (Figure 3-19). It enables one to distinguish between original extracts and fraudulent materials [16].

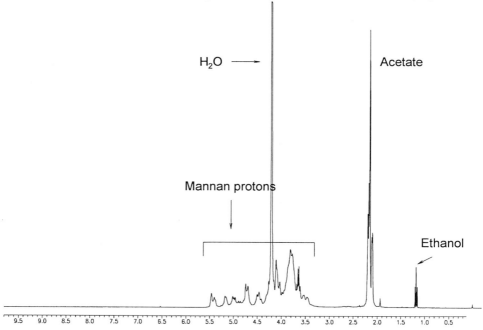

Figure 3-17: High-temperature ^{1}H NMR spectrum of acetylated polymannose (Acemannan®) from Aloe Vera gel, 300 MHz, solvent D_2O at 353K, "SSL".

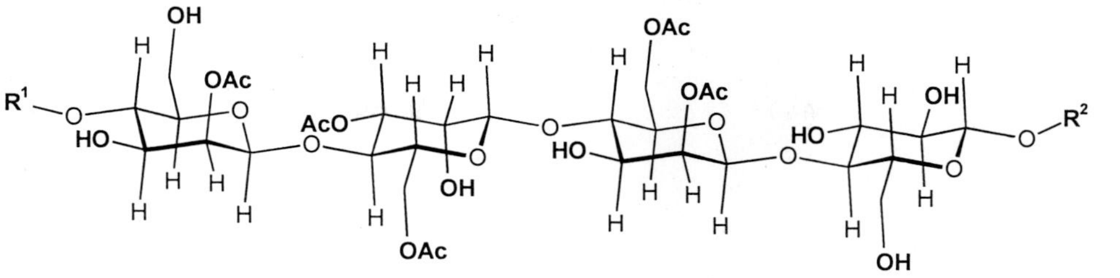

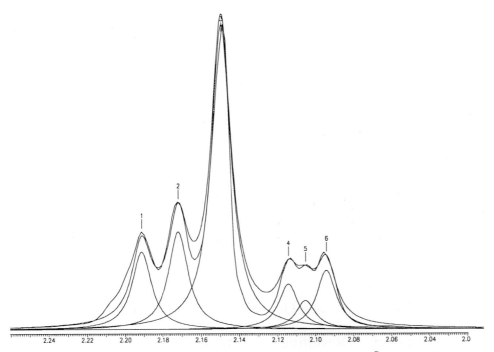

Figure 3-18 Chemical structure of acetylated polymannose (Acemannan®)

Figure 3-19: High-temperature ^{1}H NMR spectrum of Acemannan®, acetate fingerprint, 300 MHz, solvent D$_2$O at 353K, "SSL".

In addition, the chemical and enzymatic degradation of the Aloe Vera gel can be seen by ^{1}H NMR spectroscopy. Typically, degradation products, e.g. lactic and acetic acid, are easy to detect and to quantify (Figure 3-20).

32 *Analysis of Drugs*

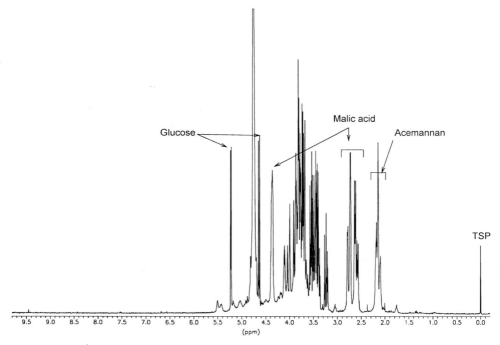

Figure 3-20: ^{1}H NMR spectrum of freeze dried Aloe Vera gel, 300 MHz at 353K, "SSL".

As can be concluded from the examples described above, further applications of NMR spectroscopy can be expected to be introduced to the Ph. Eur. and national European pharmacopoeias as well as the JP and the USP in the future. The high potential of NMR spectroscopy in terms of identification and quantification of drugs and their impurities resulting from the synthesis pathway or degradation will be demonstrated in the following sections.

3.3   Identification and Quantification of Impurities in Drugs

3.3.1   Impurities Resulting from the Synthesis Pathway

There is a long-standing tradition of using TLC and HPLC for the quantification of impurities arising from the synthesis or degradation of drugs. In recent years, the pharmacopoeia committees have required the structures of impurities to be described in a transparency statement at the end of each drug substance monograph. The HPLC and TLC analyses are optimized and validated with respect to these impurities. However, in the meantime many new manufacturers (often from Eastern Europe or Asia) provide drugs on the market which have been synthesized in other ways than registered. Consequently, the pattern of impurities has changed and the HPLC method used in the pharmacopoeias may no longer be suitable. In this context, an NMR method can be superior to HPLC methods, because a new impurity can be easily found and identified in NMR spectra. This will be demonstrated by some representative examples.

The impurity pattern of dequalinium chloride varies depending on synthesis conditions (see Figure 3-22 and Figure 3-23). The drug is usually obtained by conversion of 4-aminoquinaldine with 1,10-dihalodecane. Apart from residual aminoquinaldine (also a product of photolysis [17]) and a product consisting of 3 quinaldine molecules connected by two decanes, as well as another overalkylation product, which were found by HPLC, the 400-MHz ^{1}H NMR analysis exhibits a further impurity, 4-amino-1-[10-[(2-methylquinolin-4-yl)amino]decyl]-2-methylquinolinium chloride (see Figure 3-23) [18], which has a defined pharmacological activity. It is used in practice to treat trypanosomiasis in cattle.

Thus, this impurity has to be quantified and limited in a pharmacopoeia monograph. ^{1}H NMR is an effective method, because examination of the methyl region (2.6 to 2.9 ppm) reveals distinct signals for at least one of the methyl groups of each compound. Accurate integration of these methyl signals allows the determination of the (molar or mass) composition of the samples. Some commercially available samples were found to consist of 75.7% dequalinium chloride, 18.2% overalkylation impurity, 5.8% of the new impurity and 0.3% 4-aminoquinaldine [18], [19]. Interestingly the formerly DAB9 restricted the content of non-quaternary amines only. NMR analyses of two dequalinium chloride samples provided by Ravensberg GmbH and by Prof. Dr. K. Eger [20], University of Leipzig, FRG, were performed by Spectral Service [2] (see Figure 3-21). The methyl group as well as the aromatic CH-signals (6.3–7.12 ppm) show the same impurity signals. The molar content of the impurities was calculated from the methyl signal region as well as from the aromatic signal region. The results are displayed in Table 1.

Table 1: Molar distribution of dequalinium chloride and by-products

Compound		Old commercial sample [18]	Synthesis sample	Sample provided by Ravensberg
Dequalinium chloride		75.7	96.9	96.3
Overalkylation product	(a)	18.2	1.5	1.9
Antitrypanosomiasis	(b)	5.8	1.2	1.4
Aminoquinaldine	(c)	0.3	0.4	0.4

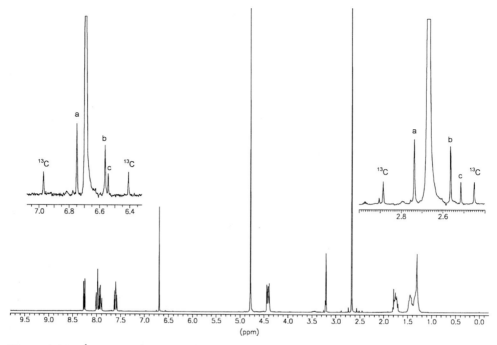

Figure 3-21: ¹H NMR spectrum of dequalinium chloride, 300 MHz, solvent D_2O, "SSL".

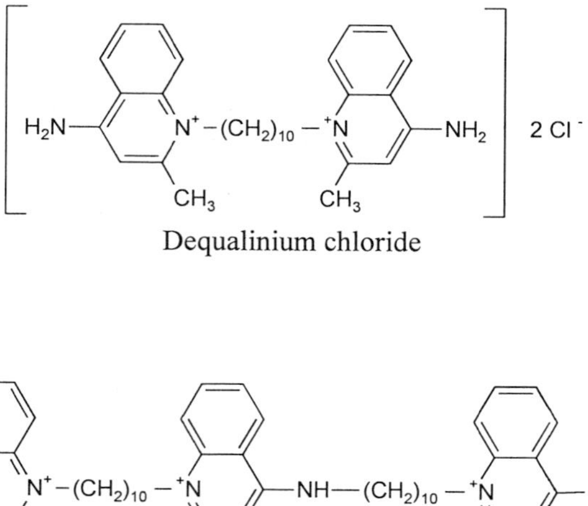

Dequalinium chloride

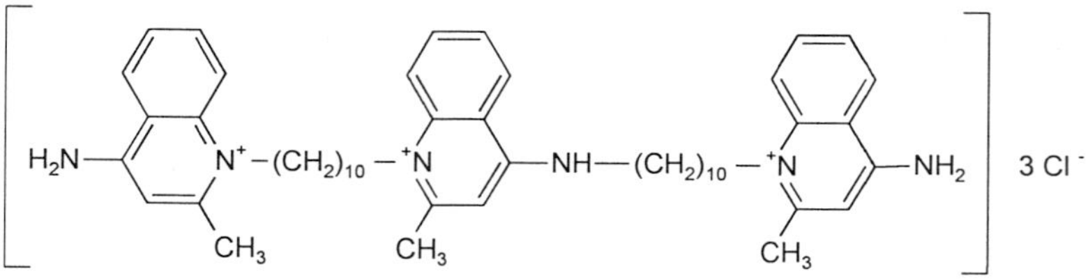

Compound a

Figure 3-22: Dequalinium chloride and by-product a.

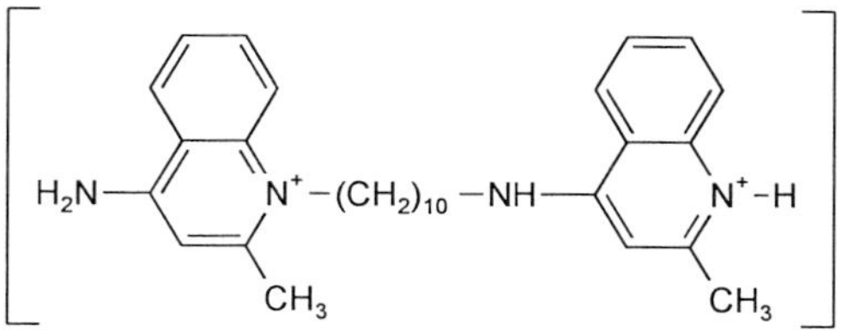

Compound b 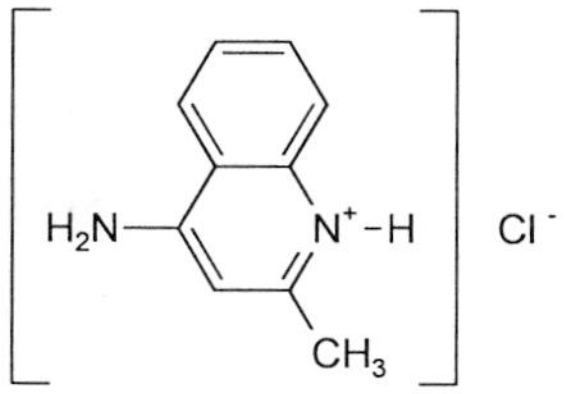   Compound c

Figure 3-23: Dequalinium chloride by-products b and c.

Thus, the new samples meet the demand of the content described as "not less than 95.0% and not more than the equivalent of 101.0%" in the DAB1998 and Pharmeuropa [21].

Captopril, an antihypertensive agent, is known to contain the disulfide analog, a metabolite, the RS-diastereomer (epicaptopril) and methylmercaptopropionic acid as impurities (see Figure 3-23) [22], [23]. Although the 400-MHz ^{1}H NMR spectrum recorded in DMSO-d_6 exhibits two amido conformers, the diastereomers and the disulfide can be clearly distinguished by signals between 4.2 and 4.7 ppm, representing the methine hydrogen attached to C-2 of the pyrrolidine ring [24]. These signals can be used for quantification.

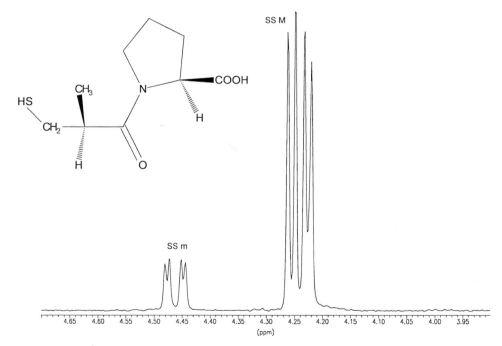

Figure 3-24: ^{1}H NMR spectra of captopril, 300 MHz, solvent DMSO-d_6; m = minor, M = major, "SSL".

Vitamin E, ± α-tocopherol as free phenol, acetate or succinate ester, contains small amounts of β-, γ-, and/or δ-tocopherol. Owing to the sometimes poor resolution and the requirement for an authentic reference standard, HPLC and GC analysis turned out to be unsatisfactory. Baker and Myers [25] could prove that both methods, ^{1}H and ^{13}C NMR, are suitable to quantify the composition of vitamin E by integration of corresponding signals.

Heteronuclear correlation experiments can be useful in terms of identification and quantification of traces in complex mixtures such as soil or water. The detection of trace amounts of organophosphorus compounds related to the chemical weapons convention impressively demonstrates how the advantages of each nucleus were combined, namely the sensitivity of the ^{1}H NMR spectra and the freedom from background noise for ^{31}P NMR spectra [26]. This was achieved by measuring inverse ^{1}H-^{31}P NMR spectra utilizing the HSQC experiments. These techniques will be introduced to drug analysis in body fluids.

3.3.2 Decomposition Reactions

3.3.2.1 Hydrolysis Reactions

Benzodiazepines, which attract interest as anxiolytics and sedatives, are known to hydrolyze in a physiological medium, giving benzophenone derivatives by ring opening (Figure 3-25). Dawson et al. tried to observe the degradation process of flurazepam dihydrochloride in different media at various temperatures, using ^{1}H, ^{13}C and ^{19}F NMR spectroscopy [27], [28].

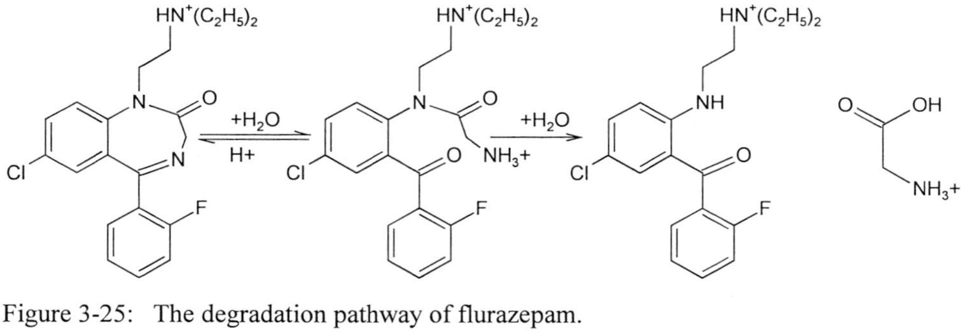

Figure 3-25: The degradation pathway of flurazepam.

Since ^{19}F NMR allowed detection of initial trace amounts (<1%) of the ring-opened product, it also afforded the best means of detection and quantifying the various entities in solution. Over a period of 24 h the equilibrium between flurazepam and the ring-opened form in D$_2$O was found to amount to a 44 : 56 ratio. The equilibrium did not influence the degradation reaction resulting in the benzophenone. Some aged samples were investigated; the percentage of benzophenone ranged from 5 to 10, no further degradation was observed in solution over a period of 24 h at 0 and 27°C.

The chemistry of penicillin-related drugs in different media can also be observed by ^{1}H and ^{13}C NMR spectra. Since the spectra are rather simple, it is possible to observe epimerizations as well as the decompositions resulting in penicilloic acid and related products. Details have been already reviewed by Branch et al. [29], [30]. In addition, the conversion of 6-aminopenicillanic acid to 8-hydroxypenillic acid accelerated by carbonate can be followed by ^{1}H NMR using the signal of the methyl hydrogens in position 2 [31]. The alkaline hydrolysis of cefotaxime was also elucidated, and the kinetics of the reaction observed by means of ^{1}H NMR spectroscopy [32].

The investigation of the ester hydrolysis reaction can take advantage of the suppression of the water signal by means of the WATR technique (water attenuation by T_2 relaxation). Using this method, signals in the region of $\delta = 4$–5 ppm can easily be observed. In this manner, the kinetics of hydrolysis of neostigmine bromide in weak acid aqueous solution [33] as well as the shelf life of aqueous solutions of acetylcholine, carbachol and atropine [34] were determined.

The decomposition process of bispyridinium aldoximes TMB-4, by acetylcholine esterase reactivators, were observed in D_2O solutions of different pD values and various temperatures [35]. Whereas the aldoximes turned out to be rather stable, corresponding ether and cyano derivatives were found to convert to the pyridone in alkaline medium at high temperatures.

3.3.2.2 Photodecompositions

The process of photodecomposition of the calcium channel blocker nifedipine (see Figure 3-26) can be observed by ^{1}H NMR spectroscopy [36] looking at the singlets between 2.3 and 2.6 ppm representing the methyl groups and 3.3 and 3.7 which belong to the methoxy groups and are well separated for each compound. Nifedipine accurately weighed was taken from a capsule, dissolved in a mixture of 40% $CDCl_3$ in CCl_4, and *tert*-butanol was added as an internal standard. The spectrum was recorded and the content of photodecomposition calculated from the integral of the methoxy signals in comparison with the *tert*-butyl signal. Additionally, a spectrum was taken after exposure of the solution in the NMR tube to diffused sunlight (see Figure 3-28). Nearly 100% of the 4-(2'-nitrophenyl)pyridine compound was found after 8 h. The method was as accurate and precise as the HPLC method described in the USP. In this case, the NMR method is superior to the HPLC procedure with respect to rapidity (about 5 min), simplicity and specificity.

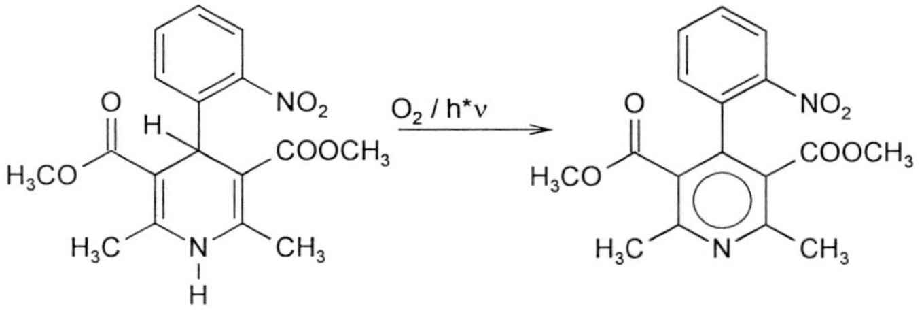

Figure 3-26: Nifedipine and its photodecomposition product.

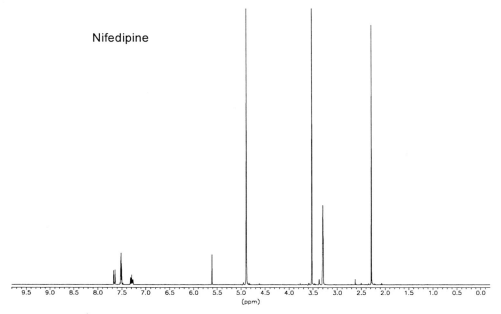

Figure 3-27: ^{1}H NMR spectrum of nifedipine, 300 MHz, solvent methanol-d$_4$, "SSL".

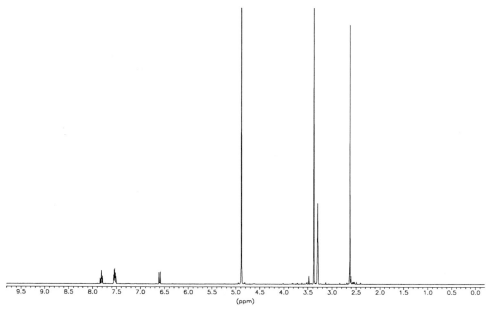

Figure 3-28: ^{1}H NMR spectrum of nifedipine after photo-oxidation, 300 MHz, solvent methanol-d$_4$, "SSL"

Fluoroquinolone compounds bearing a fluorine atom in position 6 and mostly a piperazine ring in position 7 attract interest as potent antibacterials. They are known to decompose upon irradiation, resulting in products which do not have any antibacterial activity, but show photosensitization properties in vivo. Hence, many of studies utilizing especially different hyphenated chromatographic-spectroscopic techniques were performed in order to elucidate the structure of the photo-products. In the case of ciprofloxacin heteronuclear ^{1}H-^{13}C NMR correlations such as HMQC and HMBC provide enough information to assign the spectra to a quinolone derivative which has replaced the piperazine ring with an amino group [37]. By making use of the sensitive C-F coupling constants, the carbon spectra could be assigned; subsequently, the correlation experiments were used to assign the corresponding ^{1}H NMR spectra. Highly fluorinated fluoroquinolones tend to loose a fluorine atom upon radiation. Martinez et al. [38] were able to observe the reaction by measuring time-dependent ^{19}F NMR spectra: the intensity of the signal of the additional fluorine in position 8 decreased, whereas a signal caused by a fluorine ion began to increase during the photolysis.

3.3.2.3 Miscellaneous

Some drugs are not stable in aqueous solutions and undergo equilibrium reactions with further ingredients in the drug formulation. These equilibration reactions can be observed by NMR spectroscopy. Madopar®, used in treatment of Parkinson's disease, consists of levodopa and benserazide (BZ). Serylhydrazide (SZ) is a degradation product of benserazide always present in tablets. The amount of serylhydrazide increases during storage due to a reaction with water. The same reaction takes place when a sample is prepared for analysis in an aqueous solution [2]. The equilibrium reaction is shown in Figure 3-29.

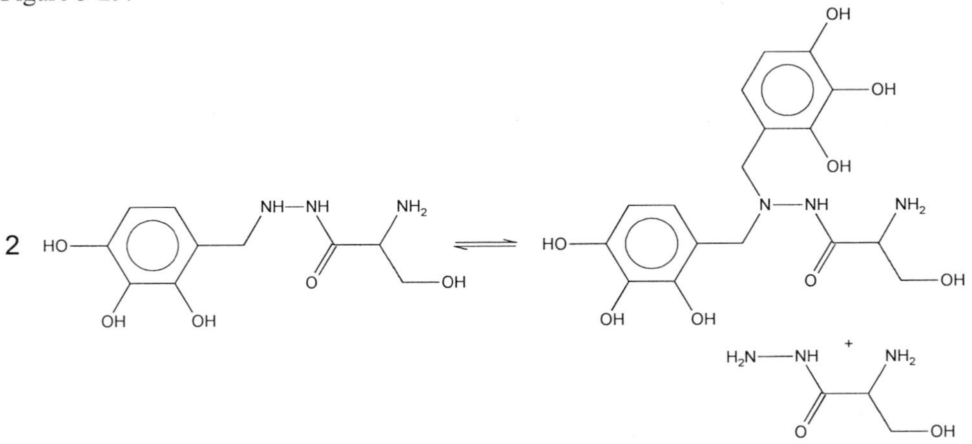

Figure 3-29: Equilibrium reaction of benserazide.

In ^{1}H NMR spectra, all three molecules are detectable. The ratio of the aromatic products can be taken from the integrals of the benzylic protons, since no response factor has to be taken into account. This ^{1}H NMR technique does not need standards and can be used as an ab initio method. Figure 3-30 - Figure 3-32 show the ^{1}H NMR spectra of a freshly prepared MeOD/D$_2$O/DCl extract of Madopar® and of an extract which is heated for 30 min to 80°C. The equilibrium ratio is about 1:1. It is not possible to isolate the dibenserazide (DBZ) to produce a standard for HPLC analysis.

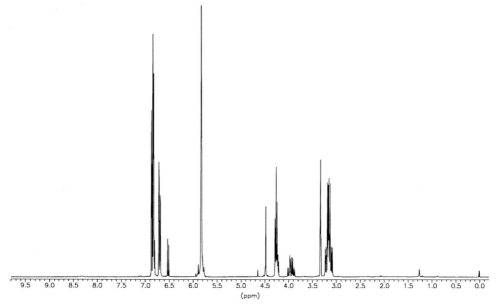

Figure 3-30: ^{1}H NMR spectrum of Madopar®, 300 MHz, solvent methanol-d$_4$, "SSL".

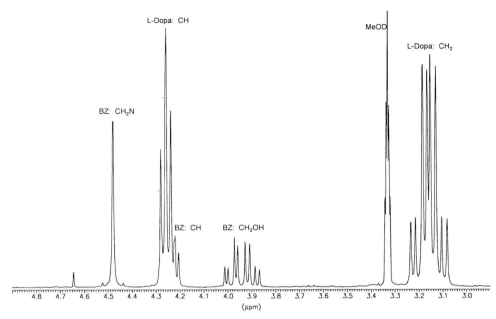

Figure 3-31: ^{1}H NMR spectrum of Madopar®, CH$_2$OH region, 300 MHz, solvent methanol-d$_4$, "SSL".

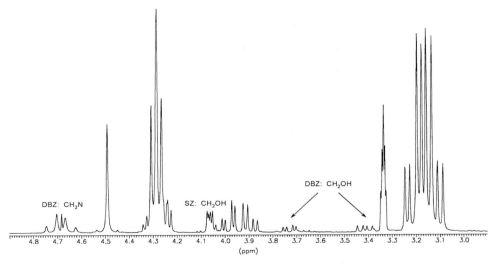

Figure 3-32: ^{1}H NMR spectrum of Madopar® after 30 min at 80°C, CH$_2$OH region, 300 MHz, solvent methanol-d$_4$, "SSL".

Cyanoacrylates are used to close cuts or wounds, even blood vessels can be glued together. These materials contain stabilizers such as hydroquinone in amounts between 100 and 500 ppm. Because of the high reactivity with respect to polymerization, chromatographic methods are not practicable and cannot be recommended to quantify the amount of hydroquinone. In a CDCl$_3$ solution the acrylate is stable for some hours and quantification was possible by means of ^{1}H NMR spectroscopy [2].

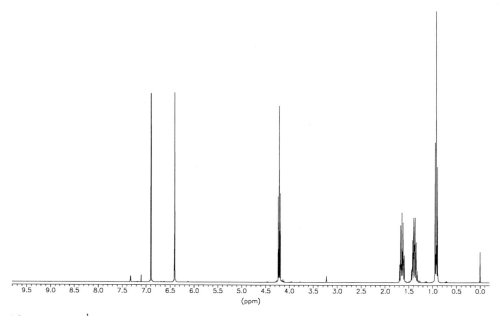

Figure 3-33: ^{1}H NMR spectrum of cyanoacrylate and hydroquinone as antioxidant, 300 MHz, solvent CDCl$_3$/benzene-d$_6$ (5:1) "SSL".

The ^{13}C satellites of the acrylate signals are used for calibration of the hydroquinone signal. To facilitate integration, the addition of some benzene-d$_6$ is necessary to shift the hydroquinone signal between the ^{13}C satellites of the acrylate signal. Because of the natural abundance of the ^{13}C nucleus, the area of a ^{13}C satellite is defined to be 5600 ppm in relation to the main signal. The amount of hydroquinone can be calculated from the expression, where M represents molecular mass:

$$\text{Hydroquinone [ppm]} = \frac{5600 \times \text{area}_{\text{hydroquinone}} \times M_{\text{hydroquinone}}}{2 \times (\text{area}_{\text{H1}} + \text{area}_{\text{H2}}) \times M_{\text{acrylate}}}$$

The method was validated in the range 30–1000 ppm. In Figure 3-34 the ^{1}H NMR spectrum of a cyanoacrylate containing 450 ppm hydroquinone is shown.

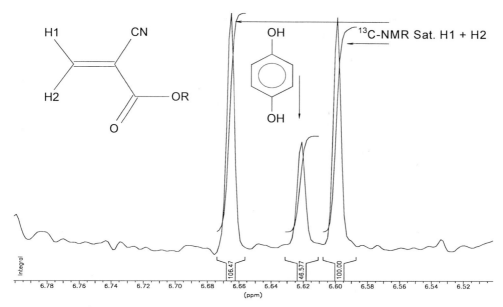

Figure 3-34: ¹H NMR spectrum (detail) of cyanoacrylate stabilized with hydroquinone, 300 MHz, solvent CDCl$_3$/benzene-d$_6$ (5:1), "SSL".

There are countless examples which describe the structure elucidation of drug impurities by means of NMR spectroscopy. Mostly, the impurities were collected and purified, and an NMR spectrum was recorded and assigned, e.g., the recently reported isomers of betamethasone [39]. Since no quantitative analyses have been performed, these examples are outside the scope of this book.

Quantitative analysis could be improved by the introduction of the joint application of HPLC and NMR spectroscopy in drug impurity profiling. This is demonstrated by Görög et al. for some examples, such as the 17α and 17β isomers of an ethinylestradiol and impurities of enalapril maleate [40]. The details of this technique and further applications will be discussed in the section NMR in body fluids.

3.4 Quantification of Drugs in Dosage Forms

3.4.1 Analysis of Single Drugs in Dosage Forms

The characterization of drugs in dosage forms by means of NMR spectroscopy is an interesting goal, because the identity and quantity can be determined simultaneously. Hanna et al. [41] describe a method for analysis of chlorpheniramine maleate in tablets and injection. After extraction of the drug, an internal standard (*tert*-butanol) is added and a 90-MHz spectrum is recorded. Using the integrals of the signals, which appear in a range between 1.95 to 2.70 ppm, representing the aliphatic hydrogens of chlorpheniramine, and the integrals of the signal at 1.25 ppm belonging to the methyl groups of the internal standard, the quantity of the drug can be calculated. Even using only a 90-MHz spectrometer, the method was found to be as accurate as an HPLC analysis.

Fardella et al. [42] reported on the quantitative analysis of fluoroquinolones, i.e. pefloxacin, norfloxacin and ofloxacin, in various pharmaceutical forms using both ^{1}H and ^{19}F NMR spectroscopy. The powder of the drug tablets, accurately weighed, and the internal standard, 2-amino-3,5-dibromo-6-fluorobenzoic acid and 2-fluoro-5-aminobenzoic acid, respectively, were extracted with DMSO-d$_6$, and the suspensions so obtained were centrifuged, filtered and subjected to the NMR measurements. The instrumental conditions had to be optimized especially, in terms of full T_1 relaxation: an extra delay of 6 to 8 s between the pulses was applied. The quantity of the substances was determined using intregrals of well-separated signals; each integral was the average of five readings. The quantity was calculated from the following formula:

$$\text{quantity (mg)} = (I_p/I_{is}) \times (EW_p/EW_{is}) \times C$$

where I_p is the average integral value for the drug, I_{is} the average integral value for the internal standard (is), EW_p the equivalent weight of the drug (molecular weight), EW_{is} the equivalent weight of the internal standard and C the weight (in mg) of the internal standard. Recovery experiments resulted in average recovery of 99.8% in the ^{1}H analysis and of 99.7% in ^{19}F, which confirms that the method is accurate and precise.

The following table shows that NMR spectroscopy is a simple and reliable means of quantifying a substance, no matter whether it exists in a drug or in dosage forms.

Table 3-1:   Quantification of drugs by NMR spectroscopy

Drug compound	Dosage form	MHz	Internal standard	Average recovery	Ref.
antazoline	Tablets	60	HMCTS	99.80%	[43]
Methimazole	Tablets	60	Benzoic acid	99.12%	[44]
Clofibrate	Capsules	60	HMCTS	92.45%	[45]
Tolazoline	Tablets	60	HMCTS	100.06%	[46]
Carbamazepine	Tablets	60	HMCTS	96.88%	[47]
Vitamins B$_1$ + B$_6$, simultanously	Tablets	60	Maleic acid	97.57% 97.42%	[48]
Phenytoin	Capsules, tablets	90	Acetamide	99.90%	[49]
Diatrizoate	Inj. solution	90	Sodium acetate	100.30%	[50]
Carbachol	Ophthalmic solution	90	Acetamide	100.00%	[51]
Bethanechol	Tablets	90	Acetamide	99.90%	[52]
Dicyclomine	Tablets, capsules, injections	90	Maleic acid	99.98%	[53]
Metoclopramide	Tablets, injections	90	Acetamide	-	[54]

HMCTS = hexamethylcyclotrisilazane

3.4.2 Analysis of Drug Mixtures in Dosage Forms

It is also possible to identify and quantify a mixture of ingredients in a dosage form. Capsules of Extra Strength Excedrin®, consisting of 250 mg acetylsalicylic acid, 250 mg paracetamol and 65 mg caffeine, can be easily examined after dissolution in Unisol® (a mixture of DMSO-d_6, CDCl$_3$, CD$_2$Cl$_2$), filtration, and recording an inverse gated decoupled ^{13}C NMR spectrum (after optimization of the relaxation time), using acetophenone as an internal standard [55]. The power of ^{1}H NMR spectroscopy in the analysis of pharmaceutical formulations containing different concentrations of drugs is demonstrated in the ^{1}H NMR spectra of Grippostad C®; the method is similar to that applied to Extra Strength Excedrin®. The amounts of the four drugs paracetamol, caffeine, ascorbic acid and chlorphenamine hydrogen maleate including the maleic acid can be analyzed in one spectrum (see Figure 3-35). Details of this spectrum are shown in Figure 3-36 and Figure 3-37.

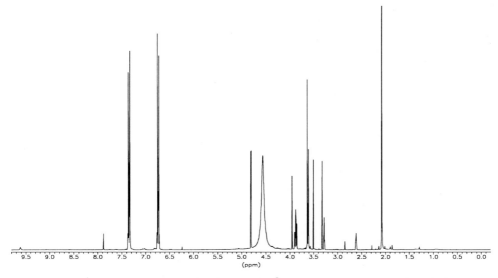

Figure 3-35: ^{1}H NMR spectrum of Grippostad C®, 300 MHz, solvent methanol-d_4, "SSL".

Simultaneous analysis of different drug compounds is difficult by classical HPLC analysis when the responses differ very greatly. NMR spectroscopy can be used in many of these cases. 1,1,1-Trichlor-2-methyl-2-propanol is used as a preservative in aqueous procaine formulations. In Figure 3-38 such an aqueous procain solution is measured as neat liquid [2]. A difficult sample preparation is not necessary. A benzene-d_6 filled capillary is used as the external lock. The determination of the amounts of the two ingredients can be done in a second measurement after adding the appropriate reference material (standard addition Figure 3-39). In the corresponding case the amount of 1,1,1-trichlor-2-methyl-2-propanol can be calculated by comparing the integral areas before and after standard addition. The standard deviation is within the range ± 1% or less.

46 Analysis of Drugs

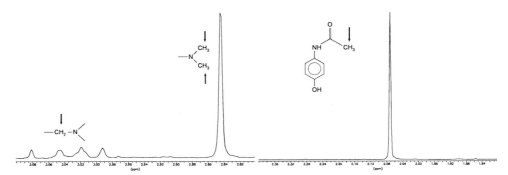

Figure 3-36: ^{1}H NMR spectrum (expansion; 1.82 to 3.10 ppm) of Grippostad C®, 300 MHz, solvent methanol-d$_4$, "SSL".

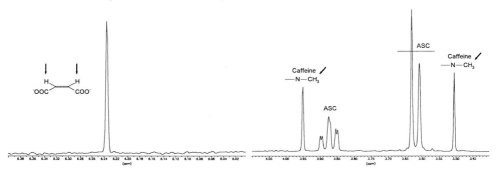

Figure 3-37: ^{1}H NMR spectrum (expansion; 3.4 to 8.1 ppm) of Grippostad C®, 300 MHz, solvent methanol-d$_4$, "SSL".

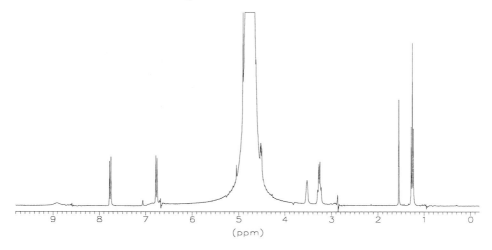

Figure 3-38: ^{1}H NMR spectrum of a procaine solution, "SSL".

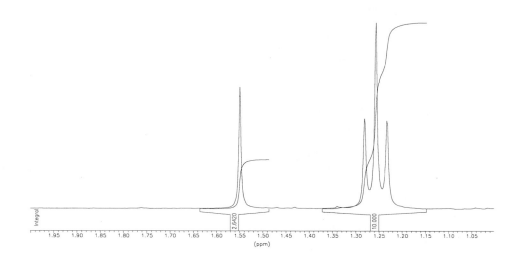
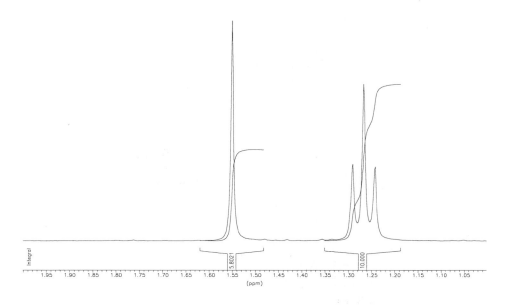

Figure 3-39: ¹H NMR spectrum of a procaine solution, detail (*top*) after addition of 0,4% 1,1,1-trichlor-2-methyl-2-propanol (*bottom*), "SSL".

Often natural products are accompanied by related compounds which may have a different pharmacological profile. For example, quinidine sulfate contains varying amounts of dihydroquinidine, which exhibits a greater antiarrhythmic activity and intrinsic toxicity than quinidine itself. Thus, dihydroquinidine is limited to 15% (Ph. Eur. 1998). Using HMCTS as an internal standard it is possible to determine dihydroquinidine in quinidine sulfate by integration of the methine proton (in neighborhood of the OH group) and the vinyl protons [56]. The average recoveries were found to be 98.9%. Most commercial tablets in this study contained 8% dihydroquinidine.

3.4.3 Analysis of Drugs and Decomposition or Isomer Traces in Dosage Forms

Hanna and Lau-Cam reported on the quantification of diphenhydramine in bulk material, capsules and freeze-dried solutions for injection. Using *tert*-butanol as the internal standard, they obtained mean recoveries of 100.0%. In addition they were able to quantify the decomposition products, benzhydrol, diphenylchloromethane and 2-(dimethylamino)ethanol, to a minimum of about 2% of the parent drug, by utilizing the integral of the corresponding benzylic protons [57].

A similar study has been performed for the diuretic drug furosemide, which tends to hydrolytically degrade to 4-chloro-5-sulfamoylanthranilic acid (CSA) [58]. Again, *tert*-butanol was chosen as a standard. Here the decomposition product CSA was easy to determine to a minimum of 1% of furosemide. Although the results of both studies were obtained on a low field 90-MHz NMR spectrometer, they were not much worse than the HPLC methods applied in the Ph. Eur. 1997 and do not need a reference sample. The following example show the superiority of NMR spectroscopy using a 300-MHz machine over an HPLC method. S-Adenosyl-L-methionine, a ubiquitous enzyme cofactor which is clinically effective in treatment of intrahepatic cholestasis and mental depression, often shows inactivation caused by an epimerization occuring at the chiral sulfur atom. Since the S-methyl group shows clearly separated signals for the mixture of diastereomers in the NMR spectrum, the ratio of the isomers can be easily determined by the integration of these signals [59]. The HPLC method cannot reach the accuracy of the NMR methods, because the external standard solutions are not stable enough.

Ephedra alkaloids, namely ephedrine, norephedrine and pseudoephedrine, are attracting interest as sympathomimetic agents. They can be quantitated either singly or in mixtures with each other in D_2O using acetamide as an internal standard and a 90-MHz spectrometer [60]. The determination of diastereomeric cross-contamination of pseudoephedrine and ephedrine in the presence of norephedrine was feasible using the integrals of the benzylic proton after addition of small amounts of DCl.

1,3-Dihalo-5,5-dimethylhydantoins used in disinfectant and molluscicides are an isomeric mixture of 1,3-dichloro-, 3-bromo-1-chloro-, 1-bromo-3-chloro- and 1,3-dibromo-dimethylhydantoin. Since the methyl groups of each compound show singlets of slightly different chemical shift, the isomeric ratio can be determined and the interconversion of the isomers easily observed by ^1H NMR spectroscopy [61].

3.5 Analysis of Complex Mixtures, e.g. Excipients

Phytopharmaca often are very complex mixtures of different chemical substances. In many cases these compounds are sensitive to biochemical or chemical reactions. If such a phytopharmacon contains a common component, an origin test and a standardization is possible by using NMR methods, e.g. formulations containing garlic powder. After reconstitution by adding water to powdered garlic containing dragees, the enzyme system produces allicin from alliin. The thermally unstable allicin can be extracted by organic solvents such as $CDCl_3$. A quantitation is possible when an internal standard is used [2]. The following figures show the ^{1}H NMR spectra of allicin extracted from fresh garlic (Figure 3-40) and from a reconstituted garlic dragee (Figure 3-41).

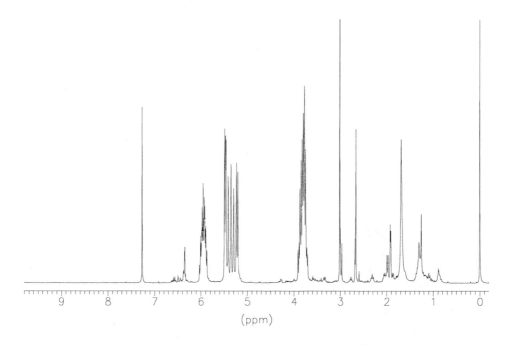

Figure 3-40: ^{1}H NMR spectrum of an organic extract of fresh garlic, 300 MHz, solvent $CDCl_3$, "SSL".

Depending on the amount of specific plant ingredients and a suitable sample preparation, NMR spectroscopy can be an alternative method in the standardization of phytopharmaca.

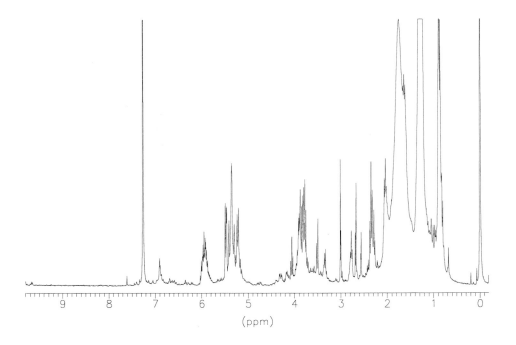

Figure 3-41: ^{1}H NMR spectrum organic extract of reconstituted garlic dragee, 300 MHz, solvent CDCl$_3$, "SSL"

3.5.1 Phospholipids

Many HPLC methods for phospholipids have been developed, but chromatographic resolution and dynamics of detection are not always satisfactory. For each source of phospholipids, special standards are needed due to the different distribution of fatty acids. These standards are expensive and in some cases are not available. Another problem is represented by the analysis of phospholipids in complex matrices. In many cases, separation is impossible or very difficult, not least due to the surface activity, which is desired in the application of phospholipids, but which complicates the analysis of these compounds. Therefore, a method is needed which is selective in the detection of phospholipids in order to avoid a separation from the matrix. The ^{31}P NMR spectroscopy of phospholipids meets these requirements. The I.L.P.S. (International Lecithin and Phospholipid Society) has chosen the ^{31}P NMR method as the reference method [62],[63],[64]. It has been tested world-wide by round robin tests in comparison to various HPLC and TLC methods. With triphenylphosphate as internal standard, a pulse angle of 15°, 10-s relaxation delay, and 32–256 accumulations, the method has a precision of <0,5%.

The ^{31}P NMR spectroscopy differentiates between the various phospholipids and has high dynamics in quantification. Only phosphorus-containing substances are detected by this method; the analysis is not disturbed by other non-phosphorus components. The resonance frequency of phosphorus depends on the chemical environment within the molecule. Phospholipids with different chemical structures are therefore recorded at distinct frequencies. Frequency can be measured very precisely, even small differences in the chemical structure of phospholipids being detected easily. Each phospholipid is represented by a single signal. Separation of the various phospholipids is not necessary as the phospholipids are characterized by their different resonance frequencies. Figure 3-42 shows a ^{31}P NMR spectrum of soy lecithin.

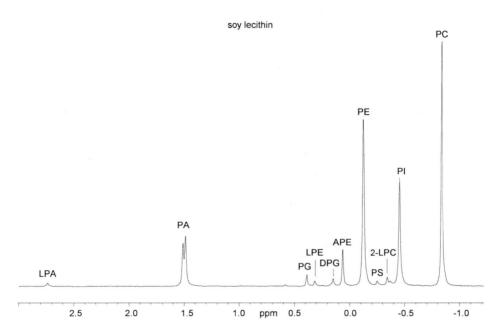

Figure 3-42: ^{31}P NMR spectrum of soy bean lecithin, "SSL";
121 MHz, solvent CDCl$_3$/methanol (2:1)
(L)PA= (lyso)phosphatidylic acid, (D)PG = diacyl-phosphatiylglycerol,
(L)PE= (lyso)phosphatidylethanolamine,
APE= acylphosphatidylethanolamine, PS = phosphatidylserine,
PI = phosphatidylinositol, (L)PC= (lyso)phosphatidylcholine.

The selective quantification of the PC degradation products 1-LPC, 2-LPC and GPC (Figure 3-43) makes ^{31}P NMR spectroscopy the ideal method for testing the stability of liposome containing formulations in pharmaceutical products. The amount of the different phospholipids can be calculated from their integral areas.

52 Analysis of Drugs

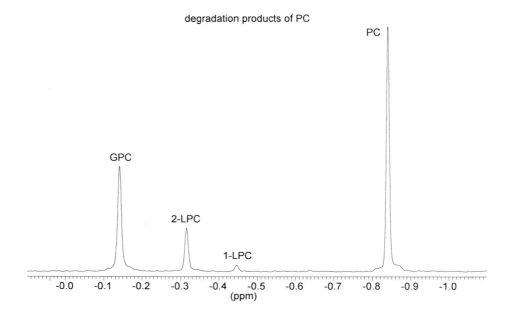

Figure 3-43: ^{31}P NMR spectrum of phospholipid degradation products, "SSL":
121 MHz, solvent CDCl$_3$/methanol (2:1)
1-LPC, 2-LPC and GPC = glycerophosphatidylcholine.

AL721 is a 7:2:1 mixture of neutral glyceride, phosphatidylcholine and phosphatidylethanolamine, which is considered to be valuable in HIV therapy. In order to determine the ratio of phosphatidylcholine and phosphatidylethanolamine based on the comparison of signal intensities, a pulse sequence has to be developed which minimizes the influence of T_1 relaxation times and NOE differences [65].

3.5.2 Silicones

The analysis of silicone products such as simethicone can be performed with good selectivity, using ^{1}H NMR spectroscopy. The NMR-method was cross-validated [2] against the official IR-method [66].

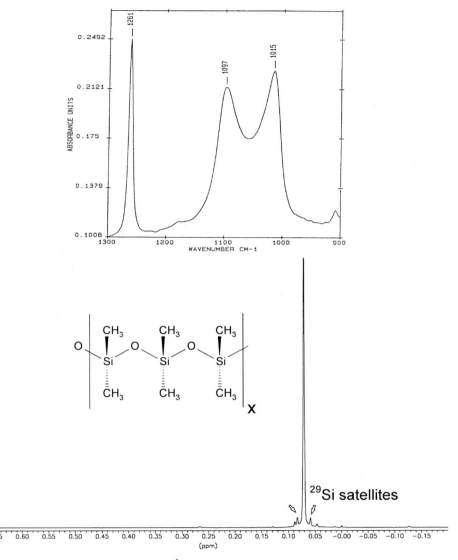

Figure 3-44: IR spectrum (*top*) and ^{1}H NMR spectrum (*bottom*) of simethicone, 300 MHz, solvent $CDCl_3$, "SSL"

The comparison of IR and ^{1}H NMR spectra (Figure 3-44) demonstrates the superiority of the NMR method; it provides more precise quantitative results and more information about structural details of silicones, as shown in the following expanded spectra Figure 3-45. Quantification can be done from the integral area by using an internal standard which does not interfere the silicone signals, e.g. 2-(methyldiphenylsilyl)-ethanol. Structural data, e.g. chain length (and with that the viscosity), end groups (OH, TMS, OCH_3 etc.) or functional groups within the silicone chains, can be detected and quantified by means of ^{1}H NMR spectroscopy. As an example, the expansion of an ^{1}H NMR spectrum of silicone is shown

54   *Analysis of Drugs*

in Figure 3-46. The end groups Si(CH$_3$)$_2$–OH and three more links within the chain are well separated from the inner chain signals. The integrals can be used for the determination of the total chain length [2].

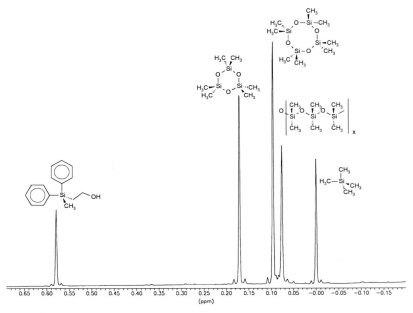

Figure 3-45: ^{1}H NMR spectrum of a mixture of siloxanes, 300 MHz, solvent CDCl$_3$, "SSL".

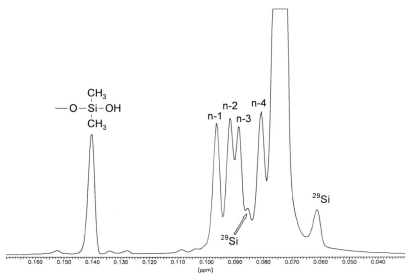

Figure 3-46: Expansion of an ^{1}H NMR spectrum of silicone, 300 MHz, solvent CDCl$_3$, "SSL".

3.5.3 Fluorine Containing Substances

Fluorocarbons such as perfluorodecalin, perfluorooctane and perfluorobromooctane are expected to have an increasing number of medical applications. Because of the toxicity of oxidized metabolites, the amount of proton containing molecules is a critical parameter. In this case, ^{1}H NMR spectroscopy is used to show the absence of protons and partly protonated fluorocarbons (Figure 3-47) [2].

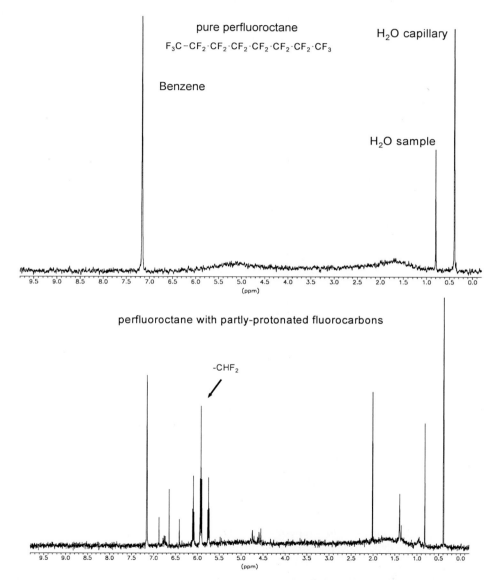

Figure 3-47: Analysis of fluorocarbons, 300 MHz, neat liquid, external lock, "SSL".

The measurement is done in neat liquid, and a capillary filled with benzene-d_6 is used as an external lock. The signals of protonated benzene and water are used for qualitative and quantitative calibration. The broad signals at 5.0 and 1.8 ppm are probe characteristics.

Especially in the field of dental care products, inorganic phosphates and fluorides are used. The combination of ^{19}F NMR spectroscopy and ^{31}P NMR spectroscopy makes it possible to determine simultaneously the content of fluoride and monofluorophosphate in the presence of ortho-, meta- and pyrophosphate [2].

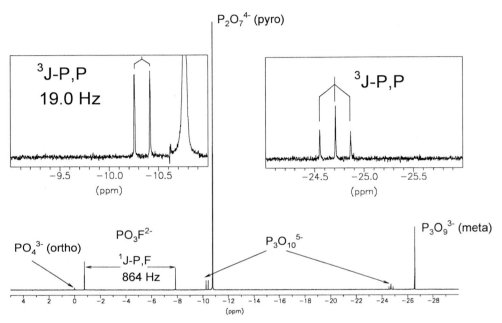

Figure 3-48: ^{31}P NMR spectrum of a toothpaste, 121 MHz, solvent D_2O, "SSL".

The standard addition method is used for quantitation. After determination of the phosphates with ^{31}P NMR (Figure 3-48) a calibration of the ^{19}F NMR spectrum is not necessary. The amount of monofluorophosphate is known from the ^{31}P NMR analysis. The fluoride content is determined from the integral areas of the fluoride and the monofluorophosphate signals in the ^{19}F NMR spectrum. Differences in the response of both nuclei are determined by measurement of standardized solutions.

The sample preparation of dental care products depends on the kind of formulation. Organic components such as emulsifiers, surfactants and thickeners have to be removed as well as the water-insoluble abrasives. In the case of very low fluoride and phosphate content, it is recommended to evaporate the aqueous extract. The chemical shift of the phosphate signals depends considerably on the pH of the solution, therefore a buffer of pH of 8–9 should be used. With strong acidic or basic medium, hydrolysis of meta- and pyrophosphate occurs. The ring opening of the metaphosphate producing the linear triphosphate is easily detected in the ^{31}P NMR spectrum (Figure 3-48).

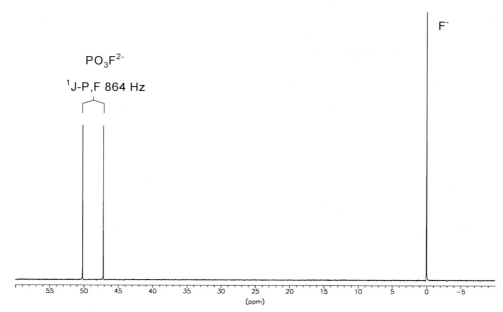

Figure 3-49: ^{19}F NMR spectrum of a toothpaste, 297 MHz, solvent D$_2$O, "SSL".

The outer P atoms show a doublet at –10 ppm, the inner P atom is represented by a triplet at –25 ppm. The large coupling constant (864 Hz) between the fluorine and the phosphorus nucleus (Figure 3-49) is worth mentioning, which is of course found in the ^{31}P NMR spectrum as well as in the ^{19}F NMR spectrum.

3.6 Concluding Remarks

These examples demonstrate the high potential of NMR spectroscopy in quantitative analysis. On the one hand, a drawback of NMR in comparison to MS is the fact that the sample size required for structure elucidation of an impurity is much higher. On the other hand, there are several advantages:

1. In addition to the structural information, an NMR spectrum can simultaneously provide information on the quantity of an impurity.
2. In most cases, the integration of signals used for quantification is more precise and accurate than the HPLC analysis.
3. Normally, no isolation of the impurity is necessary.
4. No expensive authentic reference samples are necessary.
5. In most cases, NMR spectroscopy is quicker (e.g. no equilibration of HPLC column), easy to perform and more specific.

References

1. P P Lankhorst, M M Poot, M P A de Lange, *Pharmacopeial Forum* **22** No 3, 2414-2422 (1996)
2. Method developed by Spectral Service GmbH Cologne, Germany, not published previously
3. A E Tonelli, The NMR spectra and the microstructures of polymers 55-95, (1996), in Polymer Spectroscopy edited by Allan H Fawcett, John Wiley & Sons, Chichester
4. K A M Thakur, R T Kean, E S Hall, J J Kolstad, T Lindgren, M A Doscotch, J I Siepmann, E J Munson, *Macromolecules* **30**, 2422-2428 (1997)
5. K A M Thakur, R T Kean, E S Hall, M A Doscotch, E J Munson, *Anal Chem* **69**, 4303-4309 (1997)
6. G A Neville, T J Racey, R N Rej, A S Perlin, *J Pharm Sci* **79**, 425-427 (1990)
7. G A Neville, F Mori, T J Racey, P Rochon, K R Holme, A S Perlin, *J Pharm Sci* **79**, 339-343 (1990)
8. G A Neville, F Mori, K R Holme, A S Perlin, *J Pharm Sci* **78**, 101-104 (1989)
9. U R Desai, R J Linhardt, *J Pharm Sci* **84**, 212-215 (1995)
10. S Tachibana, S Nishiura, S Ishida, K Kakehi, S Honda, *Chem Pharm Bull* **38**, 2503-2506 (1990)
11. B Casu, M Guerrini, A Naggi, G Torri, L De-Ambrosi, G Boveri, S Gonella, A Cedro, L Ferro, E Lanzarotti, M Paterno, M Attolini, M G Valle, *Arzneim Forsch/Drug Res* **46**, 472-477(1996)
12. F Nachtmann, G Atzl, W D Roth, *Anal Profiles Drug Subst* **12**, 215-276 (1983)
13. K G Ludwig-Baxter, R N Rej, A S , Perlin G A Neville, *J Pharm Sci* **80**, 655-660 (1992)
14. G A Neville, P Rochon, R N Rej, A S Perlin, *J Pharm Sci* **80**, 239-244 (1991)
15. G L Silvey, *J Pharm Sci* **81**, 471-474 (1992)
16. A Yagi, S Orndorff, P Toothman, B W K Diehl, *Japan Res Inst Indust Sci* **9**, 53-65 (1998)
17. A K Parkash, J K Sudgen, Z Yucan, W Mingxin, *Int J Pharmaceutics* **92**, 151-155 (1993)
18. C Boschetti, G Fronza, C Fuganti, P Grasselli, A G Magnone, A Mele, C Pellegatta, *Arzneim -Forsch /Drug Res* **45**, 1217-1221 (1995)
19. G Grübler, W Voelter, H Henke, H Mayer, *Pharm Ind* **52**, 794-800 (1990)
20. T Leopold, Ph D Thesis, Leipzig (1996)
21. *Pharmeuropa* **8**, 521-523 (1996)
22. F S El-Feraly, H Y Aboul-Enein, *Spectr Lett* **16**, 425-430 (1983); H Y Aboul-Enein, L Maat, J A Peters, *Spectr Lett* **18**, 419-424 (1985)
23. *Pharmeuropa* **6**, 291 (1994)
24. A F Casy, G H Dewar, *J Pharm Biomed Anal* **12**, 855-861 (1994)
25. J K Baker, C W Myers, *Pharm Res* **8**, 761-770 (1991)
26. C Albaret, D Loeillet, P Auge, P -L Fortier, *Anal Chem* **69**, 2694-2700 (1997)
27. B A Dawson, D B Black, G A Neville, *J Pharm Biomed Anal* **13**, 395-407 (1995)
28. G A Neville, H D Beckstead, D B Black, B A Dawson, H F Shurvell, *J Pharm Sci* **83**, 1274-1279 (1994)
29. S K Branch, A F Casy, E M A Ominde, *J Pharm Biomed Anal* **5**, 73-103 (1987)

30. S K Branch, A F Casy, A Lipczynski, E M A Ominde, *Magn Res Chem* **24**, 465-479 (1986)
31. A F Casy, A Lipczynski, *J Pharm Pharmacol* **46**, 533-534 (1994)
32. B Vilanova, F Munoz, J Donoso, J Frau, F G Blanco, *J Pharm Sci* **83**, 322-327 (1994)
33. A J Ferdous, R D Waigh, *J Pharm Pharmacol* **45**, 559-562 (1993)
34. A J Ferdous, N A Dickinson, R D Waigh, *J Pharm Pharmacol* 43, 860-862 (1991)
35. E Inkmann, U Holzgrabe, K F Hesse, *Pharmazie* **52**, 764-774 (1997)
36. G S Sadane, A B Ghogare, *J Pharm Sci* **80**, 895-898 (1991
37. K Torniainen, J Mattinen, C -P Askolin, S Tammilehto, *J Pharm Biomed Anal* **15**, 887-894 (1997)
38. L J Martinez, G Li, C F Chignell, *Photochem Photobiol* **65**, 599-602 (1997)
39. TM Chan, A Evans, K A Belsky, E A Ditnfeld, D J S Tsai, A T McPhail, *Magn Res Chem* **34**, 1025-1030 (1996)
40. S Görög, G Balogh, M Gazdag, *J Pharm Biomed Anal* **9**, 829-833 (1991)
41. G M Hanna, C A Lau-Cam, *J Pharm Biomed Anal* **11**, 855 - 859 (1993)
42. G Fardella, P Barbetti, I Chiappini, G Grandolini, *Int J Pharmaceut* **121**, 123-127 (1995)
43. H Y Aboul-Enein, A I Jado, M S El-Din Rashid, *Spectroscopy Lett* **11**, 931-938 (1978)
44. H Y Aboul-Enein, *J Pharm Pharmacol* **31**, 196 (1979)
45. H M El-Fatatry, H Y Aboul-Enein, *Spectroscopy Lett* **11**, 921-930 (1978)
46. H Y Aboul-Enein, H M El-Fatatry, M S Rashed, *Spectroscopy Lett* **12**, 187-197 (1979)
47. H M El-Fatatry, H Y Aboul-Enein, E A Lotfi, *Anal Lett* **12**, 951-961 (1979)
48. H Y Aboul-Enein, M A Loutfy, H M El-Fatatry, *Chem Biomed Environ Instrumentation* **11**, 69-76 (1981)
49. G M Hanna, Drug Develop *Ind Pharm* **10**, 341-354 (1984)
50. G M Hanna, C A Lau-Cam, *J AOAC Int* **79**, 833-838 (1996)
51. G M Hanna, C A Lau-Cam, *Drug Develop Ind Pharm* **14**, 43-51 (1988)
52. G M Hanna, C A Lau-Cam, *J AOAC* **70**, 557-559 (1987)
53. G M Hanna, *J AOAC* **67**, 222-224 (1984)
54. G M Hanna, C A Lau-Cam, Drug Develop *Ind Pharm* **17**, 975-984 (1991)
55. T A Schmedake, L E Welch, *J Chem Educ* **73**, 1045-1048 (1996)
56. G M Hanna, C A Lau-Cam, *J AOAC* **71**, 1118-1121 (1988)
57. G M Hanna, C A Lau-Cam, *Pharmazie* **39**, 816-818 (1984)
58. G M Hanna, C A Lau-Cam, *J AOAC Int* **76**, 526-530 (1993); previuos study by H Y Aboul-Enein, A A Al-Badr, M S El-Din Rashed, *Spectroscopy Lett* **12**, 323-331 (1979)
59. L K Revelle, D A d'Avignon, J C Reepmeyer, R C Zerfing, *J AOAC Int* **78**, 353-358 (1995)
60. G M Hanna, *J AOAC Int* **78**, 946-954 (1995)
61. J Beihoffer, E Bour, J H Lowry, J J Reschl, J L Seidel, C P Gibson, *J AOAC Int* **79**, 823-828 (1996)

62. B W K Diehl, W Ockels, Quantitative Analysis of Phospholipids In: *Proceedings of the 6th International Colloquium Phospholipids Characterization, Metabolism and Novel Biological Applications*, 29-32 Edited by G Cevc & F Paltauf Champaign: AOCS Press (1995)
   B W K Diehl, W Ockels, H Herling, R Unger, S Winkler, (1996) Quantitative Determination of Egg-Lecithin in a Liposome-Preparation, Validation of the ^{31}P NMR Method 7th International Congress on Phospholipids Brussels, Belgium, September (1996)
63. J De Kock, (1993) The European Analytical Subgroup of I L P S - A Joint Effort to Clarify Lecithin and Phospholipid Analysis *Fat Sci Technol* **95** (9), 352-355
64. J De Kock, (1995) The European Analytical Subgroup of I L P S In *Characterization, Metabolism and Novel Biological Applications*, pp 22-28 Edited by G Cevc & F Paltauf Champaign: AOCS Press
65. L L Olson, A P Cheung, *J Pharm Biomed Anal* **8**, 725-728 (1990)
66. *Pharmeuropa* **8**, 232-234 (1986)

Chapter 4

G. Hägele and U. Holzgrabe

4 pH-Dependent NMR Measurements

4.1 Introduction

Drugs in clinical use tend to be weak bases or acids and may, therefore, carry positive or negative charges depending on the pH of the liquid or tissue they have to pass through on their way to the target area or at the target molecule. Mostly, the non-ionized form of a drug is lipid soluble and can migrate through the lipid membranes of e. g. the gastric mucosa. In turn, the ionized molecules are water soluble and not able to penetrate the lipid membranes. Thus, the penetration of a drug through a membrane and the pharmacokinetics are determined by the drug's pK_a value [1]. Most drugs interact with a defined target structure, such as a receptor protein. The affinity of the drug to this protein is determined by the strength of the interaction. Ion-ion interactions play a decisive role because it is through their electrostatic field that two species start to recognize each other. From the pharmacodynamic point of view it is, therefore, important to know whether and at which atom a drug is ionized or non-ionized under physiological conditions in the presence of the target protein [2].

The commonly used methods of pK_a determination, such as potentiometric or spectrophotometric titrations, do not generally give information on the sites of protonation. Since it is well known that the chemicals shifts of all atoms, e.g. hydrogen, fluorine, carbon or phosphorus, are related to the amount of charge present at the atom [3], they can be used to determine both the pK_a values and the locus of protonation.
As early as 1973 Breitmaier [4] reported on the dependence of the ^{13}C NMR chemical shifts in pyridine and analogs on the pH: On increasing the pH from 0 to 8 he observed a considerable upfield shift of the α carbon atoms and a downfield shift of the β atoms in pyridine. Evaluation of the data resulted in titration curves, which can be used to calculate the pK_a value of pyridine:

$$pH = pK + \log (\delta_{max}-\delta)/(\delta-\delta_{min})$$

with δ_{max} = pH-independent chemical shift in acid medium and δ_{min} = pH-independent chemical shift in basic medium. The pK_a values determined by the NMR method were found to be in the same range as those reported in the literature [4]. To be precise, when using this method the influence of the solvent on the pH has to be considered.

4.2 Determination of the Site of Protonation

The change in the chemical shift upon addition of NaOD to a neutral solution of the acetylcholinesterase (AChE) reactivator TMB-4, which is a carbon analog of the clinically used obidoxime, is demonstrated in Figure 4-1. TMB-4, a symmetrical molecule, is characterized by two acidic oxime functions which can be deprotonated by NaOD. Since the exchange of the protons is very fast, an average spectrum of the three species in equilibrium was observed in each pH case, which results in titration curves. Due to mesomerism the deprotonation of both groups with NaOD results in a neutral molecule and, thus, an upfield shift of the pyridine and methine hydrogen atoms [5]. Interestingly, after 2 days of storing of the solution with 2 equivalents of NaOD, an exchange of the α-pyridinium hydrogens with deuterium occured. This observation was also made by Lin and Klayman for the analogous AChE reactivator 1,1′-methylenebis(4-hydroximino-methylpyridinium)dichloride [6].

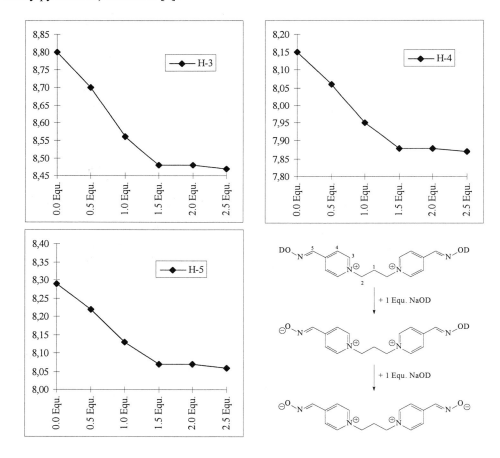

Figure 4-1: Titration curves of the TMB-4 pyridinium and methine hydrogens. Ordinate: δ values in ppm; abscissa: equivalents of NaOD.

The protonation equilibria of quinolone antibacterials, which consist of a β-keto-carboxylic group in position 3/4 and a piperazine ring in position 7, was also investigated by means of NMR spectroscopy. Buckingham et al. [7] titrated norfloxacin in a pD range of 5.5–10.9 and found that the chemical shifts of H5 and H8 remaining constant while those of H2 and the piperazine hydrogens underwent upfield shifts with increasing pD. Evaluation of the data of H2 and the piperazine hydrogens by means of titration curves gave the pK_a values of 6.90 and 9.25 with respect to the carboxyl group and the piperazine ring, respectively. In addition, the external piperazine ring was assumed to be the site of protonation. The latter result is confirmed by HMBC experiments performed by other authors [8], [9] for a variety of quinolone compounds. Further data for quinolones and the corresponding metabolites in acidic, neutral and basic media can be found in [10].

X=CH, R=H	norfloxacin
X=CH, R=CH₃	pefloxacin
X=N, R=H	enoxacin

R = H	ciprofloxacin
R = CH₃	N-methylciprofloxacin
R = C₂H₅	enrofloxacin

Table 4-1:   Some representative quinolone antibacterials.

Whenever a drug has more than one basic or acidic functional group, a conventional determination of the pK_a value does not provide the order of protonation or deprotonation upon variation of the pH of a solution. The usefulness of ^1H and ^{13}C NMR measurements can be impressively demonstrated for itraconazole, an azole-type antifungal drug. The molecule consists of a piperazine and two triazole rings which are all candidates for protonation. The pK_a is reported to be 3.7 [11].

Itraconazole was dissolved in a neutral solution of $CDCl_3:CH_3OD:D_2O$ = 16:8:1, which was acidified with 1, 2, 3, and 4 equivalents of DCl. The assignment was established by HETCOR and COLOC experiments [12] and compared with the assignment of the structurally related ketoconazole [13]. Adding one equivalent of DCl results in a strong downfield shift of the aromatic hydrogens H22/24 and the piperazine hydrogens which can be attributed to a protonation of nitrogen 26 (see Figure 4-2). In addition, the ^{13}C NMR spectrum shows a strong upfield shift of the α carbon atom C23 and a downfield shift of the β-atoms 22/24, which supports the assumed protonation of nitrogen 26 (see Figure 4-3). Further addition of DCl causes a very strong downfield shift of the triazole hydrogens 10 and 12, which belong to the left hand heterocyclus. The carbon atoms C12 and C10 show an upfield shift; the Δδ value of C12 is more pronounced than that of C10.

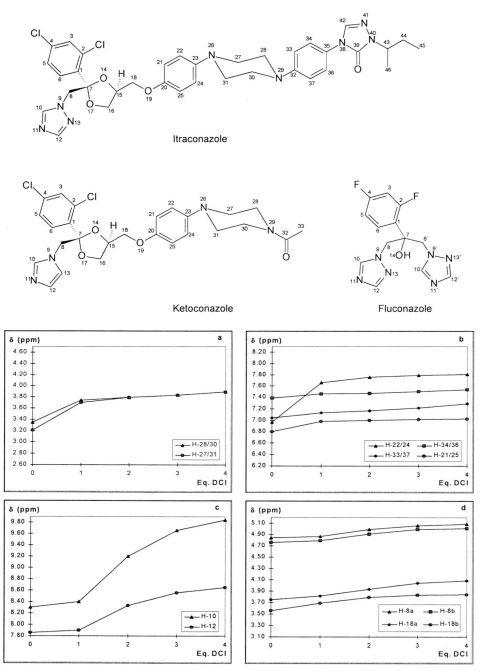

Figure 4-2: Chemical shifts of the itraconazole hydrogens of the piperazine ring *a*, the phenyl rings attached to the piperazine *b*, the triazole ring *c* and the methylene groups *d* as a function of the number of DCl added.

In order to find out which of the three nitrogens of this triazole ring is protonated, semiempirical calculations by the AM1 method were performed. Comparison of the chemical shifts of the carbon atoms and the theoretically obtained electronic charges indicates that nitrogen 11 happened to be protonated [12]. Taken together, the above mentioned pK_a value can be attributed to nitrogen 26.

Similar investigations were performed for fluconazole [12], which lacks a piperazine ring. Logically, the triazole rings were protonated upon addition of DCl. Since only one set of signals was observed upon addition of 1 and 2 equivalents of DCl, and the $\Delta\delta$ values were found to be smaller then those observed with itraconazole, it may be concluded that both triazole rings were protonated simultaneously.

The usefulness of pH-dependent NMR measurements for the determination of pK_a values, the order and locus of protonation, and information about structural changes upon protonation has often been shown. Instead of the time-consuming pK_a and NMR measurements in a series of samples of varying pH, the automated NMR-controlled titration method using ^{31}P, ^{113}Cd, ^{13}C, ^{1}H and ^{19}F atoms can be employed.

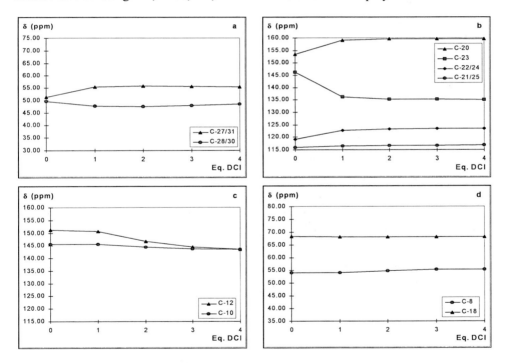

Figure 4-3: Chemical shifts of the itraconazole carbon atoms of the piperazine ring **a**, the phenyl rings attached to the piperazine **b**, the triazole ring **c** and the methylene groups **d** as a function of the number of DCl added.

4.3 Determination of Dissociation Constants and Stability Constants by NMR-controlled Titrations

First some fundamental principles will be described. The protonation equilibrium of an *n*-valent base is described by eq. 1-3.

$$i H^+ + L^{-a} = H_i L^{i-a} \quad (i=1-n) \tag{1}$$

$$\beta_i = \frac{c_{H_i L^{i-a}}}{c_{H^+}^i \cdot c_{L^{a-}}} \quad (i=0-n) \tag{2}$$

$$x_i = \frac{10^{(\lg\beta_i - i \cdot pcH)}}{\sum_{k=0}^{n} 10^{(\lg\beta_k - k \cdot pcH)}} \tag{3}$$

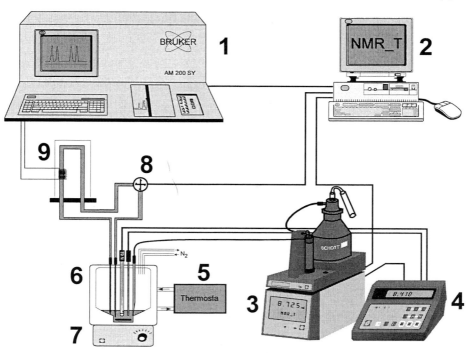

Figure 4-4: Hardware set up for NMR-controlled titrations: [1]NMR spectrometer, 2 PC, 3 burette, 4 glass electrode and potentiometer, 5 thermostat, 6 thermostated titration vessel, 7 stirrer, 8 pump, 9 NMR probe head.

The component L^{-a} and each of the species H_iL^{i-a} present in the equilibrium contribute "ion-specific" NMR-parameters $\delta_{H_iL}^{i-a}$ in an exchange reaction which is rapid on the NMR time scale. An averaged chemical shift $<\delta>$ appears, given by eq. 4.

$$<\delta> = \sum_{i=0}^{n} x_i \delta_{H_iL^{i-a}} \qquad (4)$$

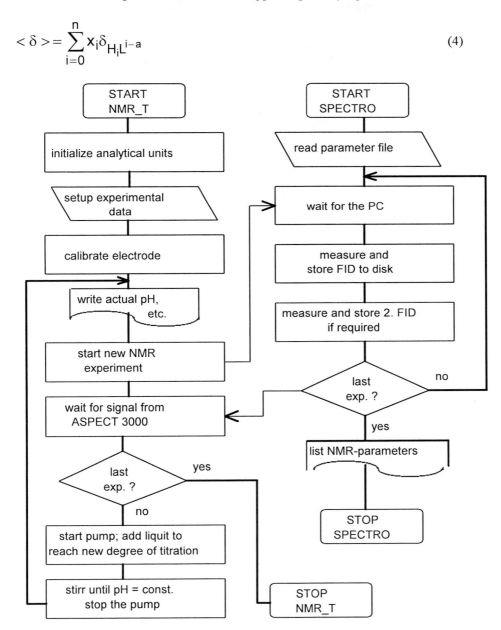

Figure 4-5: Software set-up for NMR-controlled titrations.

Consequently the dynamic chemical shift $<\delta>$ is a function of intrinsic "ion-specific" chemical shifts and molar fractions of protolytic species, where the latter depend simply on stability constants and concentrations of reagents used. Obviously the molar fractions depend on the state of titration and the actual pH in solution. $<\delta>$ may be plotted as a function of the state of titration (δ, τ spectra) or as a function of pH (δ, pH spectra) as shown below. The hardware and software set-up for PC-guided NMR-controlled titrations or alternatively titration-dependent NMR spectroscopy is shown in Figure 4-4 and Figure 4-5 below.

4.4 Applications in Pharmacy

4.4.1 Macroscopic Dissociation Equilibria

The macroscopic protonation and dissociation constants of leucine are obtained from potentiometric and $^{13}C\{^1H\}$-NMR controlled titrations of L-leucine + 1 HNO_3 vs NaOH: lg$b$1 = 9.85, lg$b$2 = 12.29, pK_1 = 2.34 and pK_2 = 9.85. Ion-specific chemical shifts δ_C [ppm] for carbon C1 in L-leucine were calculated as 175.70 (H_2L^+), 177.87 (HL) and 186.62 (L^-) using numerical data from Figure 4-6 and Figure 4-7.

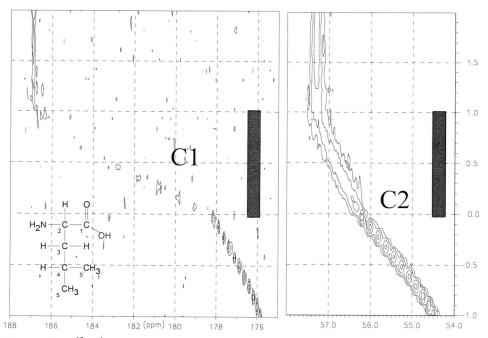

Figure 4-6: $^{13}C\{^1H\}$-NMR-controlled titration of L-leucine + 1 HNO_3 vs NaOH for C_1 and C_2. δ_C [ppm] (x axis) vs τ, the degree of titration (y axis).

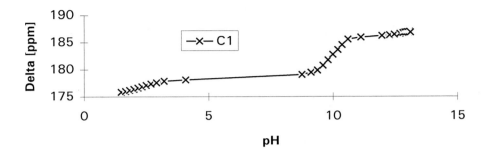

Figure 4-7: $^{13}C\{^1H\}$-NMR-controlled titration of *L*-leucine vs NaOH for C1. δ_C [ppm] (y axis) vs pH (x axis).

4.4.2 Microscopic Dissociation Equilibria

Phenylephrine

Potentiometric titration of phenylephrine vs NaOH (Figure 4-8) yields the macroscopic dissociation constants while the microscopic dissociation constants are accessible from simultaneous observation of potentiometric and $^{13}C\{^1H\}$-NMR-controlled titrations (Figure 4-9 and Figure 4-10). The microscopic dissociation species defined in Table 4-1 are involved in the microscopic dissociation equilibrium shown in Figure 4-6.

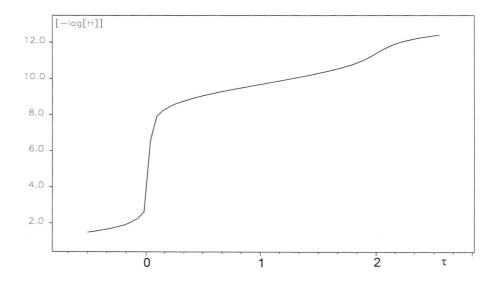

Figure 4-8: Titration curve pH = f(τ); phenylephrine + HNO$_3$ vs NaOH; macroscopic dissociation constants: pK_1 = 9.17 and pK_2 = 10.22.

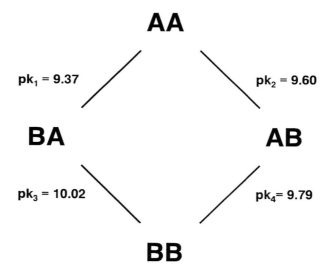

Figure 4-9: Microscopic dissociation scheme and dissociation constants of phenylephrine derived from potentiometric and ^{13}C{^{1}H}-NMR-controlled titrations.

microscopic species	symbol
HO-C₆H₄-CH(OH)-CH₂-NH₂⁺-CH₃	AA
⁻O-C₆H₄-CH(OH)-CH₂-NH₂⁺-CH₃	BA
HO-C₆H₄-CH(OH)-CH₂-NH-CH₃	AB
⁻O-C₆H₄-CH(OH)-CH₂-NH-CH₃	BB

Table 4-2: Microscopic dissociation species of phenylephrine.

The microscopic and macroscopic dissociation constants are correlated by:

$$K_1 = k_1 + k_2 \qquad (5)$$

$$K_2^{-1} = k_3^{-1} + k_4^{-1} \qquad (6)$$

$$K_1 * K_2 = k_1 * k_3 = k_2 * k_4 \qquad (7)$$

Microscopic molar fractions are given by eq. 8–12.

$$x_{AA} = \frac{1}{\Sigma} \qquad (8)$$

$$x_{BA} = \frac{\dfrac{k_1}{c_{H^+}}}{\Sigma} \quad (9)$$

$$x_{AB} = \frac{\dfrac{K_1 - k_1}{c_{H^+}}}{\Sigma} \quad (10)$$

$$x_{BB} = \frac{\dfrac{K_1 K_1}{c_{H^+}^2}}{\Sigma} \quad (11)$$

$$\Sigma = 1 + \frac{K_1}{c_{H^+}} + \frac{K_1 K_2}{c_{H^+}^2} \quad (12)$$

The dynamic chemical shift $<\delta>$ observed by NMR-controlled titrations follows from eq. 13:

$$<\delta> = \sum_{i=AA}^{BB} x_i \delta_i \quad (13)$$

Under specific restriction and for given values of macroscopic constants K1 and K2 the microscopic constant k_1 and henceforth all k_i can be extracted from $<\delta>$. The restriction is: the indicator spin to be monitored is sensitive to one type of protonation process only, long-range protonation effects have to be zero or negligible. For example, the chemical shift of C1 in phenylephrine is identical in the pairs of microscopic dissociation species (AA and AB) and in (BA and BB) respectively. Analogous reasoning has to hold for other indicator spins.

Results from $^{13}C\{^1H\}$-NMR-controlled titration of phenylephrine hydrochloride + 1 HNO$_3$ vs NaOH shown below in Figure 4-10 – Figure 4-12 were used to evaluate the microscopic dissociation constants.

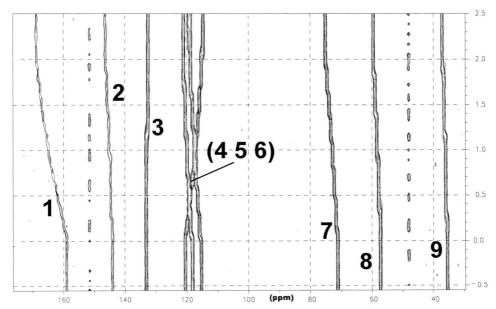

Figure 4-10: $^{13}C\{^1H\}$-NMR controlled titration of phenylephrine hydrochloride + 1 HNO$_3$ vs NaOH; δ_C vs TMS; overview for all carbon atoms C1-C9; spin labeling given in Figure 4-8.

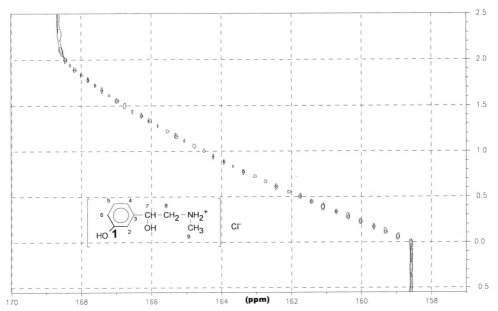

Figure 4-11: $^{13}C\{^1H\}$-NMR of phenylephrine hydrochloride + 1 HNO$_3$ vs NaOH; δ_C of C1 (*ipso* to phenolic hydroxy group) vs TMS.

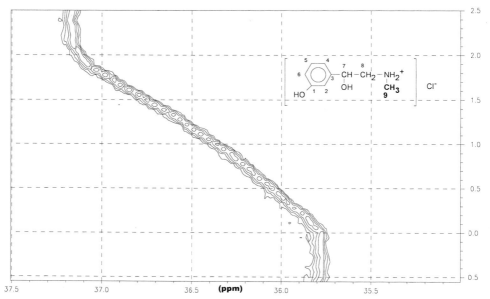

Figure 4-12: $^{13}C\{^1H\}$-NMR of phenylephrine hydrochloride + 1 HNO_3 vs NaOH; δ_C of C9 (*N*-methyl group) vs TMS.

Table 4-3 lists the microscopic dissociation constants as obtained from carbon atoms C1 to C3 and C7 to C9. Suitable calculations showed, that C1 and C9 are good indicators holding the restrictions as required above.

	C1	C9	C1 / C9	C2	C3	C7	C8
pk1	9.369	9.366	9.368	9.310	9.454	9.395	9.390
pk2	9.600	9.604	9.602	9.722	9.485	9.559	9.566
pk3	10.019	10.022	10.020	10.078	9.934	9.993	9.998
pk4	9.788	9.784	9.786	9.666	9.903	9.829	9.822

Table 4-3: Micro dissociation constants of phenylephrine evaluating carbon atoms C1 to C3, C7, C8 and C9. Best solutions are obtained from C1 and C9.

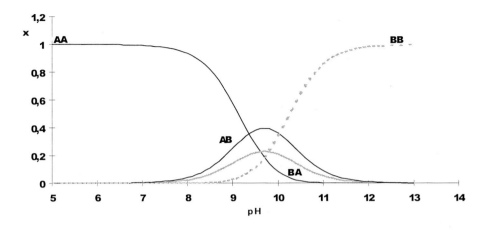

Figure 4-13: Microscopic molar fractions for protonation species of phenylephrine vs pH.

Figure 4-13 shows the molar fractions of phenylephrine as functions of pH. Obviously in phenylephrine deprotonation of the ammonium cationic site is preferred to deprotonation of the phenolic function. Details of basic principles, hardware and software concepts and applications of NMR-controlled titrations are described in [14], [15], [16], [17], [18], [19], [20], [21], [22]. Results from more laborious UV/Vis [23], [24] and now the easily accessible automated NMR-controlled titrations of phenylephrine [22] are consistent. The method of NMR-controlled titration is a powerful tool in analytical and structural chemistry, and was recently applied successfully to a series of biorelevant phosphorus compounds [18],[19],[20],[21].

References

1. L Z Benet, J R Mitchell, L B Sheiner, Pharmacokinetics: the Dynamics of Drug Absorption, Distribution, and Elimination, 1-32 (1992), in Goodman and Gilman's The Pharmacological Basis of Therapeutics, edited by A Goodman Gilman, T W Rall, A S Nies, P Tayler McGraw-Hill, New York
2. E M Ross, Pharmacodynamics: Mechanism of Drug Action and Relationship between Drug Concentration and Effect, 33-48 (1992), in Goodman and Gilman's The Pharmacological Basis of Therapeutics, edited by A Goodman Gilman, T W Rall, A S Nies, P Tayler, McGraw-Hill, New York
3. For review: D Farcasiu, A Ghenciu, *J Progr NMR* **29**, 129-168 (1996)
4. E Breitmaier, K -H Spohn, *Tetrahedron* **29**, 1114 - 1152 (1973)
5. E Inkmann, U Holzgrabe, K -F Hesse, *Pharmazie* **52**, 764-774 (1997)
6. A I Lin, D L Klayman, *J Pharm Sci* **75**, 797-799 (1986)
7. D A Buckingham, C R Clark, A Nangia, *Aust J Chem* **43**, 301-309 (1990)

8. K Takacs-Novak, B Noszal, I Hermecz, G Kereszturi, B Podanyi, G Szasz, *J Pharm Sci* **79**, 1023-1028 (1990)
9. C M Riley, D L Ross, D vander Velde, F Takusagawa, *J Pharm Biomed Anal* **11**, 49-59 (1993)
10. U Holzgrabe, S K Branch, *Magn Res Chem* **32**, 192-197 (1994)
11. The Merck Index, 12th edtn, (1996)
12. E Inkmann, U Holzgrabe, *J Pharm Biomed Anal, J Pharm Anal* **20**, 297 - 307 (1999)
13. B A Dawson, *Can J Spectros* **35**, 27-30 (1990)
14. (a) 1 M Grzonka and G Hägele; *Workshop "Computer in der Chemie", Software-Entwicklung in der Chemie* **2** 229 (1988), Springer Verlag, Berlin (ed J Gasteiger)
    (b) G Hägele, M Grzonka, H -W Kropp, J Ollig, H Spiegl, *Phosph Sulf Silic* **77**, 85 - 88 (1993)
    (c) G Hägele, S Varbanov, J Ollig, H -W Kropp, *Z Allg Anorg Chem* **620**, 914 - 920 (1994)
    (d) G Hägele; Review in "Phosphorus-^{31}P-NMR Spectral Properties in Compound Characterisation and Structural Analysis" (Edit L D Quin und J G Verkade) VCh
15. G Hägele, M Grzonka, *Workshop "Computer in der Chemie", Software-Entwicklung in der Chemie* **3**, 181 (1989) Springer-Verlag, Berlin (ed J Gasteiger)
16. G Hägele, M Grzonka, H -W Kropp, J Ollig und H Spiegl; *Phosph Sulf and Silic* **77**, 85 - 88 (1993)
17. G Hägele, S Varbanov, J Ollig and H -W Kropp; *Z Allg Anorg Chem* **620**, 914 - 920 (1994)
18. G Hägele; Review in "Phosphorus-^{31}P-NMR Spectral Properties in Compound Characterisation and Structural Analysis" (ed L D Quin, J G Verkade) VCH, 395–409, (1994)
19. J Ollig, G Hägele, *Comp Chem* **19/3**, 287 - 294 (1995)
20. G Hägele, C Arendt, H W Kropp, J Ollig, *Phosph Sulf Silic* **109-110**, 205 - 208 (1996)
21. J Ollig, M Morbach, G Hägele, E Breuer, *Phosph Sulf Silic* **111**, 55 (1996)
22. J Ollig, Dissertation, Heinrich-Heine-Universität Düsseldorf (1996)
23. a) G Hägele, H -J Majer and F Macco; *GIT* **9**, 922–929 (1992)
    b) H -J Majer, Dissertation Heinrich-Heine-Universität Düsseldorf (1993)
    c) C Arendt, Diplomarbeit Heinrich-Heine-Universität Düsseldorf (1993)
24. C Arendt, G Hägele, *Comp Chem* **19/3**, 263 - 268 (1995)

Chapter 5

B.W.K. Diehl and U Holzgrabe

5 Complexation Behavior of Drugs Studied by NMR

5.1 Selfassociation of Drugs and Association with Other Components of a Formulation

For a long time the usefulness of NMR spectroscopy in the characterization of complexes has been well known. As early as 1969 Stamm [1] described the behavior of caffeine alone and in combination with sodium benzoate in $CHCl_3$. Looking especially at the changes of the ^1H NMR chemical shift of the methyl groups in caffeine in various concentrations and upon addition of increasing amounts of benzoate he could observe the break of the self association of the dimeric and tetrameric caffeine superstructures and subsequently the heteroassociation with the benzoate. Due to the fact that only the N-methyl groups and the carbon C8 carry hydrogen atoms which are all located on the periphery of the molecule ^1H NMR spectra can give only poor information about the complex geometry [2]; ^{13}C NMR spectra are a much better tool to elucidate the dimeric structure of caffeine. The concentration-dependent measurement of ^{13}C NMR spectra (0.16% - 4.48%) revealed a strong upfield shift of all carbon atoms belonging to the dioxopyrimidine ring, whereas the imidazole atoms were found to be almost unaffected. The geometry of the vertical stacking derived from these findings is displayed in Figure 5-1 [3]. The dimeric complex is stabilized by electrostatic and charge-transfer interactions. From corresponding concentration-dependent ^1H NMR measurements [4] dimerization constants were calculated, but are somewhat ambiguous, because the hydrogens do not participate in the complexation [3] (see above) and, in addition, the simultaneous formation of higher oligomers leads to misinterpretations of the data [5].

As can be seen from the differences in chemical shifts in concentration-dependent measurements, theophylline, lacking a methyl group in position 7, also tend to form dimeric complexes. These are characterized by two NH···O=C hydrogen bondings [6],[7] and by a vertical stacking as found in the case of caffeine [8]. Both kinds of complexation were recently found in crystals of anhydrous theophylline [9].

The tendency to form complexes strongly influences the water solubility of the xanthine derivatives, which plays a pivotal role for a high bioavailability. Destroying of the superstructure by addition of e.g. benzoate goes along with an increase in solubility and the change of properties important in pharmacokinetics and in addition in pharmaceutics.

The example of xanthine derivatives demonstrates the power of NMR spectroscopy in the elucidation of superstructures formed by the self-association of molecules. The development of solid state NMR spectroscopy enables the investigation of the structure of drug powders. The knowledge of a powder structure can help to standardize the solubility of a powder, the suitability for drug formulations etc.

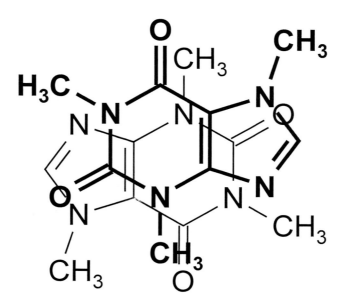

Figure 5-1: Structure of the dimeric complex of caffeine.

5.2 Complexation with Cations

Some drugs tend to bind metal ions which often results in a varied bioavailability and pharmacology. Especially the fluoroquinolones and tetracyclines, both antimicrobial drugs are known to form chelate complexes with divalent and trivalent ion, such as calcium, magnesium, bismuth and aluminum ions, using the β-keto carboxylic moiety [10]. Besides the reduced oral bioavailability, the antimicrobial activity of the fluoroquinolones is diminished. Since these ions are components of antiacid and multivitamin drugs, a co-administration of the antimicrobials and these drugs should be avoided. In order to find out whether a drug is able to complex with metal ions, NMR spectroscopy can help to elucidate the stoichiometry and the geometry of the chelate complexes.

As early as 1988 Mendoza-Diaz et al. [11] described the complexation of nalidixic acid, a fluoroquinolone of the first generation, with Cu^{2+} ions via the 4-oxo and 3-carboxylate groups. ^{13}C NMR spectra showed a dramatic change in the chemical shift and a broadening of the carbon atoms of this moiety upon complexation with the Cu^{2+} ions. In 1992, Shimada and coworkers [12] observed a reduced absorption of lomefloxacin orally administered concomitantly with aluminum- and magnesium-containing antacids. Again, ^{13}C NMR measurements revealed the complexation: whereas Al^{3+} ions produced a second set of signals of the keto-carboxylic carbon atoms, Mg^{2+} ions caused a broadening of these signals. More detailed studies of lomefloxacin and norfloxacin [13] utilizing the ^{1}H NMR spectroscopy clearly showed the formation of two different complexes upon addition of increasing amounts of Al^{3+} ions. Three sets of signals for H5 and H2 (between $\delta = 7.5$ and 9.3 ppm) appeared at a drug:metal ratio of 2:1. They belong to the free drug and complexes of 2:1 and 3:1 stoichiometry. This indicates a slow exchange between free and bound drug molecules on the NMR time scale. Studies with increasing amounts of Mg^{2+} showed a shift of the NMR signals. Due to the fast exchange on the NMR time-scale between the free and complexed quinolones, single average signals were observed which do not give any information about the stoichiometry.

Lecomte at al. studied magnesium complexes with ciprofloxacin, pefloxacin and norfloxacin by means of ^{19}F NMR spectroscopy [14], [15]. The fluorine signals were observed between 46 and 50 ppm with respect to trifluoroacetic acid. Upon addition of 10 mM Mg^{2+} a second peak belonging to the complex appeared. Data obtained with increasing ratios of Mg^{2+}/quinolone (0-150) were fitted with a 1:1 and a 2:1 stoichiometry for the complex. The 1:1 stoichiometry yielded affinity constants between 10 and 21×10^2 M^{-1}. Sakai et al. [16] intensively explored the complexation of lomefloxacin, levofloxacin and ciprofloxacin with Al^{3+}, Mg^{2+} and Ca^{2+} by means of ^1H and ^{13}C NMR spectroscopy. Upon addition of 0.2 mol Al^{3+} they found a different extent of complexation: 56% of levofloxacin, 28% of ciprofloxacin and 10% of lomefloxacin. In turn, the binding affinity of the metal ions towards the drugs was found to be $Al^{3+} \gg Ca^{2+} > Mg^{2+}$.

Even though all studies revealed a different complexation behavior of the drugs on the one hand and the metal ions on the other hand, it can be stated that fluoroquinolones should not be co-administered with antacid or multivitamin drugs in order to avoid the complexation and, as a result, a decreased antimicrobial activity.

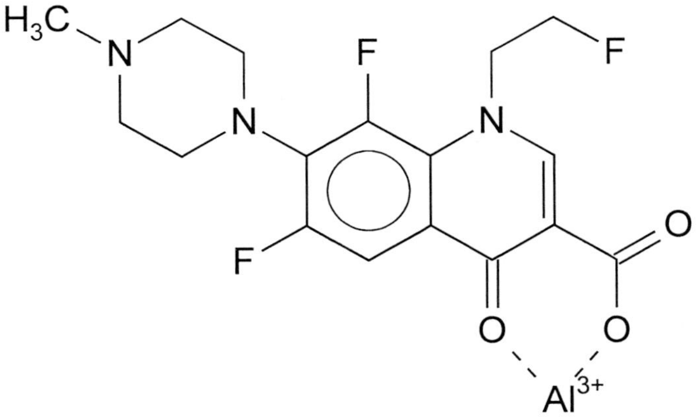

Figure 5-2: One possible solution structure of fleroxacin in presence of Al^{3+}.

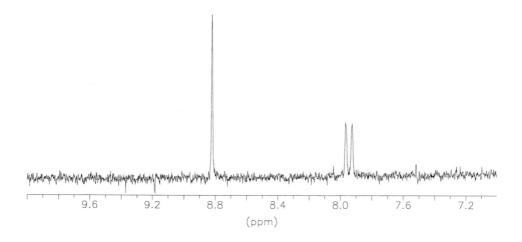

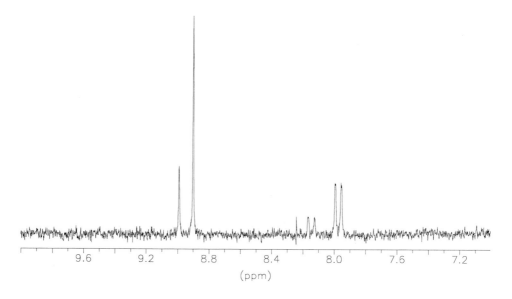

Figure 5-3: ^{1}H-NMR spectra of fleroxacin before (*top*) and after addition of AlCl$_3$ (*bottom*) at 37°C, "SSL".

References

1. H Stamm, *Arch Pharm* (Weinheim) **302**, 174-184 (1969)
2. A L Thakkar, L G Tensmeyer, R B Hermann, W L Wilham, *Chem Commun*, 524-525 (1970)
3. R Matusch, Habilitationsschrift, Marburg 1977
4. I Horman, B Dreux, *Helv Chim Acta* **67**, 754-764 (1984)
5. H Stamm, F Timeus, *Helv Chim Acta* **67**, 2161-2163 (1984)
6. A L Thakkar, L G Tensmeyer, W L Williams, *J Pharm Sci* **60**, 1267-1269 (1971)
7. D Guttman, T Higuchi, *J Pharm Sci* **60**, 1267-1270 (1971)
8. J Nishijo, I Yonetani, K Tagahara, Y Suzuta, E Iwamoto, *Chem Pharm Bull* **34**, 4451-4456 (1986)
9. Y Ebisuzaki, P D Boyle, J A Smith, *Acta Cryst* **C53**, 777-779 (1997)
10. F Sörgel, M Kinzig, *Am J Med* **94**, 56-69 (1994)
11. G Mendoza-Diaz, K H Pannell, *Inorg Chim Acta* **152**, 77-79 (1988)
12. J Shimada, K Shiba, T Oguma, H Miwa, Y Yoshimura, T Nishikawa, Y Okabayashi, T Kitagawa, S Yamamoto, *Antimicrob Agents Chemother* **36**, 1219-1224 (1992)
13. C Riley, D L Ross, D van der Velde, F Takusagawa, *J Pharm Biomed Anal* **11**, 49-59 (1993)
14. S Lecomte, M H Baron, M -T Chenon, C Coupry, N J Moreau, *Antimicrob Agents Chemother* **38**, 2810-2816 (1994)
15. S Lecomte, M -T Chenon, *Int J Pharm* **139**, 105-112 (1996)
16. M Sakai, A Hara, S Anjo, M Nakamura, *J Pharm Biomed Anal* **18**, 1057-1067 (1999)

Chapter 6

B.W.K. Diehl and U. Holzgrabe

6 Determination of the Isomeric Composition of Drugs

Stereochemistry is essential for biochemistry, so that it is preferable to use chiral drugs. Diastereomeres have different physically properties, which cause different NMR spectra.

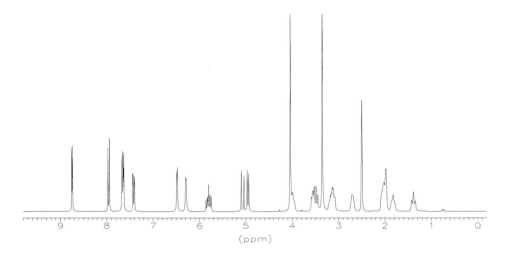

Figure 6-1: ^{1}H NMR spectrum of quinine HCl, 300 MHz, solvent DMSO-d_6, "SSL".

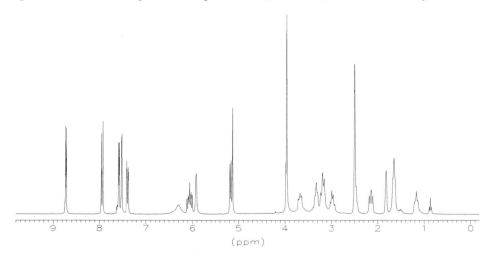

Figure 6-2: ^{1}H NMR spectrum of quinidine H$_2$SO$_4$, 300 MHz, DMSO-d_6, "SSL".

The comparison of the ¹H NMR (Figure 6-1 and Figure 6-2) and ¹³C NMR (Figure 6-3 and Figure 6-4) of quinine and quinidine is a good example, and is typical of many others.

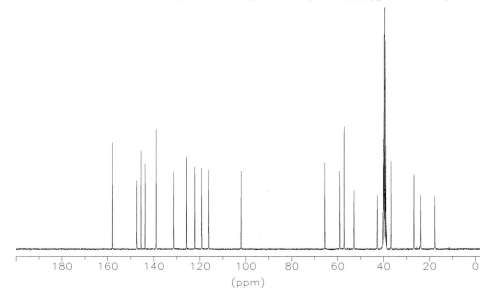

Figure 6-3: ^{13}C NMR spectrum of quinine HCl, 75 MHz, solvent DMSO-d_6, "SSL".

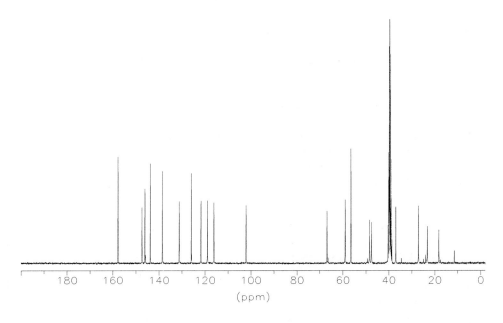

Figure 6-4: ^{13}C NMR spectrum of quinidine H_2SO_4, 75 MHz, solvent DMSO-d_6, "SSL".

Even apparently simple non chiral molecules such as citric acid (Figure 6-5) or glycerol (Figure 6-6) show complex ¹H NMR spectra caused by the diastereotopy. On the one hand these effects make ¹H NMR spectra in particular more difficult to interpret, but on the other hand much, even of three-dimensional information, can be extracted.

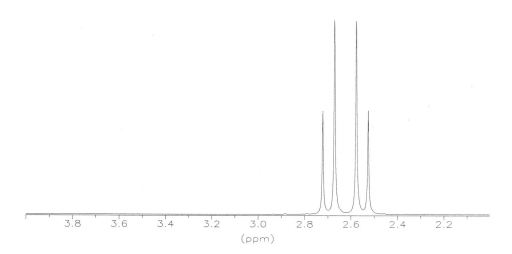

Figure 6-5:   ¹H NMR spectrum of citric acid, 300 MHz, solvent D_2O, "SSL".

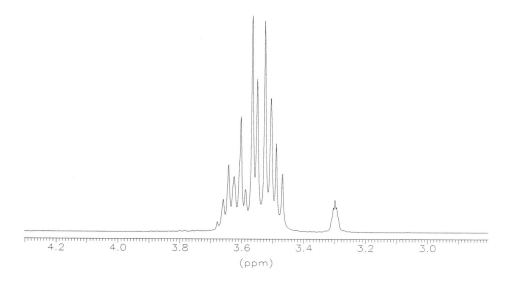

Figure 6-6:   ¹H NMR spectrum of glycerol, 300 MHz, solvent CD_3OD, "SSL".

6.1 Determination of the Enantiomeric Excess

Apart from chiral HPLC methods, ^{1}H NMR spectroscopy has often been used to determine the enantiomeric excess (ee[%] = [(R-S)/(R+S) x 100) of an asymmetric synthesis by derivatization with chiral (enantiomerically pure) reagents, e.g. Mosher's reagent, α-methoxy-α-trifluoromethyl-phenylacetic acid (MTPA) [1]. Recently, international authorities, e.g. the European Pharmacopoeia Commission, have encouraged for the introduction of NMR spectroscopy for chiral analysis.

Derivatization methods yielding diastereomeric substances strongly depend on the quantitative conversion of both enantiomers with the chiral reagent. In addition, it is necessary to prove that no isomerization occurs during the derivatization reaction, as the subsequent analysis may be falsified by this reaction [2]. Moreover, the additional step is time-consuming, which is undesirable in routine procedures. Chiral lanthanide shift reagents (CSR), such as tris[3-(trifluoromethylhydroxymethylene)-*d*-camphorato]-ytterbium [Yb(tfc)$_3$] or tris{[(heptafluoropropyl)-hydroxy-methylene]-*d*-camphorato}-europium [Eu(hfc)$_3$], also have a long-standing tradition in chiral discrimination, especially for low-field NMR instruments [3]. Because of the formation of diastereomeric complexes between the enantiopure shift reagent and the enantiomers analyzed, a double set of more or less shifted signals can be seen for the diastereomers.

The integrals of a corresponding set of hydrogens can be used to determine the optical purity of a sample. However, on high-field spectrometers the CSRs tend not to work well because of severe line-broadening.

In addition, impurities in the sample measured as well as in the solvent may create problems, e.g. the presence of water, which results in the formation of a precipitate of europium oxide [4], [5]. Therefore, most analyses under development use chiral solvating agents (CSA) for chiral discrimination, because they do not suffer from these disadvantages [6]. Again, chiral recognition is created by formation of diastereomeric complexes. Hence, CSA and solute should have common features of complementary functional groups, which are in general hydrogen-bond donors or acceptors. Thus, acids, amines, alcohols or cyclic compounds, e.g. cyclodextrins (CDs), crown ethers or peptides are suitable CSAs (see Figure 6-8) [7] .

Dawson et al. presented a study using different reagents, such as α-, β-, and γ-CDs, S-α-methylbenzylamine, (R)-*N*-3,5-dinitrobenzoyl-methylbenzylamine, (S)-1-(1-naphthyl)-ethylamine and (R)-2,2,2-trifluoro-1-(9-anthrylethanol) (R-TFAE), for enantiomeric evaluation of the dissolution of ketoprofen (Figure 6-7) formulations [8]. Samples of known amounts of drugs were dissolved in CDCl$_3$ containing the various CSAs as well as a known amount of internal standard (*o*-dimethoxybenzene) and ^{1}H NMR spectra recorded on a 400-MHz instrument. It was checked whether the CSAs produced well-resolved peaks for both compounds, whether the CSAs were stable and had no resonances close to the signals of interest in ketoprofen, and whether the signals of the enantiomers of ketoprofen were well separated at reasonably low concentrations of the CSAs. With the exception of the naphthylethylamine, all CSAs produced good separations of optical isomer resonances, and so can be used to determine the isomeric ratio upon dissolution and metabolism.

Racemic Compounds

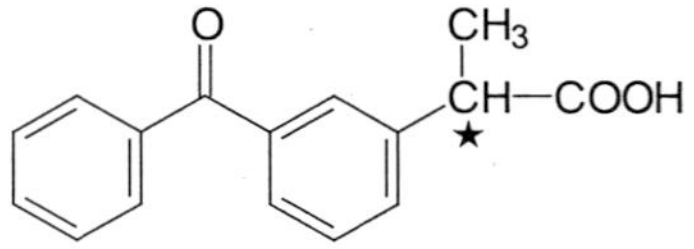

ketoprofen

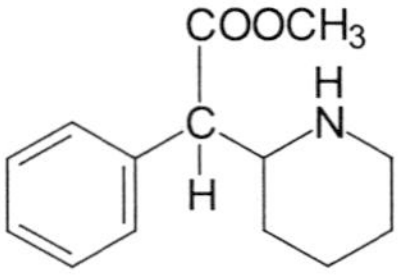

methylphenidate

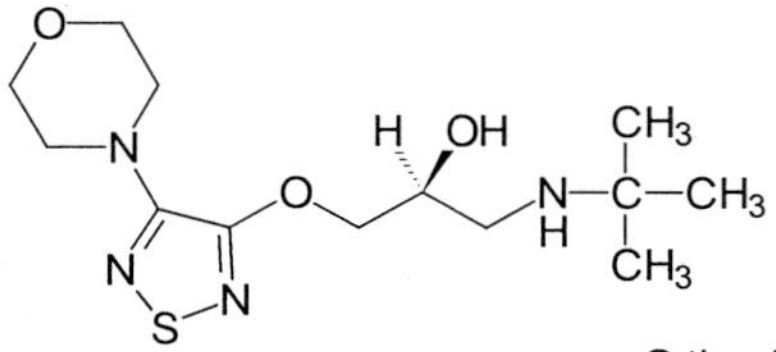

S-timolol

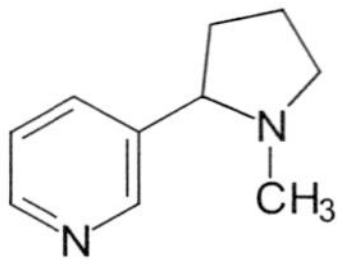

nicotine

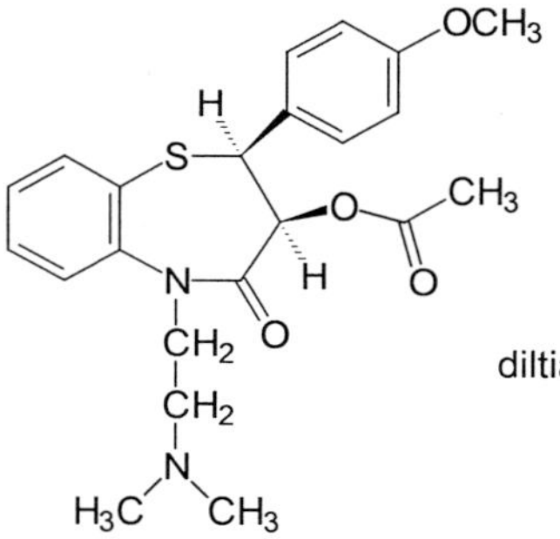

diltiazem

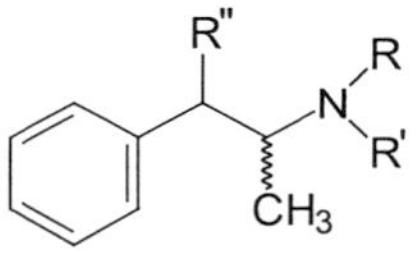

selegiline (R= CH$_3$, R'= CH$_2$−C≡CH, R"= H)
norephedrine (R= R'= H, R"= OH)
methamphetamine (R= CH$_3$, R'= H, R"= H)

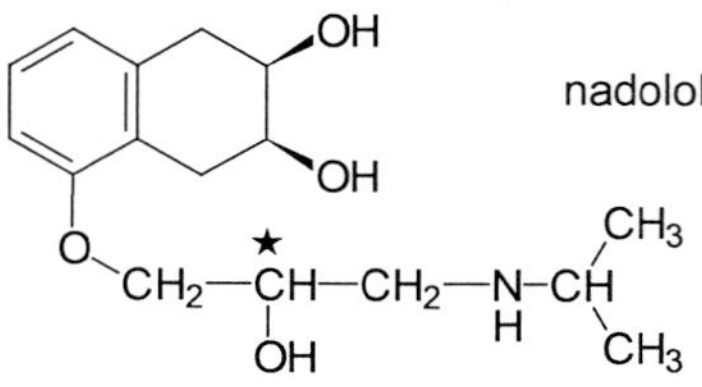

nadolol

Figure 6-7: Structural formula of the racemic compounds discussed.

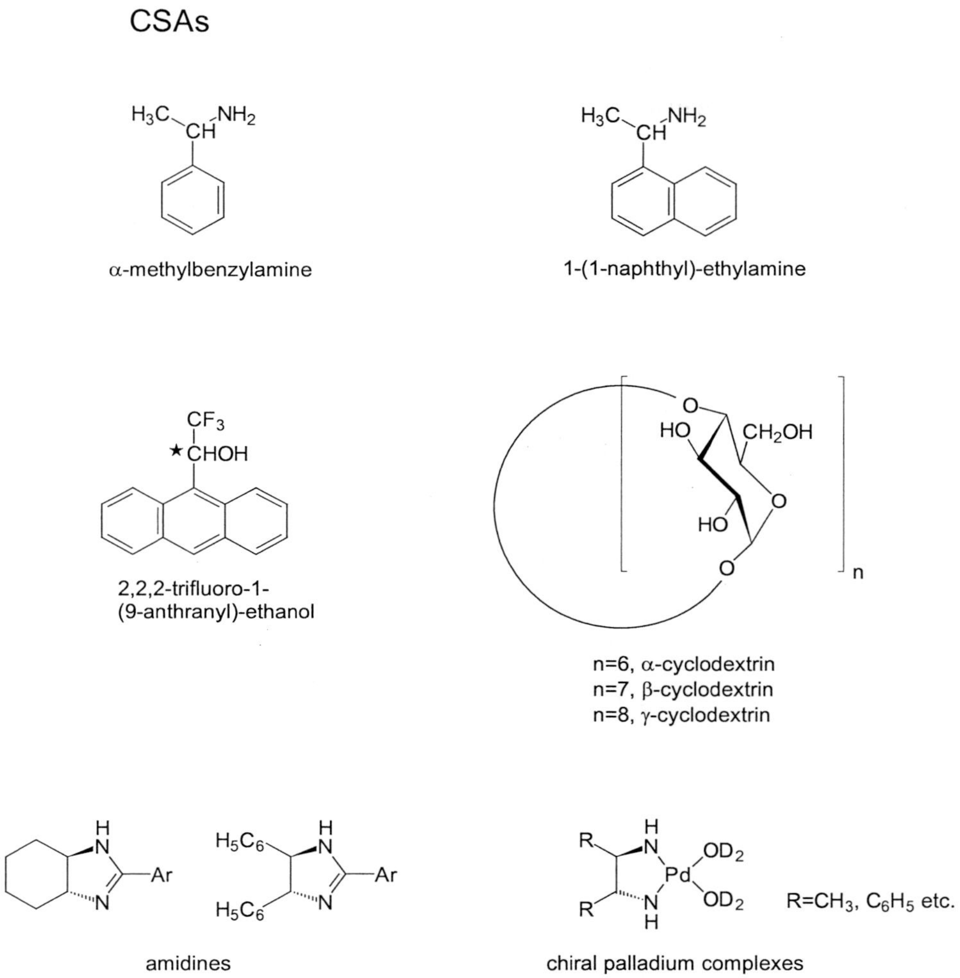

CSAs

Figure 6-8: Structural formulae of the chiral solvating agents discussed.

6.2 Comparison of Different CSAs

A comparison of different reacting CSAs can be demonstrated by the enantiomeric studies of mexiletine by 2,2,2-trifluoro,1(9-anthranyl)-ethanol (TFAE) (Figure 6-9) and tris[-3-(trifluoromethylhydroxymethylene)-D-campherato]-europium (III) (Figure 6-11 and Figure 6-13). In principle the enantiomerically pure CSAs build a diastereomeric complex with the target molecules. The diastereomeric splitting of NMR signals are caused by the magnetic interaction. The effect of europium CSAs, like that of other rare earth elements, is based on their special magnetic characteristics. At low magnetic fields, the europium CSAs were very useful, but, at modern high field strengths leading to line broadening, other CSAs with anisotropic effects are more useful.

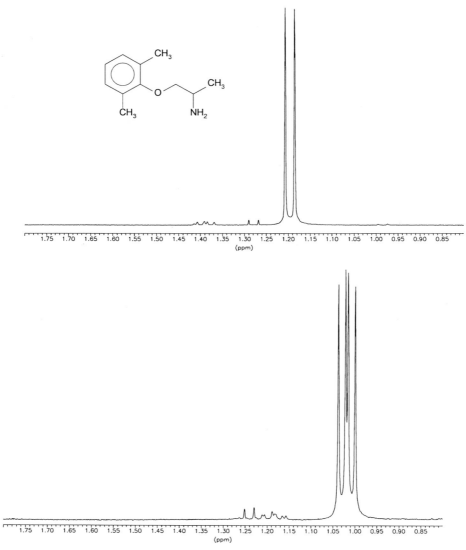

Figure 6-9: ¹H NMR spectrum of mexiletine, 300 MHz solvent CDCl₃; detail of the CH-**CH₃** group before (*top*) and after addition of TFAE (*bottom*), "SSL".

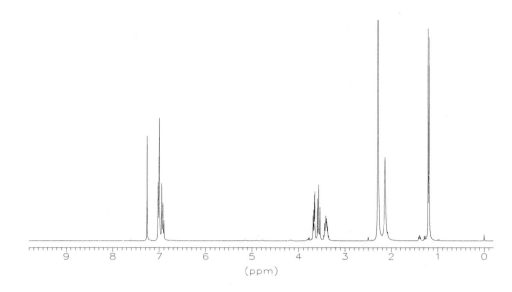

Figure 6-10: ¹H NMR spectrum of mexiletine, 300 MHz solvent CDCl$_3$, "SSL".

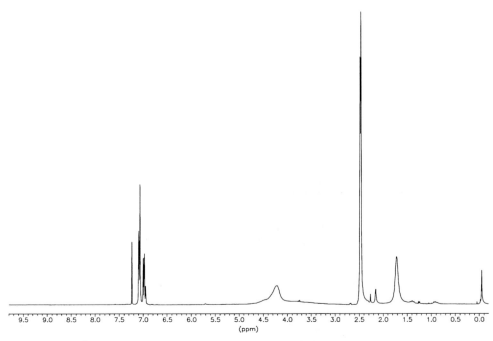

Figure 6-11: ¹H NMR spectrum of mexiletine and europium CASs, 300 MHz solvent CDCl$_3$, "SSL".

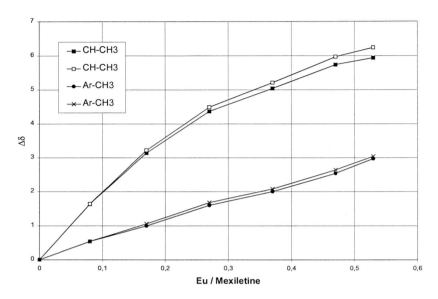

Figure 6-12: Changes of the chemical shifts of the methyl groups in mexiletine using increasing amounts of Eu(hfc)$_3$.

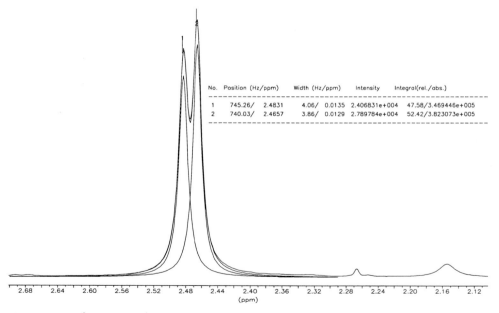

Figure 6-13: ^{1}H NMR spectrum of mexiletine and europium CASs, 300 MHz solvent CDCl$_3$; aromatically linked CH$_3$, deconvolution analysis, "SSL".

The selectivity of ³¹P NMR spectroscopy affords a very useful method of analyzing the different types of cyclic phosphamides [9] Figure 6-14 and their metabolites in urine. The pharmaco- kinetics also can be checked by ³¹P NMR spectroscopy [10]. Cyclic phosphamides (CP) are used as antitumor drug [11]. These are available as racemic mixtures of enantiomers. By the use of α-, ß- and γ-cyclodextrins as chiral shift reagents, the determination of the stereochemistry of the three most used drugs cyclophosphamide, trofosphamide and ifosphamide is possible by ³¹P NMR spectroscopy. In addition it is shown that cyclophosphamide in human urine is a racemic mixture [10].

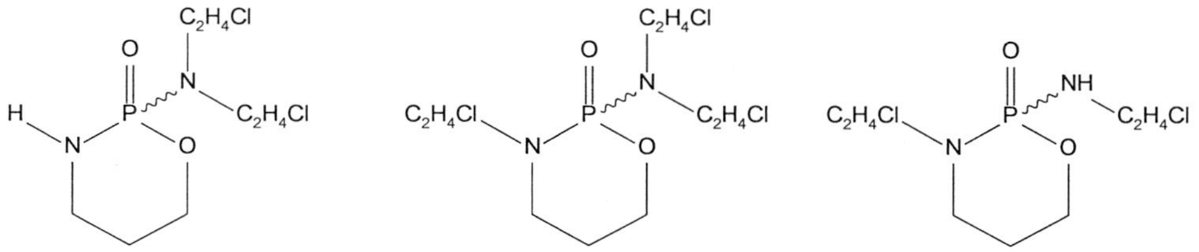

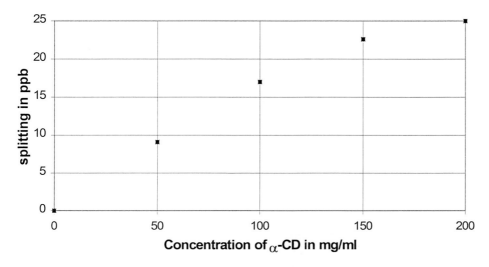

Figure 6-14: cyclophosphamide (*left*) trofosphamide (*middle*) and ifosphamide (*right*)

The resolution and splitting of the ³¹P NMR signals show an optimum at pH values between 9 and 11. The distance Δ between the signals of the two enantiomeric molecules strongly depends on the concentration of CD and more less on that of the cyclic phosphamide. Concentrated solutions of >100 mg/ml CD give baseline-separated signals.

Figure 6-15: Enantiomeric splitting of cyclophosphamide at different concentrations of α-CD.

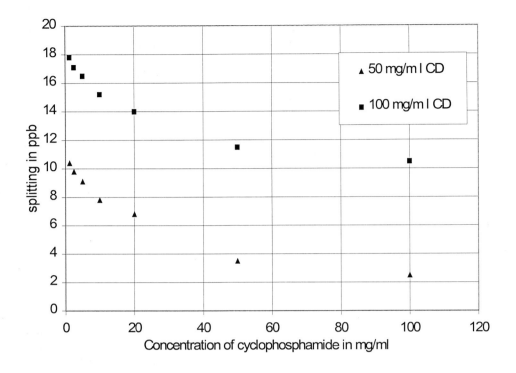

Figure 6-16: Dependence of enantiomeric splitting of cyclophosphoamide at different concentration and constant α-CD concentrations.

The interaction of the cyclic phosphamides with the three CDs is very individual, depending on the three-dimensional structure of the complex. For cyclophosphamid the splitting increases in the order α- to β- and γ-CD. The mixture of α- and β-CD (1:1) leads to a splitting of the theoretical mean value of the two single measurements. In principle, all combinations of cyclic phosphamides and CD give baseline separated signal splittings exept the combination of trofosphamide and β-CD.

Table 1: Enantiomeric splitting in ppb of different cyclophosphamides at RT, 5 mg CP and 100 mg CD-

	α-CD	β-CD	γ-CD	a/β-CD
Cyclophosphamide	12,1	62,4	93,2	36,2
Ifosphamide	64,8	35,6	15,1	-
Trophosphamide	92,6	-	294,2	-

trofosphamide and β- and γ-CD show poor solubility at room temperature at the given concentration. Therefore temperature-dependent NMR measurements were performed starting at the high temperature of 366 K and decreasing in steps of 10 K.

Table 2: Enantiomeric temperature-dependent splitting of trofosphamide.

Temperatore [K]	366	356	346	336	326	316	306	296
α-CD	30,1	35,0	42,2	51,0	61,5	71,4	82,3	94,1
γ-CD	-	102,8	130,9	169,5	197,8	237,9	264,8	294,7

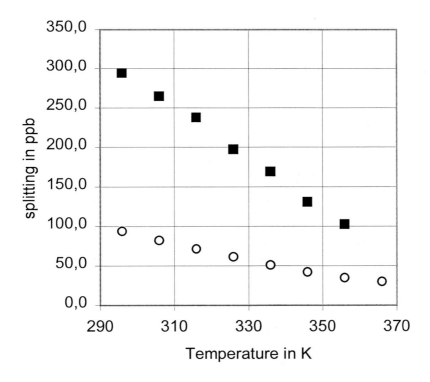

Figure 6-17: Temperature-dependent enantiomeric splitting of trofosphamide after addition of α-CD (■) and γ-CD (○).

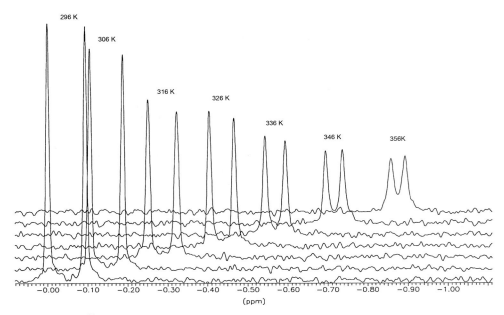

Figure 6-18: ^{31}P NMR spectra of trofosphamide, 121 MHz, "SSL".

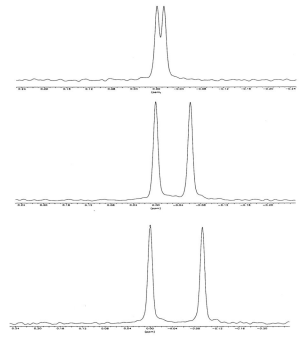

Figure 6-19: ^{31}P NMR spectra of cyclophosphamide and α- (*top*), β- (*middle*) and γ-CD (*bottom*), 121 MHz, "SSL".

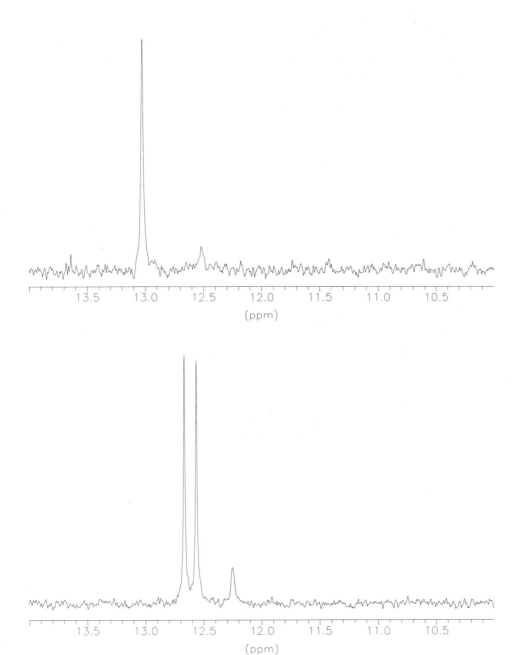

Figure 6-20: ^{31}P NMR spectrum of human urine pure (*top*) and after addition of γ-CD (*bottom*), 121 MHz, "SSL".

Apart from cyclodextrins, chemically modified, e.g. acetylated, cyclodextrins (ACDs) were checked for their utilizability to determine the enantiomeric purity of some representative phenetylamines. The results were compared with electrophoresis experiments [5]. The ^{1}H NMR spectrum of racemic adrenaline and ACD are shown in Figure 6-21. For quantification, deconvolution is often a useful data processing method.

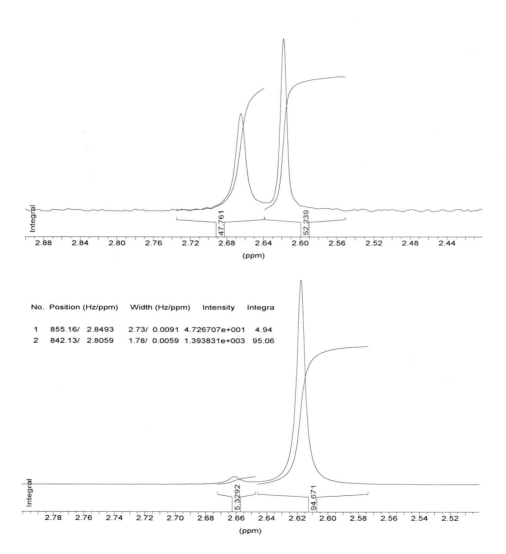

Figure 6-21: ^{1}H NMR spectra of ACD and adrenaline racemate, integration (*top*) and adrenaline 90% EE, integration and deconvolution (*bottom*), 300 MHz, "SSL".

Dauwe and Buddrus [12] tested a series of chiral amines for their ability to discriminate between the enantiomers of several acidic compounds. Strong basic amidines, as shown in Figure 6-8, turned out to be suitable for the resolution of even weak acidic compounds such as phenols, barbiturates and alcohols. The same group [13] developed chiral palladium complexes with diamines for discrimination of α-amino acids and determination of enantiomeric excess resulting from asymmetric synthesis.

Methylphenidate, an antipsychotic drug, has two centers of chirality (Figure 6-7): A comparative study of the central and peripheral stimulant effects of the four isomers has indicated that the *threo* isomers are more active than the *erythro* forms, and that the (2R,2′R)-*threo* enantiomer is more active than the corresponding antipode [14]. Thus, it is pivotal to know the isomeric composition of the drug in clinical use. Hanna and Lau-Cam first used a chiral Eu(III) shift reagent for the determination of the enantiomeric excess [15]. Although the method suffered from line-broadening, they were able to determine 98.8–100.5% of the added enantiomer (99.5 ± 0.7% of added). (R)-TFAE was found to be superior to the CSR in a 90-MHz spectrum when using a TFAE/substrate molar ratio of 4.8 [16]. Looking at the methyl resonances of methylphenidate the mean ± SD recovery value for the (2S,2′S)-*threo*-isomer amounted to 99.9 ± 0.6% of added, which is in good agreement with the CSR study. However, the TFAE approach is simpler, because it does not require reagents and solvents of high purity and anaerobic working conditions. In addition, with (R)-TFAE it was possible to determine the absolute configuration of the enantiomers.

Lacroix et al. compared an HPLC with a ^{1}H NMR method for quantification of the R-enantiomer in S-timolol maleate, a β-blocking substance (Figure 6-7) [17]. For satisfactory resolution, the HPLC required the expensive cellulose tris-3,5-dimethylphenylcarbamate column (Chiracel OD-H), a mobile phase of 0.2% dimethylamine and 4% isopropanol in hexane and ^{1}H NMR (400-MHz) R-TFAE as a CSA in CDCl$_3$. The resonance of the *tert*-butyl group between 1.0 and 1.1 ppm was used to measure the ee. With both methods, 0.2– 4.0% of the R-enantiomer in S-timolol could be determined precisely. Additionally, both methods (HPLC and NMR) were found to be superior to specific rotation, which is used in the USP XXIII for determination of optical purity. This superiority of NMR and HPLC methods over optical rotation has often been reported [2].

Diltiazem hydrochloride, a coronary vasodilatator, contains two centres of chirality in the benzothiazepine skeleton, which results in four isomers (Figure 6-7). Only the 2,3-*cis*-(+)-isomer is pharmacologically active. From a 400-MHz ^{1}H NMR spectrum of raw material in CDCl$_3$, minimum amounts of the *trans*-diastereomer could easily be detected over the range 0.5– 5.0% [18]. After addition of 10 mg (R)-TFAE to 2 mg (*cis*)-diltiazem in 500μl CDCl$_3$, the amount of the antipode could be determined using the integration of the acetyl signal. The enantiomeric impurity could be quantified down to levels of about 0.2%. The limits for the method could easily be bottomed by increasing the number of scans (here 128 scans). It is noteworthy that the method is also superior to the specific rotation applied in the USP, which allows about 2.6 % of the antipode to be present (+110– +116°).

Nadolol, another β-blocker used in the treatment of angina pectoris and hypertension, consists of two additional centres of chirality in the 1,2,3,4-tetrahydro-naphthalenediol applied in the USP, which allows about 2.6 % of the antipode to be present (+110–+116°).

Nadolol, another β-blocker used in the treatment of angina pectoris and hypertension, consists of two additional centres of chirality in the 1,2,3,4-tetrahydro-naphthalenediol moiety besides the propanol chain (Figure 6-7). Since the ring hydroxyl groups are in the *cis* position, four isomers, two racemic diastereomers, are possible. The racemic ratio was determined after derivatization with benzoic anhydride/dimethylamino pyridine to give a tribenzoate derivative [19]. Using the relative height of the signals of the *tert*-butyl groups, the racemic composition was found not to meet the USP limits. Thus, a further determination of enantiomeric composition made no sense. However, the use of CSAs, such as (R)-1,1´-bis-2-naphthol, in order to separate *tert*-butyl signals of the four isomers failed.

The S-isomer of nicotine (Figure 6-7), originally extracted from the *Nicotiana* species, is attracting pharmacological interest for its potential to aid smokers to cope with the cigarette abstinence syndrome. The R-enantiomer is also of pharmacological interest. Thus, it is important to have a method to determine the enantiomeric excess in inhalers, nasal sprays, skin absorption patches and chewing gums. Since CSR, such as Eu(tfc)$_3$ and Yt(tfc)$_3$, cause a strong line-broadening of the ^{1}H NMR signal [20],[21], ^{13}C NMR spectroscopy in conjunction with Yt(tfc)$_3$ was applied in this case [21]. Upon complexation with the chiral lanthanide reagent, signals of all aliphatic carbon atoms were resolved. The signal of the *ipso*-carbon atom C-2´ turned out to be the best for routine analysis. Using INEPT spectra with pulse delays of 0.9 s, a >100-fold excess of one of the enantiomers could be determined. For very high ees of one enantiomer an amount of 1–3 mg nicotine and an overnight acquisition was necessary. Larger amounts found in pharmaceutical formulations would be analyzed within 1–2 h.

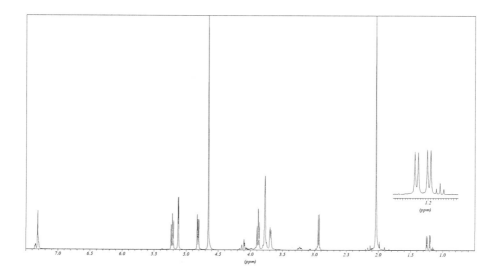

H-1 antihistamines [26], gliclazide [27], or thromboxane antagonists [28] could be well resolved. Derivatized CDs often exhibit improved chiral recognition properties [29]. *Heptakis*(2,3-di-*O*-acetyl)-β-CD is able to resolve a series of pharmacologically active phenethylamines with their structures systematically varied [30],[31],[32]. Similar results have been reported for hydroxypropyl- as well as for permethylated β-CD [33] and perphenylcarbamated β-CD to resolve atenolol [34].

The advantages of CDs as CSA are water solubility, no signal-broadening effects, and the narrow chemical-shift range of the CDs. Thus, they are suitable reagents for quantification of the chiral purity of drugs. The enantiomeric excess of the psychostimulants amphetamine and norephedrine and the antiparkinson drug selegiline (Figure 6-7) could easily be determined using *heptakis*(2,3-di-*O*-acetyl)-β-CD. For example, many signals of the racemate selegiline are resolved in the presence of the CD dissolved in aqueous buffer, pH 4.5, see Figure 6-22. The signals of the C-methyl group were used to determine the enantiomeric impurities (see Figure 6-23) [35].

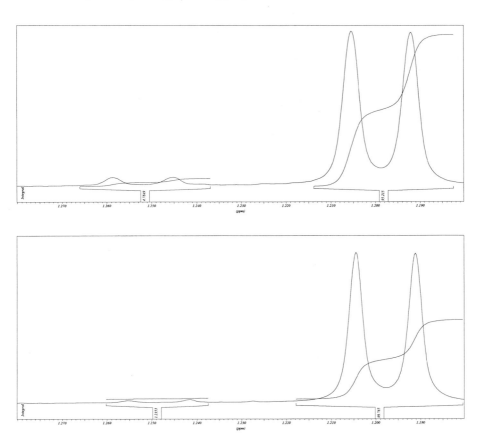

Figure 6-23: ^{1}H NMR expansions of the C-methyl region of 95 % (*top*) and 98.75% (*bottom*) R-selegiline in presence of *heptakis*(2,3-di-*O*-acetyl)-β-CD.

Taken together with regard to the determination of the isomeric composition of drugs, NMR has several advantages over HPLC:

1. The development and validation of an NMR method usually takes a few weeks, while an HPLC methodology may take several months.
2. NMR methods were found to be more robust than HPLC methods, which are subject to incalculabilities of column reproducibility and mobile phase composition; a suitability test has to be performed to eliminate the vagaries of HPLC.
3. After development, an NMR analysis can be completed within an hour, whereas an HPLC analysis may take a day or more, unless the instrument is already prepared for the job. In routine analysis both methods can be automated. Hence, the analysis time will be the same.
4. An NMR analysis will usually take less technician time than an HPLC analysis. Thus, the cost of running the NMR analysis is usually the lower of the two.

References

1. J A Dale, D L Dull, H S Mosher, *J Org Chem* **34**, 2543-2649 (1969)
2. D Parker, *Chem Rev* **91**, 1441-1457 (1991)
3. H Y Aboul-Enein, *Anal Lett* **21**, 2155-2163 (1988)
4. L M Sweeting, D C Crans, G M Whiteside, *J Org Chem* **52**, 2273-2276 (1987)
5. P E Peterson, M Stepanian, *J Org Chem* **53**, 1907-1911 (1988)
6. G R Weisman, in Asymmetric Synthesis; J D Morrison, Academic Press, New York, 1983, Vol 1, Chap 8, p 153-171
7. W H Pirkle, D J Hoover, *Top Stereochem* **13**, 264-331 (1982)
8. B A Dawson, S A, Qureshi, D B Black, 33rd Experimental Nuclear Magnetic Resonance Conference, Pacific Grove, CA, (1992)
9. D Busse, F W Busch, F Bohnenstengel, M Eichelbaum, P Fischer, J Opalinska, K Schumacher, E Schweizer and H K Kroemer, *J Clinic Oncol* **15**, 1885-1896, (1997)
10. Spectral Service, not published previously
11. G Hempel, S Krümpelmann, A May-Manke, B Hohenlöchter, G Blaschke, H Jürgens and J. Boos, *Cancer Chemother Pharmacol*, 45-50 (1997)
12. C Dauwe, J Buddrus, *GIT Fachz Lab* **5/94**, 517-518
13. B Staubach, J Buddrus, *Angew Chem* **108**, 1443-1445 (1996)
14. Arzneistoff-Profile, Eds V Dinnendahl, U Fricke, Govi-Verlag, Frankfurt a M (1982 - 1996)
15. G M Hanna, C A Lau-Cam, *Pharm Res* **7**, 726-729 (1990)
16. G M Hanna, C A Lau-Cam, *J Pharm Biomed Anal* **11**, 665-670 (1993)
17. P M Lacroix, B A Dawson, R W Sears, D B Black, *Chirality* **6**, 484-491 (1994)
18. B A Dawson, D B Black, 37th Experimental Nuclear Magnetic Resonance Conference, Pacific Grove, CA, (1996)
19. B A Dawson, D B Black, *J Pharm Biomed Anal* **13**, 39-44 (1995)
20. C G Chavdarian, E B Sanders, R L Bassfield, *J Org Chem* **47**, 1069-1073 (1982)
21. J W Jaroszewski, A Olsson, *J Pharm Biomed Anal* **12**, 295-299 (1994)

22. A F Casy, *TRAC* **12**, 185-189 (1993)
23. C A Marchant, S K Branch, *J Pharm Pharmacol* **42**, P68 (1990)
24. T T Ndou, S Mukundan Jr, I M Warner, *J Incl Phenom* **15**, 9-15 (1993)
25. M Wiese, H -P Cordes, H Chi, J K Seydel, T Backensfeld, B W Müller, *J Pharm Sci* **80**, 153-156 (1991)
26. A F Casy, A F Drake, C R Ganellin, A D Mercer, C Upton, *Chirality* **4**, 356-366 (1992)
27. J R Moyano, M J Arias-Blanco, J M Gines, A M Rabasco, J I Perez-Martinez, M Mor, F Giordano, *J Pharm Sci* **86**, 72-75 (1997)
28. A F Casy, A D Cooper, T M Jefferies, R M Gaskell, D Greatbanks, R Pickford, *J Pharm Biomed Anal* **9**, 787-792 (1991)
29. C J Easton, S F Lincoln, *Chem Soc Rev* 163-170 (1996)
30. S K Branch, U Holzgrabe, T M Jefferies, H Mallwitz, M Matchett, *J Pharm Biomed Anal* **12**, 1507-1517 (1994)
31. S K Branch, U Holzgrabe, T M Jefferies, H Mallwitz, F J R Oxley, *J Chromatogr A* **758**, 277-292 (1997)
32. U Holzgrabe, H Mallwitz, S K Branch, T M Jefferies, M Wiese , *Chirality*, in press
33. A Taylor, D A R Williams, I D Wilson, *J Pharm Biomed Anal* **9**, 493-496 (1991)
34. Y Kuroda, Y Suzuki, J He, T Kawabata, A Shibukawa, H Wada, H Fujima, Y Go-oh, E Imai, T Nakagawa, *J Chem Soc Perkin Trans II*, 1749-1759 (1995)
35. U Holzgrabe, M Thunhorst, *Magn Res Chem,* submitted

Chapter 7

K. Albert

7 On-line Coupling of HPLC or SFC with NMR Spectroscopy

7.1 Introduction

^{1}H NMR monitoring of natural products, as well as drugs and their corresponding metabolites in body fluids (see Chapter 8), suffers from signal overlap due to the complexity of the fluids studied. Therefore, extraction and separation techniques are required for the isolation of pure compounds. Structure determination is then performed by the use of spectroscopic techniques such as mass spectrometry, infrared and NMR spectroscopy. This time-consuming procedure is still employed in many research laboratories, but the need for efficient and rapid screening techniques in modern drug research favors the hyphenation of extraction and separation techniques together with a structure-related method. In the case of volatile organic compounds, the combination of gas chromatography together with mass spectrometry has proven to be of extreme value, for instance, for the investigation of complex hydrocarbon mixtures. Nonvolatile compound mixtures are effectively separated by reversed-phase high performance liquid chromatography (HPLC) employing an ultraviolet (UV) or refractive index (RI) detector. These modes of detection do not allow one to derive detailed structural information for unknown peaks, therefore the on-line combination with mass and NMR spectrometers should be the method of choice. Recently, the coupling of HPLC together with electrospray mass spectrometry (MS) has proven to be one of the most informative LC-MS methods [1], [2]. However, not all structure-related questions can be answered by the use of MS techniques: the assignment of stereochemical configuration and conformation, e.g., for stereoisomers, can only be performed by the application of NMR techniques. Especially in the case of oxidation-sensitive natural compounds, the hyphenation of HPLC together with NMR spectroscopy in a closed-loop separation-identification system should enable a rapid, effective and unambiguous assignment of unknown peaks.

NMR spectra of liquids are routinely acquired in 5-mm NMR glass tubes inserted into the coaxial radiofrequency coil of an NMR probe in the room-temperature bore of a cryomagnet (Figure 7-1).

To eliminate magnetic field inhomogeneties, the NMR tube is rotated at a speed of 20 Hz. This design cannot be used for the continuous registration of flowing chromatographic peaks. A well-proven continuous-flow NMR probe consists of a U-type glass tube, to which the r. f. coil is directly attached (Figure 7-1)[3]. The r.f. coil length with the glass internal diameter within the r. f. coil together define the NMR flow-cell detection volume, amounting to between 60 and 120 µl for current routine HPLC-NMR applications.

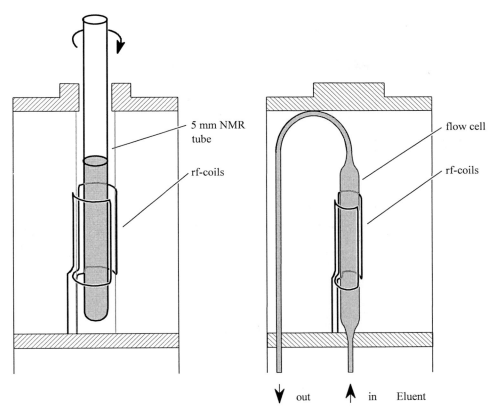

Figure 7-1: Design of NMR probes for conventional and continuous-flow detection.

Due to the fixing of the r.f. coil directly to the NMR glass tube, rotation is not possible. Thus, test spectra show a broader signal half width, but routine continuous-flow spectra are similar to conventional NMR spectra. This high detection volume is necessary to meet the current sensitivity requirements of routine NMR spectroscopy. Whereas the resolution of the chromatographic separation is degraded for early eluting peaks, the successful use of these continuous-flow probes in many on-line HPLC-NMR applications clearly proves its high practical value for drug analysis.

On-line HPLC-NMR separations can be conducted either in the continuous-flow or stopped-flow mode. In the continuous-flow mode, the chromatographic effluent resides for a distinct time within the flow-cell volume defined by the r.f. coil. This residence time τ is defined by the ratio between flow-cell volume and flow rate. It should be in the range of 3 to 5 s [3]. Otherwise severe distortions of the NMR signal line widths could occur. In the stopped-flow mode the peak maximum is transferred to the NMR flow cell and 2D NMR spectra can be recorded without any time restrictions.

The experimental arrangement for the on-line coupling of a chromatographic separation system with an NMR cryomagnet is outlined in Figure 7-2.

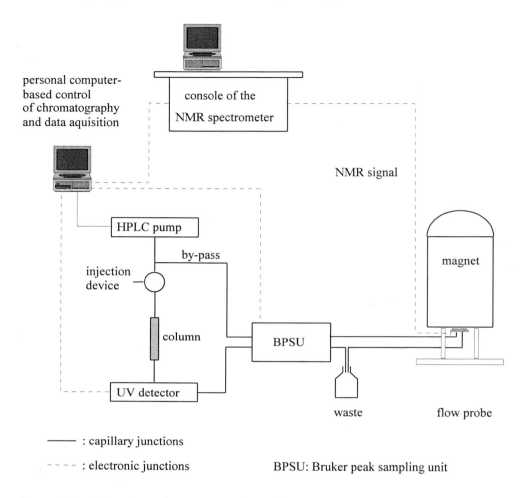

Figure 7-2:   Experimental arrangement for on-line HPLC-NMR coupling.

When a shielded cryomagnet is used, the chromatographic separation device can be located at a distance of 30 cm from the magnet. For unshielded cryomagnets, a distance of 2 m is advisable. Connection between the chromatographic separation system and the NMR flow probe is by capillary tubing (I. D. 0.25 mm). For the continuous NMR monitoring of HPLC separations, flow-rates between 0.3 and 0.5 ml/min are usually employed, resulting in sufficient residence times τ. For stopped-flow acquisitions, the insertion of the peak maximum into the NMR flow cell is triggered by the UV detector of the HPLC instrument. A switching valve positioned between the UV detector and the continuous-flow probe guarantees the capture of the peak maximum in the flow cell. Automated LC-NMR systems use peak sampling units wherein the separated peaks are stored in loops and sequentially measured in the NMR flow cell. To obtain high quality NMR spectra with a tolerable S/N value, the chromatographic column must be overloaded, yet still provides adequate chromatographic resolution.

7.2 HPLC-NMR Coupling

The main problem in recording NMR spectra of HPLC reversed-phase effluents results from the strong proton signals of acetonitrile, methanol and water. Water can be substituted by D_2O, but the price of CD_3CN is too high, at least for academic institutions. For the proper adjustment of the receiver gain of the NMR instrument, the solvent signals are suppressed by employing a presaturation or non-excitation method. In many practical experiments, the application of a NOESY-type presaturation sequence has proven to be advantageous [4], [5].

7.2.1 Continuous-flow Measurements

Provided the available amount of substances is in the upper µg range, on-line HPLC-NMR separation is a feasible method for structure identification. Figure 7-3 shows the ^{1}H NMR spectrum of a mixture of five isomers of vitamin A acetate.

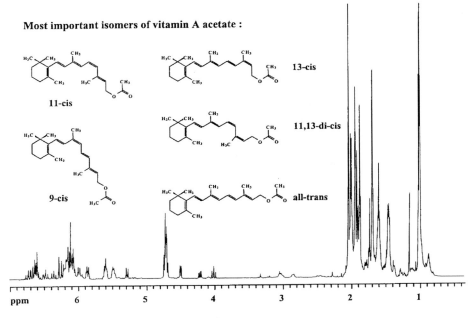

Figure 7-3: ^{1}H NMR spectrum (400 MHz) of a mixture of vitamin A acetate isomers.

Without any knowledge of chemical shifts and coupling constants for these isomers, full assignment of the ^{1}H NMR spectrum would be impossible. Employing a cyanopropyl-modified silica gel, an on-line HPLC-NMR separation was performed with two 250 x 4.6 mm HPLC columns using hexane as eluent at a flow rate of 0.3 ml/min [6]. Eight transients were coadded, defining a time resolution of 8 s.
The Fourier-transformed spectrum results in a "row" in the two-dimensional contour plot of the separation (Figure 7-4).

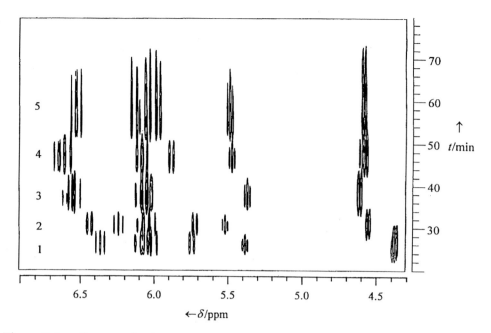

Figure 7-4: Contour plot (400 MHz) of the separation of vitamin A acetate isomers.

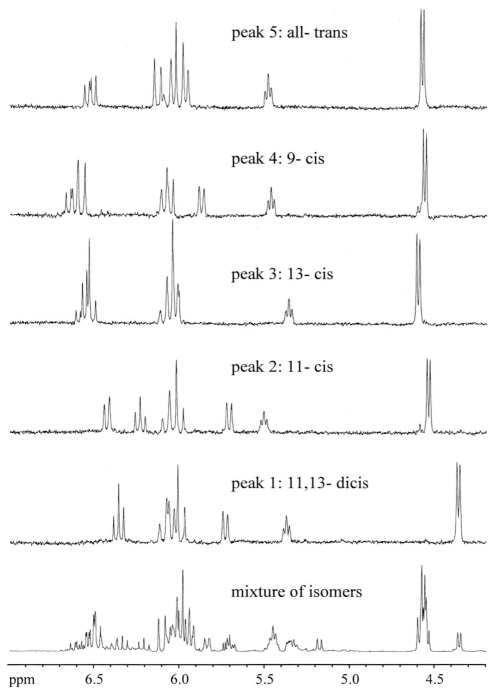

Figure 7-5: Continuous-flow ^{1}H NMR spectra (400 MHz) of vitamin A acetate isomers.

Here, the proton chemical shifts of 128 rows are displayed versus the chromatographic retention times. Similarly to the display of a topographical map, the peak summits are shown from a bird's eye view. An alternative display is a "stacked plot" of 128 ^{1}H NMR spectra versus retention time. Figure 7-5 shows the continuous-flow ^{1}H NMR spectra of the separated vitamin A acetate isomers.

The spectra were recorded at the corresponding peak maxima. Both display modes demonstrate the power of continuous-flow HPLC-NMR coupling. Whereas peaks 1 and 2 (11,13-di-*cis* and 11-*cis* vitamin A acetate) are not fully separated, full discrimination of both compounds is possible in the two-dimensional contour plot (Figure 7-4). On the other hand, the quality of the recorded continuous-flow ^{1}H NMR spectra (Fig. 5) easily allows the determination of characteristic chemical shift values, coupling constants and integration ratios.

7.2.2   Stopped-flow Measurements

NMR parameters derived from continuous-flow experiments consist of chemical shifts, coupling constants and integration ratios. These parameters provide valuable information for the classification of unknown peaks in rapid through put screening applications, but a full structure elucidation, e.g., of an unknown drug metabolite, can only be performed with data obtained from two-dimensional NMR spectra together with the molecular data derived from mass spectra. The registration of two-dimensional NMR spectra is only possible in the stopped-flow mode. Here, either the peak maximum is transferred to the continuous-flow probe and the separation is stopped, or all interesting peaks are stored in small capillaries (loops) and 2D NMR registration of these fractions is performed automatically.

Structural isomers of essential natural compounds, such as carotenoids and hop bitter acids, show different biological activities. For structure-activity correlation studies in modern medical treatment it is of extreme relevance to know the stereochemical structure of the active ingredient. The classical method of isolation, sample enrichment, purification and transfer to an NMR sample tube frequently fails because the isolated compound isomerizes and decomposes.

Two essential carotenoids, lutein and zeaxanthin, play an important role in the visual process. For the treatment of age-related macular degeneration (AMD), knowledge of the isomeric composition within the macula is of particular interest [7]. Full assignment of isomeric configuration is possible by the registration of two-dimensional proton-proton correlated NMR spectra. Figure 7-7 shows, as an example, the COSY stopped-flow NMR spectrum of all-*trans* zeaxanthin isolated from ox retina.

Full stereochemical assignment is possible via the cross peaks in the spectrum. As indicated, the correlation between H-11 and the neighboring protons H-10 and H-12 is evident, while H-15 couples with H-14.

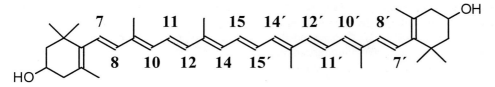

Figure 7-6: Chemical structure of all-*trans* zeaxanthin.

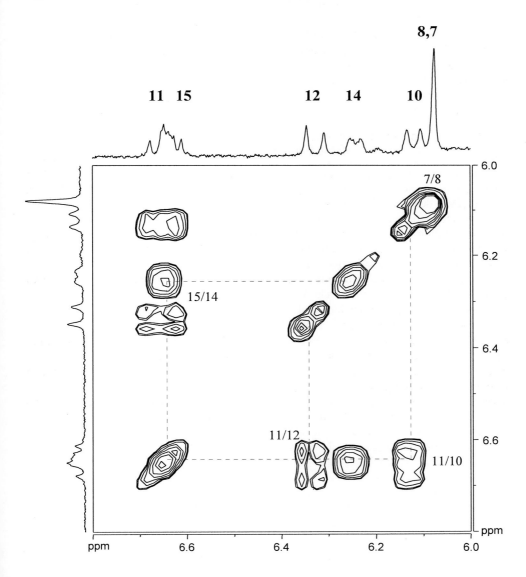

Figure 7-7: COSY stopped-flow spectrum (600 MHz) of all-*trans* zeaxanthin.

Apart from metabolism studies, which will be exclusively discussed in Chapter 8 the linking of HPLC to NMR spectroscopy provides a powerful tool for the identification of impurities in drugs. Using the stopped-flow technique, Roberts and Smith [8] were able to find minor components in a mixture of chromane compounds at a 3% level. Peng et. al. [9] Applied the LC-NMR to identify the degradation products of the protease inhibitor N-hydroxy-1,9-di-(4-ethoxybenzenesulfonyl)-5,5-dimethyl-(1,3)cyclohexyl-diazine-2-carboxamide in a dosage formulation.

Even though the sensitivity of HPLC NMR hyphenation is about 10 µg on-column with the on-flow mode and 1 µg with the stopped-flow mode [3], [10] the linking is shown to be rather effective. As soon as it is possible to routinely detect components in mixtures in the nanomolar range [11], NMR detection will surpass the MS technique because NMR detection automatically provides detailed structural information according to the direct and well-understood relationship between molecular structure and NMR chemical shifts, coupling constants and relaxation times.

7.3 SPE-HPLC-NMR Coupling

Despite the continuous improvement in NMR probe development with inherent gain in sensitivity, concentration levels of drugs in body fluids and in tissues are far below the current NMR detection limit in the nanogram range. Thus, careful sample enrichment has to be performed prior to the HPLC-NMR analysis. Solid phase extraction (SPE) employing a short HPLC column (12.5 x 4.6 mm) is one means of achieving this goal [12]. In on-line SPE-HPLC-NMR experiments, the sample loop of the injection device is substituted by a sample enrichment cartridge. After accumulating the desired compound on the column under a definite mobile phase composition, the enriched sample is desorbed from the column directly to the chromatographic column by changing the mobile phase composition. With this procedure, light and oxygen exposure are excluded and further HPLC-NMR investigation even in the continuous-flow mode is possible. Figure 7-8 shows the contour plot of the HPLC-NMR separation of napopxren, fenoprofen and ibuprofen, performed after sample enrichment of 10 ml of a solution containing 5 µg/ml of each drug with a C_{18} column. Structural elucidation of the compounds is easily derived from the different substitution patterns in the aromatic region. This is demonstrated with the ^{1}H NMR spectrum of the last eluting compound, ibuprofen, as shown in Figure 7-9. This example demonstrates that analyte detectability can be significantly increased by combining on-line trace enrichment prior to HPLC-NMR analysis.

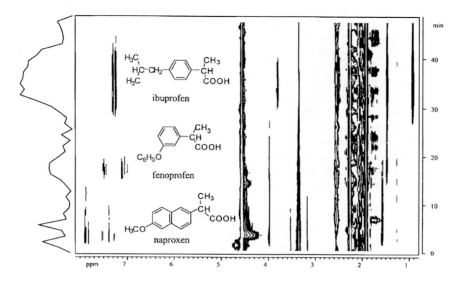

Figure 7-8: ¹H NMR chromatogram (contour plot, 400 MHz) of an SPE-HPLC separation of naproxen, fenoprofen and ibuprofen.

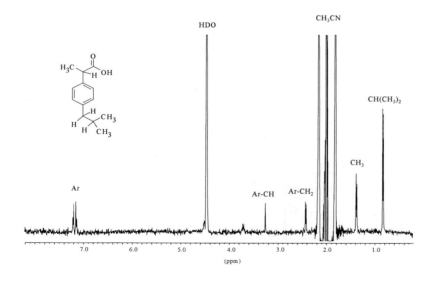

Figure 7-9: ¹H NMR spectrum (400 MHz) of ibuprofen extracted from the SPE-HPLC-NMR run depicted in Figure 7-8.

7.4 SFE-NMR and SFC-NMR Coupling

Another widely-used extraction technique is supercritical fluid extraction (SFE) with supercritical CO_2. The extraction capability of supercritical CO_2 is dependent upon the applied pressure. For pressures in the range of 40 bar, CO_2 has an extraction polarity similar to that of hexane. For pressures near 400 bar, the extraction behavior is similar to that of dichloromethane. Thus, by using a pressure gradient, effective extraction of lipophilic compounds can be performed. Figure 7-10 shows a feasible arrangement for on-line SFE-NMR coupling [13], [14].

Here, the extraction medium is loaded onto a conventional HPLC column. NMR detection is performed with a pressure-stable probe containing a sapphire tube instead of glass. As an example of supercritical fluid extraction, Figure 7-11 shows the ^{1}H NMR spectrum of piperine extracted from black pepper at a temperature of 370 K [13].

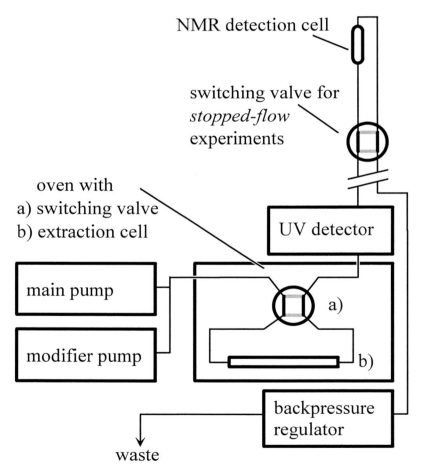

Figure 7-10: Experimental arrangement for on-line SFE-NMR coupling.

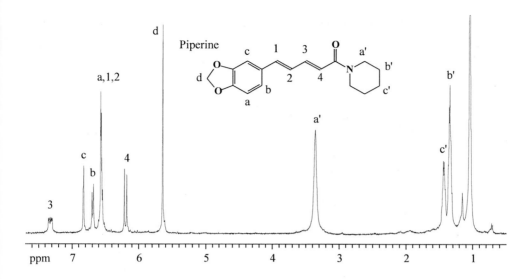

Figure 7-11: ^{1}H NMR spectrum (400 MHz) of piperine extracted from black pepper.

A supercritical separation technique employing non-protonated solvents is supercritical fluid chromatography (SFC) using CO_2. At a temperature of 323 K and a pressure of 161 bar, high-resolution, continuous-flow NMR spectra in supercritical CO_2 can be obtained [15], [16], [17], [18], [19]. This is demonstrated in Figure 7-12, which shows the ^{1}H NMR spectrum of ethylbenzene in supercritical CO_2 [16].

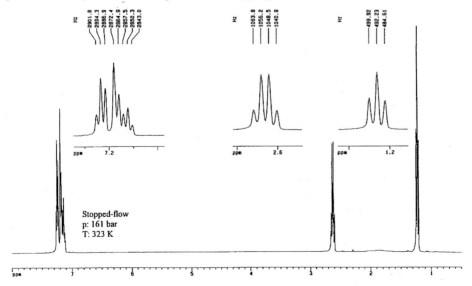

Figure 7-12: ^{1}H NMR spectrum (400 MHz) of ethylbenzene in supercritical CO_2.

Very often, SFC-NMR separations can be performed with a pressure gradient. Thus, different isomers of vitamin A acetate are easily separated (Figure 7-13) [20].

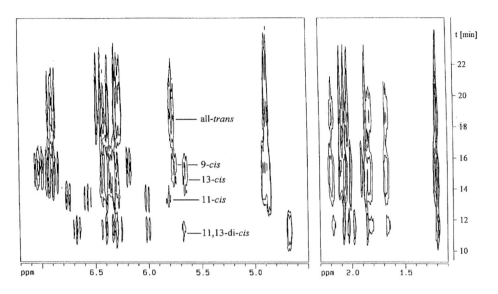

Figure 7-13: Contour plot (400 MHz) of the separation of vitamin A stereoisomers in supercritical CO_2.

7.5 Capillary Separations

In important applications in biotechnology and genetic engineering only a few nanograms of biologically active mixtures of compounds are available. A full characterization can only be accomplished by the miniaturization of the closed-loop separation-identification system. One hardware approach is the use of a solenoidal coil directly attached to a fumed-silica capillary [21], [22], [23]. The other approach is the utilization of a microprobe with a double-saddle Helmholtz micro coil [11] [24] [25] [26] [27] [28]. The capillary is fixed within the microcoil along the z-direction of the cryomagnet. This approach has several advantages and disadvantages. Because the capillary is inserted within the coil, it can be easily exchanged to meet other separation problems. This design has the inherent disadvantage that the filling factor (the ratio of the sample volume to the coil volume) is lower in comparison to a design where the coil is directly fixed to the capillary. Solenoidal coils with a perpendicular orientation to the B_o direction of the cryomagnet result in a threefold increase in sensitivity versus the Helmholtz microcoil design, but the currently developed approaches suffer from susceptibility-induced line-broadening as a result of the solenoidal coil being fixed to the fused silica capillary. This design has already been improved by the application of a susceptibility matching fluid and shows the potential for further optimization. However, the orientation of the capillary along the axis of the B_o field has the tremendous advantage that the NMR signal half width is not affected when electrophoretically driven separation techniques (capillary electrophoresis, capillary electrochromatography) are performed. Here, the magnetic field induced by the current within the capillary has no component along the z direction of the cryomagnet. Thus, a vertically oriented Helmholtz microcoil is a feasible design at least for capillary separations.

Coupling of capillary HPLC, rather than conventional separation techniques, with NMR has several distinct advantages [11], [21-28]. The reduced solvent consumption allows the use of deuterated solvents, thus rendering elaborate solvent suppression unnecessary. If only small amounts of the sample are available, higher concentrations of analyte in the detection cell are obtained when the column dimensions are small. Sensitivity levels currently obtainable for this design are in the 0.5 µg range with acquisition times of several seconds.

The structure elucidation of a vitamin A dimerization product was one of the first practical results obtained by on-line capillary HPLC-NMR coupling [11]. The structures of the dimers of vitamin A acetate, so-called kitols, were unknown for a long time because these compounds are very sensitive to UV irradiation and to air. The classical procedure of isolation, removal of the extraction solvent and redissolution in a deuterated solvent resulted in many isomerization products. By combining capillary separation with NMR microcoil detection, structure elucidation of a previously unknown kitol became possible [11]. Figure 7-14 shows the capillary ^{1}H NMR spectrum of the unknown, indicating that the resolution is sufficient to detract all coupling constants needed for the stereochemical assignment.

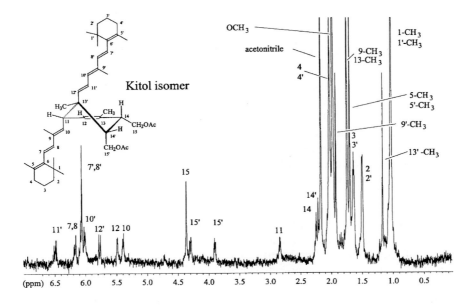

Figure 7-14: Capillary ^{1}H NMR spectrum of a dimerization product of vitamin A acetate, recorded in a detection volume of 200 nl.

The on-line capillary electrophoresis- and on-line capillary electrochromatography-NMR separation of alkylbenzoates are further examples [27], [28]. The contour plot of the separation performed in the CE and the CEC mode is shown in Figure 7-15.

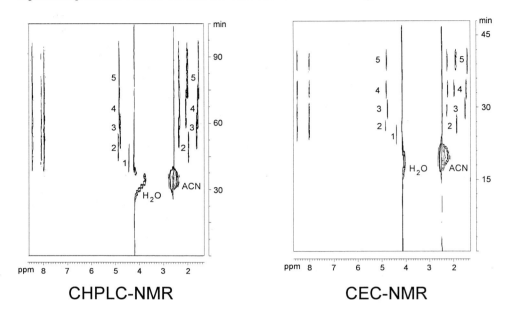

Figure 7-15: Contour plot of a CE and a CEC separation of alkylbenzoates.

The CEC-NMR contour plot indicated that all compounds are base-lined separated resulting in distinct NMR rows in the 2D display. This example shows the high application power of CE and CEC-NMR for providing unambiguous information about substances in complex organic molecules. The first steps towards a high-throughput separation system have already been made. For the successful performance of real-world applications, NMR sensitivity must be improved. NMR probes with 1-ng sensitivity should facilitate the application of capillary separations for metabolite structure elucidation and functional food analysis.

Summarizing this chapter, it can be stated that the linking of NMR spectrometers (from 400 to 750 MHz) to various separation techniques has been developed into rather a powerful tool. Stereochemical problems in metabolism studies can be easily solved if the concentration of the present compounds is above the detection threshold. The coupling of capillary separation techniques together with NMR detection will stimulate high-throughput screening applications in modern drug development.

Acknowledgment

The authors are grateful to John Pearce for reading the manuscript and to Tobias Glaser for his help in preparing the figures.

References

1. R B van Bremen, *Anal Chem* **68**, 299A-304A (1996)
2. C Rentel, S Strohschein, K Albert, E Bayer, *Anal Chem* **70**, 4394-4400 (1998)
3. K Albert, *J Chromatogr A* **703**, 123-147 (1995)
4. M Spraul, M Hofmann, P Dvortsak, J K Nicholson, I D Wilson, *Anal Chem* **65**, 327-330 (1993)
5. M Hofmann, M Spraul, R Streck, I D Wilson, A Rapp, *Labor Praxis* **10**, 36- 39 (1993)
6. K Albert, G Schlotterbeck, U Braumann, H Händel, M Spraul, G Krack, *Angew Chem Int Ed Engl* **34**, 1014-1016 (1995)
7. M Dachtler, K Kohler, K Albert, *J Chromatogr B* **720**, 211-216 (1998)
8. J K Roberts, R J Smith, *J Chromatogr A* **677**, 385-389 (1994)
9. S X Peng, B Borah, R L M Dobson, Y D Liu, S Pikul, *J Pharm Biomed Ana* **20**, 75-89, (1999)
10. M Spraul, M Hofmann, J C Lindon, J K Nicholson, I D Wilson in *Biofluid and Tissue Analysis for Drugs, Including Hypolipidaemics*; eds E Reid, H M Hill, I D Wilson, Methodology Surveys in Bioanalysis of Drug, Vol 23, Royal Society of Chemistry, 21-32 (1994)
11. G Schlotterbeck, L -H Tseng, H Händel, U Braumann, K Albert, *Anal Chem* **69**, 1421-1425 (1997)
12. J A de Koenig, A C Hogenboom, T Lacker, S Strohschein, K Albert, U A Th Brinkman, *J Chromatogr A* **813**, 55-61 (1998)
13. U Braumann, H Händel, K Albert, R Ecker, M Spraul, *Anal Chem* **67**, 930-935 (1995)
14. K Albert, *J Chromatogr A* **785**, 65-83 (1997)
15. L A Allen, T E Glass, H C Dorn, *Anal Chem* **60**, 390-394 (1988)
16. K Albert, U Braumann, L -H Tseng, G Nicholson, E Bayer, M Spraul, M Hofmann, C Dowle, M Chippendale, *Anal Chem* **66**, 3042-3046 (1994)
17. U Braumann, L -H Tseng, K Albert, M Spraul, *GIT* **38**, 77-79 (1994)
18. K Albert, U Braumann in *Frontiers in Analytical Spectroscopy*, eds D L Andrews, A M C Davies, The Royal Society of Chemistry, Cambridge, 1995, pp 86-93
19. K Albert, U Braumann, R Streck, M Spraul, R Ecker, *Fresenius J Anal Chem* **352**, 521-528 (1995)
20. U Braumann, H Händel, S Strohschein, M Spraul, G Krack, R Ecker, K Albert, *J Chromatogr A*, **761** 336-340 (1997)
21. N Wu, T L Peck, A G Webb, R L Magin, J V Sweedler, *J Am Chem Soc* 116, 7929-7930 (1994)

22. N Wu, T L Peck, A G Webb, R L Magin, J V Sweedler, *Anal Chem* **66**, 3849-3857 (1994)
23. D L Olson, T K Peck, A G Webb, R L Magin, J V Sweedler, *Science* **270**, 1967-1970 (1995)
24. K Albert, *Angew Chem Int Ed Engl* **34**, 641-642 (1995)
25. B Behnke, G Schlotterbeck, U Tallarek, S Strohschein, L -H Tseng, T Keller, K Albert, E Bayer, *Anal Chem* **68**, 1110-1115 (1996)
26. K Albert, G Schlotterbeck, L -H Tseng, U Braumann, *J Chromatogr A* **750**, 303-309 (1996)
27. K Pusecker, J Schewitz, P Gförer, L -H Tseng, K Albert, E Bayer, *Anal Chem* **70**, 3280-3285 (1998)
28. K Pusecker, J Schewitz, P Gförer, L -H Tseng, K Albert, E Bayer, I D Wilson, N J Bailey, G B Scarfe, J K Nicholson, J C Lindon, *Anal Comm* **35**, 213-215 (1998)

Chapter 8

U. Holzgrabe

8 NMR of Body Fluids

Since high-field NMR instruments have been introduced to scientific laboratories, it has been possible to directly detect and quantify drugs, metabolites of drugs, and toxic substances in urine and other body fluids. Since no assumptions have to be made prior to the analysis, more or less no working-up procedure is necessary, no artefacts have to be expected and no recovery experiments have to be performed. NMR spectroscopy is an extremely powerful tool to measure levels of metabolites in various body fluids, e.g. urine, bile, blood, or plasma, as well as to elucidate unknown metabolites and metabolism pathways. However, the NMR method suffers from the drawback that the analyte must be present at a concentration of > 50 μM in the sample. Attempts to overcome this problem will be described in the second part of this chapter.

8.1 NMR of Urine - Studies of Metabolism

8.1.1 ^{1}H NMR Spectroscopy

Since ^{1}H, ^{13}C, ^{15}N, and, with restrictions, ^{19}F, and ^{31}P are present in endogenous biomolecules and drugs, these magnetically active nuclei can be used in metabolism studies. In case of ^{1}H NMR spectra the intense water signal occurring in urine has to be suppressed by a certain pulse sequence, e. g. the application of a homo-gated secondary irradiation field at the solvent resonance frequency resulting in a selective saturation of the signal, the WEFT (water elimination Fourier Transform), WATR (water attenuation by T_2 relaxation) or the recently improved WATERGATE (water suppression by gradient-tailored excitation) method; for summary see.[1],[2],[3] An alternative method to avoid the water problem is provided by the freeze-drying of the urine. After dissolution in D_2O, the NMR spectrum can be recorded without any special pulse technique. In addition, relatively high-molecular mass substances such as proteins and lipids present in most biological samples cause a line-broadening of the signal due to high values of T_2. The Hahn spin-echo sequence can be used to overcome this problem.

Everett et al. first reported the application of NMR spectroscopy in metabolism studies. In 1984, they studied the metabolites of ampicillin in rats using spin-echo ^{1}H NMR spectroscopy [4]. Apart from the unchanged penicillin, the natural (5R)-isomer of penicilloic acid, the epimerized (5S)-isomer as well as a new metabolite, a diketopiperazine derivative, were identified in urine. Reinvestigation of ampicillin and corresponding penicillins ten years later confirmed the results; in the case of amoxycillin, a dimeric metabolite was additionally found [5].

In the 80s and 90s, it was especially the group of J.K. Nicholson who developed NMR spectroscopy into a routine method in drug metabolism studies. Some representative examples are discussed below.

In the beginning, for the most part the metabolites of drugs could only be identified in urine when a spectrum of the isolated (or independently synthesized) substance was known. For example, oxpentifylline, extensively used in the treatment of vascular diseases, and its acidic metabolite 1-(4´-carboxybutyl)-3,7-dimethylxanthine can be directly identified and quantified in a freeze-dried urine sample using a 250-MHz spectrometer [6]. The medication of lymphatic filariasis with diethylcarbamazepine has to be monitored for several reasons [7]. Since the drug lacks an absorbing chromophore (HPLC/UV-detection) and GC needs a special detector and troublesome extraction method, ^{1}H NMR spectroscopy appears to be suitable for the urine analysis: the urine samples were mixed with 10% D_2O and directly measured afterwards.

The unchanged drug and the corresponding N-oxide, a minor product of metabolism, could be quantified with high precision and accuracy which was better than 15% (obtained using a calibration graph) [7]. Often solid phase extraction (SPE) was performed before running the ^{1}H NMR spectrum. In this way, the O-desmethyl metabolite of naproxen [d-2(6-methoxy-2-naphthyl)propionic acid] could be found in human urine [8]. The main metabolites of paracetamol, i.e. sulfate and glucuronide, could be identified and quantified in the urine NMR spectrum without prior purification (Figure 8-1) [9].

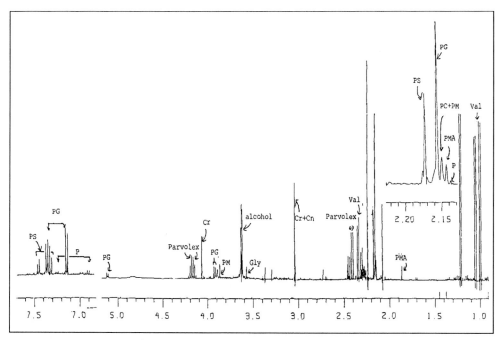

Figure 8-1: 400-MHz ^{1}H NMR spectrum of a urine sample from a patient who had taken paracetamol overdose and had been treated with parvolex (N-acetylcysteine). P = paracetamol, PG = paracetamol glucuronide, PS = paracetamol sulfate, PMA = paracetamol mercapturic acid, PC = paracetamol L-cystein, PM = methoxy-paracetamol (taken with permission from the thesis of M. L. Sitanggang, Bath, UK, 1988)

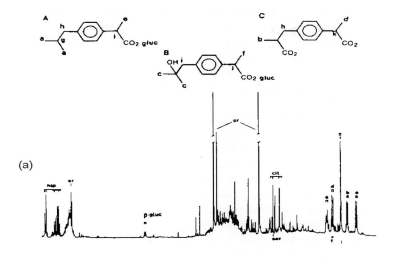
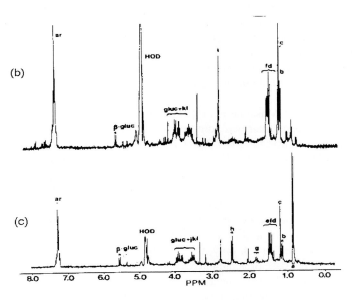

Figure 8-2: (a) 250-MHz ¹H NMR spectrum of a sample of freeze-dried urine (2 ml) obtained from a subject following oral dosing with 400 mg of ibuprofen. The sample was redissolved in D_2O for spectroscopy. (b) 200-MHz ¹H NMR spectra showing the eluates obtained with 40/60 methanol/water (248 scans) and (c) 60/40 methanol/water (66 scans) from a C18 SPE column on which 2 ml of human urine from a sample obtained for the 2-4 h after a 400-mg dose of ibuprofen (Reprinted with permission from [10], page 2831, Copyright 1997 ACS).

The main metabolites of ibuprofen, the glucuronides of ibuprofen and of the side-chain hydroxylated derivative {2-[4-(2-hydroxy-2-methylpropyl)phenyl]propionic acid} and a side-chain oxidized compound {2-[4-(2-carboxy-2-methylpropyl)phenyl]-propionic acid} could be detected after SPE using a 250-MHz instrument (Figure 8-2) [10].
However, linking NMR spectroscopy to HPLC led to the discovery of further metabolites in both cases (see Chapter 8.3.) [11]. Leo and Wu identified the glucuronide of suprofen after HPLC separation [12].
The latter examples clearly show that the ^{1}H NMR inspection of urine suffers from insensitivity due to the detection limit of ~10 nM (using a 600-MHz instrument) on the one hand, and due to chemical noise caused by considerable signal overlap on the other hand. These problems can be overcome by prior SPE or by HPLC NMR coupling which will be discussed at the end of this chapter. Increasing the field strength (600 to 750-MHz) results in an enhanced resolution of the signals of both xenobiotics and natural compounds occurring in the urine. Two-dimensional NMR techniques provide a further possibility to solve the problem of signal overlap. Whereas 2D-COSY experiments require relatively large experimental data arrays and, in addition, high usage of disk storage space and time-consuming data processing, two-dimensional J-resolved ^{1}H NMR spectroscopy (JRES) seems to offer an efficient means of reducing peak overlap [13]. Using this technique, 30 natural components of urine can be identified from the F2 projection and the additional contour plot showing the coupling pattern. In addition, the signals of paracetamol and all its metabolites are well separated. Due to different T_2 relaxation rates, the concentration of the metabolites cannot be directly determined by the measurement of the integrals of the corresponding signals. The spectra have to be carefully calibrated by means of standard addition of compounds. Taken together, the power of the JRES technique is the signal assignment in urine rather than the quantification of a component.

Applying this techniques Maschke et al. [14] made the diagnosis of an acute chloroquine poisoning: A middle aged man was admitted to the hospital after attempting suicide. 500 µL crude urine were adjusted to pH 7.0 and 300 and 600-MHz spectra were recorded after addition of 20µl D_2O containing TMSP-d_4. A selective reduction in the T_2 of the water signal was achieved by exchange with NH_4Cl and a saturation of the signal. JRES and TOCSY experiments helped to assign the signals of the urine. Beside the endogenous metabolite, signals of the unchanged chloroquine and its major metabolite monodesethylchloroquine could be identified, which were equivalent to 462 and 142^mg/L, respectively. In addition, ethanol was found in the spectra [15]. Since the measuring time amounted to 2 h only, the ^{1}H NMR spectroscopy is a rather fast tool for clinical diagnosis in the case of acute intoxication.
The 12 hours old urine of two patient with paraquat intoxication, an effective herbicide, were studied after centrifugation [16]. The comparison of simple 400-MHz spectra recorded of the urine of a healthy patient and the intoxicated patient showed a markedly increase of glucose, lactate, alanine, valine, and glutamine and a decrease of hippurate and citrate, which indicates a renal injury upon intoxication. In addition, the signals of the unchanged paraquat could be identified in the aromatic region. Quantification of the signals was made against internal TSP (compared also with formate) [17].
In order to speed up the assignment of the urine and to find new signals or signals of a changed intensity, the group of Nicholson suggested methods of pattern recognition, such as principal component analysis [18].

8.1.2 ^{19}F NMR Spectroscopy

There are several reasons why ^{19}F NMR spectroscopy is especially suitable for following the pathway of metabolism: firstly, it is almost as sensitive as ^{1}H NMR spectroscopy (83% of ^{1}H) and has a wide range (~200 ppm); secondly, it does not have any problem with the huge water signal and thirdly, there are no interfering signals with abundant urinary components of endogenous origin. The drawback that only a single atom in a drug can be monitored in the metabolism studies is partially neutralized by the fact that the chemical shift of fluorine atoms is rather sensitive to structural changes in the molecule which are 6 to 8 atoms apart from the fluorine. Thus, metabolic conversion of drugs is mostly mirrored in the change of the chemical shift.

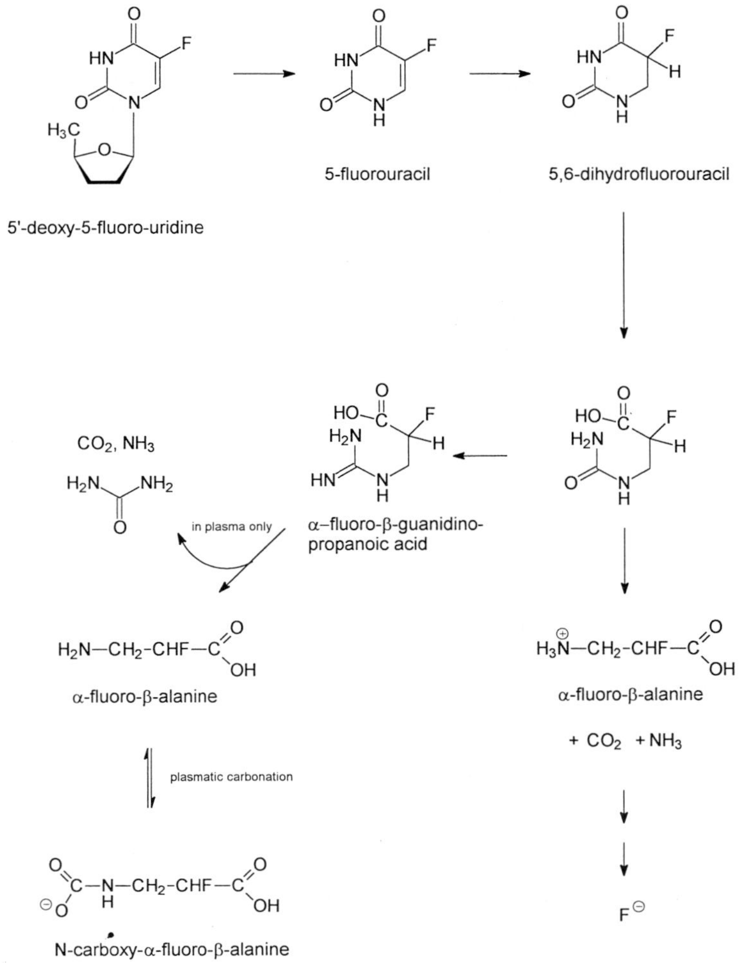

Figure 8-3:   Metabolic pathway of 5´-deoxy-5-fluoro-uridine and 5-fluorouracil, respectively.

However, the structural information derived from the spectra is poor in comparison with ^{1}H and ^{13}C NMR spectra. Thus, ^{19}F NMR spectroscopy can only be exploited for monitoring catabolic pathways of fluorinated drugs when the structures and the corresponding spectra of the metabolites are already known. As early as in 1984, Malet-Martino et al. reported the detection and quantification of five metabolites of 5´-deoxy-5-fluorouridine, an antitumor drug, in plasma, blood and urine without any sample pre-treatment [19]. Unmetabolized 5´-deoxy-5-fluorouridine, 5-fluorouracil, 5,6-dihydrofluorouracil and α-fluoro-β-alanine (Figure 8-3), the latter previously not reported, were found in blood, whereas the unchanged drug and α-fluoro-β-alanine appeared to be the major metabolites in urine. Monitoring of plasma and urine levels of the unchanged drug and each metabolite gave a complete description of the metabolic profile as displayed in Figure 8-3 [20].

As expected, the metabolism of 5-fluorouracil, the first metabolite of 5´-deoxy-5-fluorouridine, turned out to be similar [21]. Interestingly, the intermediate, α-fluoro-β-guanidinopropanoic acid, could not be detected in this study, which utilized ^{19}F NMR and HPLC methods. Martino et al. also developed a ^{19}F NMR assay method to determine the extent of protein binding of 5´-deoxy-5-fluorouridine, which was as valid as equilibrium dialysis [22].

In addition to the ampicillin studies using ^{1}H NMR spectroscopy, the group of Everett monitored the metabolism of a fluorinated penicillin, flucloxacillin, by means of ^{19}F NMR spectra [23]. Since the chemical shifts in ^{19}F NMR spectra are extremely dependent on the pH of the solution, flucloxacillin and its metabolites, the corresponding 5S- and 5R-penicilloic acid as well as 5´-hydroxymethylflucloxacillin, could be detected and quantified upon spiking the urine with the authentic sample. ^{1}H,^{19}F heteronuclear shift-correlated 2D spectra confirmed the findings [24]. In addition, the results obtained from the direct ^{19}F NMR spectroscopy were comparable to those obtained from HPLC and microbiological studies.

In a pilot scheme for xenobiotics, the metabolism of *ortho*-, *meta*- and *para*-trifluoromethylbenzoic acid (TFMBA) was studied in rats [25]. In ^{19}F NMR spectra recorded after addition of D_2O to the post-dose urine, an ester glucuronide and a transacylated ester glucuronide could be easily detected in the case of *o*-TFMBA, an ester glucuronide and a glycine conjugate in the case of *m*-TFMBA as well as transacylated ester glucuronide in the case of *p*-TFMBA. For structure elucidation of the metabolites, the urine samples were subjected to a solid-phase extraction chromatography followed by ^{1}H NMR detection. In a subsequent extended study [26], quantitative ^{19}F NMR spectroscopy was used to derive quantitative structure-metabolism relationships for xenobiotics.

The metabolism of the entire class of fluoroquinolones, bactericide gyrase inhibitors, is suitable for monitoring by ^{19}F NMR spectroscopy because metabolic conversions occur mainly at the piperazine ring which is closely located to the fluorine substituent. Unexpectedly, the fluorine signals of the fluoroquinolones were found to be broad. Upon addition of EDTA, the shape of the signals became narrow indicating that Ca^{2+} and/or Mg^{2+} caused the broadening by chelation [27]. Since it is rather difficult to dissolve the isolated metabolites in aqueous solutions, it is almost impossible to spike the urine in order to assign the different fluorine signals in the spectrum. However, signals for pefloxacin and its metabolites, norfloxacin, pefloxacin-*N*-oxide and oxopefloxacin, could be identified in human urine upon addition of a small amount of NaOD [28].

The metabolism of the anti-inflammatory drug flurbiprofen is characterized by hydroxylation of the second phenyl ring in the *para*- and *meta*-position. ^{19}F NMR using continuous broad-band ^{1}H irradiation to enhance the sensitivity and ^{19}F-^{1}H 2D shift correlated spectra revealed a total of 10 metabolites (two major and 8 minor), which were assumed to be glucuronide and sulfate conjugates of flurbiprofen and its hydroxy metabolites [29]. In a second study the urine of a volunteer treated with 200 mg flurbiprofen was subjected to an HPLC NMR analysis [30]. The two major metabolites, namely the glucuronides of flurbiprofen and 4´-hydroxy-flurbiprofen, were found to be diastereomers which were formed by in vivo conjugation of the racemic drug and its metabolite with D-glucuronic acid. The diastereomeric proportion could be evaluated.

8.1.3 ^{15}N NMR Spectroscopy

The metabolism of hydrazine, a starting product of several important industrial chemicals as well as a metabolite of drugs, e.g. isoniazide and hydralazine, is important to know. Even though the sensitivity of ^{15}N NMR spectroscopy proved to be poor, a number of metabolites could be identified in lyophilized urine, which was reconstituted in a reduced volume of water [31]. By spiking the urine with authentic samples, the signals of carbazic acid, urea, acetyl- and diacetylhydrazine could be assigned. In addition, a doublet centred at 150 ppm and a singulet at 294 ppm were assigned to the cyclization product of hydrazine and oxoglutarate and a singlet detected at 316 ppm to a pyruvate hydrazone. Although this study using ^{15}N NMR spectroscopy was fairly successful, this NMR technique has only a small chance of becoming a routine method because a huge number of accumulations (> 10,000 scans) and the optimisation of the relaxation time, both time-consuming processes, are necessary in order to be able to observe ^{15}N signals of a drug and its metabolites.

8.1.4 ^{31}P NMR Spectroscopy

The bioactivation of cyclophosphamide, a highly potent alkylating chemotherapeutic agent, can easily be observed by ^{31}P NMR spectroscopy. The phosphorus signals of the different metabolites, e. g. phosphoramide mustard (the active metabolite), dechloroethylcyclophosphamide, ketocyclo-phosphamide and carboxyphosphamide, are clearly separated when measuring the buffered urine of patients. The comparison of the ^{31}P NMR spectra of urine samples (2 ml of a 0 - 8 h fraction) obtained from a patient with breast cancer during a conventional-dose (500 mg/m^2 body surface) and high-dose (100 mg/kg body weight) adjuvant chemotherapy with cyclophosphamide revealed the change in the relative contribution of the two primary metabolic steps, e.g. formation of 4-hydroxy-cyclophosphamide, an intermediate in the bioactivation, and the side-chain oxidation. During dose escalation the inactivating pathways are favored, thereby indicating the saturation of bioactivating enzymes [32].

8.2 NMR of Bile, Blood Plasma, Cerebrospinal and Seminal Fluids

Apart from the appearance of drugs and their metabolites in urine, the concentration levels in other body fluids are often important to know in order to titrate the exact dosage necessary for a pharmacological or antimicrobial effect, or in order to find routes of excretion of a drug other than urine. In this context, rat bile was examined for a catecholic cephalosporin and its metabolites using ^{1}H NMR spectroscopy [33]. As expected from the experiments performed with urine, the cephalosporin could easily be identified in the bile; in addition, a methoxylated metabolite was found whose structure was elucidated after purification. The antibiotic cefoperazone is not metabolized by the liver. Therefore, the biliary excretion was monitored by means of ^{1}H NMR spectroscopy [34]: 43 % of the drug was found to be unchanged. The biliary excretion of several other xenobiotics, such as benzyl chloride [34] and 4-cyano-N,N-dimethyl aniline [35], were investigated after complete assignment of the natural components of the bile.

^{1}H NMR spectra of blood plasma suffer from the same problem of signal overlapping as the spectra of urine. Again, the already mentioned JRES technique can help to simplify the spectra in order to assign the intact fluid. This is especially necessary because broad resonances caused by macromolecules such as lipids and proteins (in particular albumin and immunoglobulins) can hide minor metabolites of drugs.

Although drugs and their analysis are the focus of this book, a brief overview (without going into detail) of endogenous substances and their quantitative distribution will be given here, because the distribution pattern can help to make a diagnosis of a disease or can be used to control the therapy. Using the JRES technique, Foxall et al. [13] were able to fully assign the complex spectral region between 3 and 4 ppm of human plasma consisting of signals of α- and β-glucose, several amino acids, glycerol, trimethylamine-N-oxide, choline and phosphorylcholine. The increase of the field frequency from 600 to 750-MHz results in an increase of the sensitivity now approaching 1 nM/ml. This high frequency enabled Foxall et al. [36] to detect e. g. abnormal metabolites associated with chronic renal failure, "uremic toxins" as methylamine, dimethylamine and the already reported trimethylamine-N-oxide, which are believed to cause nausea, itching, headache etc. In an extended study, Nicholson et al.37 could assign almost all components of the complex blood plasma, e. g. the different fractions of the lipoproteins (HDL, LDL, VLDL etc.), albumin, α_1-acid glycoprotein etc. by means of several pulse sequences, COSY-45, WEFT, JRES, ^{1}H-^{13}C HMQC, NOESYPRESAT, and TOCSY [38] using a 750-MHz instrument. In addition, the group of Nicholson developed two-dimensional diffusion-edited total-correlation NMR spectroscopy [39] in order to be able to determine the molecular diffusion coefficient of the components of biofluids, e.g. glucose, amino acids and lipoproteins. These coefficients can gain insight into the molecular size, molecular transport processes and molecular interactions, such as drug-protein binding or ion exchange.

Ala-Korpela et al. tried to evaluate the quantification of the lipoprotein lipid profile, because a huge number of diseases, such as diabetes mellitus, liver dysfunction and cancer, are associated with the disturbance of the balance of this system. Lineshape fitting analysis [40] and even better, utilizing artificial neural networks [41] enables to control the heavy signal overlapping such that quantification of VLDL, IDL, LDL, and HDL is possible [42]. Taken together these studies concerning the elucidation of the complex composition of the biofluids, fulfil the prerequisite to recognise a changed pattern of metabolites, and in line with this, a disease.

Inborn errors of metabolism, such as the 2-hydroxyglutaric aciduria and maple syrup urine disease, can be detected by a similar characterisation of the metabolic pattern of the urine. Whereas 250-MHz field strength turned out to be too small to find the differences in the hydroxyl acid pattern in the urine [43], 750 and 800-MHz enable one to make a differential diagnosis of the inborn diseases [44].

^{1}H and ^{31}P NMR spectroscopy have been extensively used to study the change of the phospholipid metabolite pattern of patients suffering from cancer. It is known that the level of phosphomonoesters consisting of phosphoethanolamine and phosphocholine is enhanced in malignant tissue. On the one hand it is possible to explore the levels of phospholipid metabolites by extraction of carcinoma cells with perchloric acid and performing JRES ^{1}H NMR experiments [45] which also give information on the content of carbohydrates and other components. Along with an increased level of phosphocholine, high concentrations of lactate, taurine and succinate and low levels of glucose were measured in tumor cells. On the other hand ^{31}P NMR spectra can be directly obtained from plasma and, thus, easily used to determine the level of phospholipids in the tumor and the healthy patient [46],[47],[48],[49]. Moreover, the method was evaluated to monitor the course of a therapy.

Apart from monitoring blood plasma, bile and urine, it is also possible to characterize the natural composition of a fluid such as amniotic fluid. Deviations from the normal pattern of natural metabolites may give a hint of a certain disease, e.g. diabetes, preclampsia or foetal malfunction[.]50 The amniotic fluid was centrifuged to remove cells etc. and freeze-dried for concentration. After re-suspension of the lyophilized sample with D_2O, the components of the fluid were identified and quantified upon addition of authentic samples. Citrate, valine, alanine, glucose, lactate and acetate as well as indoxyl sulphate, histidine and formate could be easily identified in a concentration range of 16 to 40 µM. With the exception of alanine and valine (due to signal overlap with HDL, VDL and LDL), the quantitative analysis by means of NMR is in good agreement with results reported from other methods.

Cerebrospinal fluid [51] was subjected to similar investigations. 46 molecules could be assigned using the JRES, COSY-45, and double-quantum filtered COSY techniques. Freeze-drying and reconstitution into water resulted in sharpening of many signals, especially those of glutamine/glutamate and other amino acids, and a loss of the volatile components.

An even greater challenge is the investigation of seminal as well as prostatic fluids, because only small amounts of liquid with low concentrations of natural metabolites are available and the ^{1}H NMR spectrum shows regions of overlapping resonances of the substance mixture. Nevertheless the group of Nicholson was able to assign the spectra measured from samples of healthy persons by means of several one- and two-dimensional techniques (see above) and, secondly, by identification of abnormal metabolites or abnormal levels of natural metabolites, both indicating malfunctions or diseases (Figure 8-4) [52],[53].

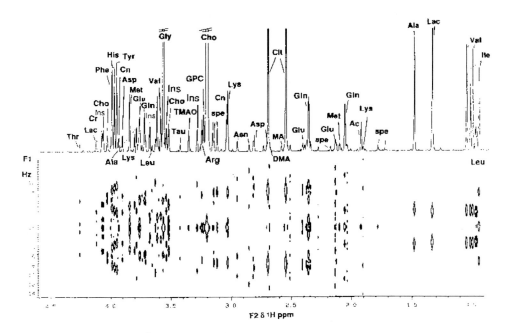

Figure 8-4: 600-MHz ^{1}H NMR spectrum JRES of the aliphatic region of control human seminal fluid from $\delta = 0.9 - 4.6$ ppm, showing contour plot and skyline F2 projection. Key: Ac = acetate, Ala = alanine, Arg = arginine, Asn = asparagine, Asp = aspartate, Cho = choline, Cit = citrate, Cr = creatine, DMA = dimethylamine, Gln = glutamine, Glu = glutamate, Gly = glycine, GPC = glycerophosphorylcholine, His = histidine, Ile = isoleucine, Ins = *myo*-inositol, Lac = lactate, Leu = leucine, Lys = lysine, MA = methylamine, Met = methionine, Phe = phenylalanine, Spe = spermine, Tau = Taurine, Thr = threonine, TMAO = trimethylamine-*N*-oxide, Trp = tryptophan, Tyr = tyrosine, Uri = uridine, Val = valine (Reprinted from [52], p 10, Copyright 1997) with kind permission of Elsevier Science -NL, Sara Burgerhartstraat 25, 1055 KV Amsterdam, The Netherlands)

Further investigations with boar seminal fluid performed by Kalic et al. [54] using a 600-MHz NMR instrument revealed the identification of the amino acids hypotaurine and carnitine. Semiselective HSQC 2D experiments on an 800-MHz spectrometer [55] enable one to differentiate between the structurally similar polyamines, spermidine, spermine and their metabolic precursor putrescine, all attracting interest for tumor therapy, in seminal fluid. Recently, Tomlins et al. [56] studied dynamic biochemical processes in incubated human seminal fluid samples: The enzymatic hydrolysis of phosphorylcholine to choline and the conversion of uridine-5´-monophosphate to uridine were found to be very fast, whereas the slow polypeptide hydrolysis to amino acids can be inhibited by the addition of EDTA. It remains to be seen which role the biochemical changes play in the reproductive function.

However, instead of inspection of the whole blood plasma, urine or other biofluids, it can be advantageous to insert a separation step before the NMR measurement, which is realized in the LC NMR hyphenation.

8.3 LC NMR Hyphenation

As described in the last sections, the monitoring of natural components as well as of drugs and their corresponding metabolites in body fluids suffers from signal overlapping due to the complexity of the fluids studied and low sensitivity. Typically, freeze-drying of the sample or solid-phase extractions were performed prior to NMR spectroscopy to overcome these problems. In this context, the hyphenation of the highly potent HPLC separation technique to NMR spectroscopy, the best technique for structure elucidation, was often discussed, but had several drawbacks owing to the need for expensive deuterated solvents, inadequate solvent suppression techniques and low sensitivity. The development of high field strength instruments (500 to 750-MHz) in the last 10 years provided greater sensitivity, new probe designs and better solvent suppression methods (for details see Chapter 7) [57], [58].

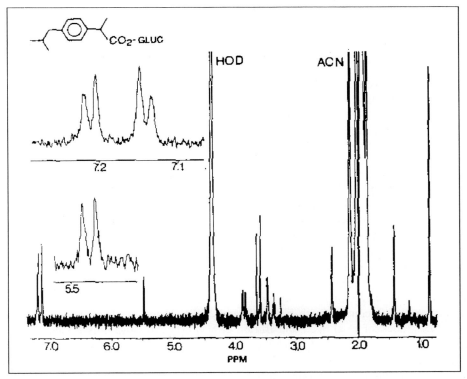

Figure 8-5: 500-MHz stop-flow ^{1}H NMR spectrum obtained for ibuprofen glucuronide; 128 scans, NOESY pulse sequence with presaturation, 1-Hz line-broadening. (Reprinted with permission from [59], p 329, Copyright 1997, ACS).

In order to prove the advantages of HPLC NMR hyphenation, urine samples of rats and man containing the drugs and their corresponding metabolites, e. g. ibuprofen, flurbiprofen, antipyrine or paracetamol (see section 8.3.), were reinvestigated. Using both continuous-flow and stopped-flow (Figure 8-5), apart from the already reported metabolism pattern [10], the non-conjugated side-chain oxidized di-acid metabolite of ibuprofen could be identified [59].

HPLC NMR measurements of flurbiprofen metabolites in human urine [60] confirmed the previous study [30]. In the case of antipyrine and its metabolites excreted in human urine, gradient RP-HPLC with stopped-flow ^{1}H NMR gave the following pattern: the 4-hydroxylated antipyrine and the corresponding ether glucuronide, the nor-antipyrine tautomerizing to give the enol form which is also conjugated to gluronic acid, and a small amount of the 3-hydroxy-antipyrine ether glucuronide [61].
Utilizing the stopped-flow operation again, an additional minor metabolite of paracetamol, N-acetylcysteinyl paracetamol, could be identified in the urine of male rats apart from the major metabolites, the sulfate and glucuronide [62],[63]. Even more interesting is the discovery of a futile deacetylation of phenacetin. In this study [64], [65] animals were dosed with either phenacetin-C^2H_3 or phenacetin and the urine was collected over a period of 24 h. Using solid-phase extraction prior to the ^{1}H HPLC NMR spectroscopy gave identification of the metabolites, the paracetamol glucuronide and sulfate and N-hydroxyparacetamol in addition to 30% futile deacetylation of the phenacetin-C^2H_3 giving the nephrotoxic aminophenol. This might be an explanation of the nephrotoxicity of the phenacetin.
HPLC NMR, especially the continuous-flow technique in connection with a 750-MHz instrument, was used to observe the acyl migration of fluorobenzoyl-glucuronides, resulting in a mixture of 1-, 2-, 3- and 4-O-acylglucuronides and corresponding α- and β-anomers [66], [67]. Similar studies concerning the acyl migration were performed with the non-steroidal antiinflammatory drug 6,11-dihydro-11-oxodibenzo[b,e]oxepin-2-acetic acid [68] [69] because the positional isomers of related drugs are believed to cause allergic reactions.
It is still a challenge to find low-level metabolites by means of HPLC NMR. Beside the improvement of the stop-flow HPLC-NMR mode using ultra-high field NMR machines (e.g. 750-MHz) [70], the additional hyphenation with mass spectroscopy is able to overcome the NMR problem of low sensitivity. Mutlib et al. [71] used the HPLC NMR MS combination to elucidate the metabolism of the new antipsychotic drug iloperidone, which is metabolized extensively for renal and biliary elimination. Thus, the identity of known metabolites could be established rather quickly by means of LC/MS and, in addition, the structures of minor unknown metabolites could be elucidated using semi-preparative HPLC coupled with both MS and NMR. Using the hyphenation of HPLC NMR MS, Shockcor et al. [72] and Burton et al. [73] reinvestigated the metabolism of paracetamol and found phenylacetylglutamine, an endogenous metabolite, in addition to the metabolites of paracetamol in urine. These HPLC MS NMR applications recently reported in the literature clearly demonstrate the power of this method.

Taking the studies of the metabolism of drugs and the exploration of various body fluids together, it can be stated that NMR spectroscopy is a powerful tool for the analysis of endogenous and exogenous components in biofluids. The most important advantages are:

1. NMR spectroscopy is non-selective; thus, it is possible to monitor the levels of all small-sized components in free solution above the detection level.
2. The technique requires only low sample volume and usually minimal sample preparation.
3. NMR measurements are mostly quite fast.
4. The content of information, especially of structural information, obtained from the spectra is very high.

However, it has to be stressed that very high resonance frequencies, 600 - 750-MHz, are necessary for this purpose. Additionally, special pulse sequences have to be applied in order to assign all signals to the components. Since such high-field instruments are only very rarely available in clinical laboratories at the moment, NMR spectroscopy is not a matter of routine examinations.

Linking of NMR spectroscopy (600 and 750-MHz) to various separation techniques (HPLC, CE, SFC) developed into a rather powerful tool, which can speed up the diagnosis of diseases associated with changes in the composition of body fluids or the elucidation of the metabolism of new drugs. It remains to be seen which analysis field will be undertaken by the new NMR techniques.

References

1. J K Nicholson, I D Wilson, *Prog Drug Res* **31**, 428-479 (1987)
2. S K Branch, A F Casy, *Prog Med Chem* **26**, 355-436 (1989)
3. M Liu Xi-an Mao, C Ye, H Huang, J K Nicholson, J C Lindon, *J Magn Res* **132**, 125-129 (1998)
4. J R Everett, K R Jennings, G Woodnutt, M J Buckingham, *J Chem Soc Chem Commun*, 894-895 (1984)
5. S C Connor, J R Everett, K R Jennings, J K Nicholson, G Woodnutt, *J Pharm Pharmacol* **46**, 128-134 (1994)
6. I D Wilson, J Fromson, I M Ismail, J K Nicholson, *J Pharm Biomed Anal* **5**, 157-163 (1987)
7. J W Jaroszewski, D Berenstein, F A Slok, P E Simonsen, M K Agger, *J Pharm Biomed Anal* **14**, 543-549 (1996)
8. I D Wilson, I M Ismail, *J Pharm Biomed Anal* **4**, 663-665 (1986)
9. J R Bales, D P Higham, I Howe, J K Nicholson, P J Sadler, *Clin Chem* **30**, 1631-1636 (1984);
S K Branch, L J Notorianni, M L Sitanggang, *J Pharm Pharmacol* **38**, 115P (1986);
M L Sitanggang, Thesis, Bath, UK, 1988; T D Spurway, K P R Gartland, A Warrander, R Pickford, J K Nicholson, I D Wilson, *J Pharm Biomed Anal* **8**, 969-973 (1990)

10. I D Wilson, J K Nicholson, *Anal Chem* **59**, 2830-2832 (1987)
11. M Spraul, M Hofmann, P Dvortsak, J K Nicholson, I D Wilson, *J Pharm Biomed Anal* **10**, 601-605 (1992)
12. G C Leo, W-N Wu, *J Pharm Biomed Anal* **10**, 607-613 (1992)
13. P J D Foxall, J A Parkinson, I H Sadler, J C Lindon, J K Nicholson, *J Pharm Biomed Anal* **11**, 21-31 (1993)
14. S Maschke, N Azaroual, J -M Wieruszeski, G Lippens, M Imbenotte, D Mathieu, G Vermeersch, M Lhermitte, *Clin Chem* **43**, 698-699 (1997)
15. S Maschke, N Azaroual, J -M Wieruszeski, G Lippens, M Imbenotte, D Mathieu, G Vermeersch, M Lhermitte, *NMR Biomed* **10**, 277-284 (1997)
16. E Bairaktari, K Katopodis, K C Siamopoulos, O Tsolas, *Clin Chem* **44**, 1256-1261 (1998)
17. M Kriat, S Confort-Gouny, J Vion-Dury, M Sciaky, P Viout, P J Cozzone, *NMR Biomed* **5**, 179-184 (1992)
18. E Holmes, J K Nicholson, F W Bonner, B C Sweatman, C R Beddell, J C Lindon, E Rahr, *NMR Biomed* 5, 368-372 (1992)
19. M -C Malet-Martino, R Martino, A Lopez, J -P Beteille, M Bon, J Bernadou, J -P Armand, *Cancer Chemoth Pharmacol* **13**, 31-35 (1984)
20. M -C Malet-Martino, J -P Armand, A Lopez, J Bernadou, J -P Beteille, M Bon, R Martino, *Cancer Res* **46**; 2105-2112 (1986)
21. W E Hull, R E Port, R Herrmann, B Britsch, W Kunz, *Cancer Res* **48**, 1680-1688 (1988)
22. D Meynial, A Lopez, M -C Malet-Martino, L S Hoffmann, R Martino, *J Pharm Biomed Anal* **6**, 47-59 (1988)
23. J R Everett, K Jennings, G Woodnutt, *J Pharm Pharmacol* **37**, 869-873 (1985)
24. J R Everett, J W Taylor, G Woodnutt, *J Pharm Biomed Anal* **7**, 397-403 (1989)
25. F Y K Ghauri, C A Blackledge, I D Wilson, J K Nicholson, *J Pharm Biomed Anal* **8**, 939-944 (1990)
26. E Holmes, B C Sweatman, M E Bollard, C A Blackledge, C R Beddell, I D Wilson, J C Lindon, J K Nicholson, *Xenobiotics* **25**, 1269-1281 (1995)
27. M Tugnait, F Y Ghauri, J K Nicholson, K Borner, I D Wilson, *Bioanalytical Approaches* Vol **22**, Chapt 48, 291-296 (1992)
28. U Holzgrabe, S K Branch, *Magn Res Chem* **31**, 192-196 (1994)
29. K E Wade, I D Wilson, J A Troke, J K Nicholson, *J Pharm Biomed Anal* **8**, 401-410 (1990)
30. M Spraul, M Hofmann, I D Wilson, E Lenz, J K Nicholson, J C Lindon, *J Pharm Biomed Anal* **11**, 1009-1015 (1993)
31. N E Preece, J K Nicholson, J A Timbrell, *Biochem Pharmacol* **41**, 1319-1324 (1991)
32. D Busse, F W Busch, F Bohnenstengel, M Eichbaum, P Fischer, J Opalinska, K Schumacher, E Schweizer, H K Kroemer, *J Clin Oncology* **15**, 1885-1896 (1997)
33. M J Basker, S C Finch, J W Tyler, *J Pharm Biomed Anal* **8**, 573-576 (1990)
34. D A Ryan, J K M Sanders, G C Curtis, H Hughes, *J Pharm Biomed Anal* **13**, 723-734 (1995)
35. D A Ryan, J K M Sanders, G C Curtis, H Hughes, *J Pharm Biomed Anal* **13**, 735-745 (1995)

36. P J D Foxall, M Spraul, R D Farrant, J C Lindon, G H Neild, J K Nicholson, *J Pharm Biomed Anal* **11**, 267-276 (1993)
37. J K Nicholson, P J D Foxall, M Spraul, R D Farrant, J C Lindon, *Anal Chem* **67**, 793-811 (1995)
38. M Liu, J K Nicholson, J C Lindon, *Anal Chem* **68**, 3370-3376 (1996)
39. M Liu, J K Nicholson, J A Parkinson, J C Lindon, *Anal Chem* **69**, 1504-1509 (1997)
40. M Ala-Korpela, Y Hiltunen, J Jokisaari, S Eskelinen, K Kiviniitty, M J Savolainen, Y A Kesäniemi, *NMR Biomed* **6**, 225-233 (1993)
41. M Ala-Korpela, Y Hiltunen, *J D Bell, NMR Biomed* **8**, 235-24 (1995)
42. M Ala-Korpela, *Progr NMR Spectr* **27**, 475-554 (1995)
43. R A Iles, A J Hind, R A Chalmer, *Clin Chem* **31**, 1795-1801 (1985)
44. E Holmes, P J D Foxall, M Spraul, R D Farrant, J K Nicholson, J C Lindon, *J Pharm Biomed Anal* **15**, 1647-1659
45. I S Gribbestad, S B Petersen, H E Fjosne, S Kvinnsland, J Krane, *NMR Biomed* **7**, 181-194 (1994)
46. M Kliszkiewicz-Janus, S Baczynski, *NMR Biomed* **8**, 127-132 (1995)
47. M Kliszkiewicz-Janus, S Baczynski, *Magn Res Med* **35**, 449-456 (1996)
48. M Kliszkiewicz-Janus, W Janus, S Baczynski, *Anticancer Res* **16**, 1587-1594 (1996)
49. M Kliszkiewicz-Janus, S Baczynski, *Biochim Biophys Acta* **1360**, 71-83 (1997)
50. P E McGowan, J Reglinski, R Wilson, J J Walker, S Wisdom, J H McKillop, *J Pharm Biomed Anal* **11**, 629-632 (1993)
51. B C Sweatman, R D Farrant, E Holmes, F Y Ghauri, J K Nicholson, J C Lindon, *J Pharm Biomed Anal* **11**, 651-664 (1993)
52. M J Lynch, J Masters, J P Pryor, J C Lindon, M Spraul, P J D Foxall, J K Nicholson, *J Pharm Biomed Anal* **12**, 5-19 (1994)
53. M Spraul, J K Nicholson, M J Lynch, J C Lindon, *J Pharm Biomed Anal* **12**, 613-618 (1994)
54. M Kalic, G Kamp, J Lauterwein, *NMR Biomed* **10**, 341-347 (1997)
55. W Willker, U Flögel, D Leibfritz, *NMR Biomed* **11**, 47-54 (1998)
56. A M Tomlins, P J Foxall, M J Lynch, J Parkinson, J R Everett, J K Nicholson, *Biochim Biophys Acta* **1379**, 367-380 (1998)
57. J C Lindon, J K Nicholson, I D Wilson, *Adv Chromatogr* **36**, 315-382 (1995)
58. J C Lindon, J K Nicholson, I D Wilson, *Prog Nucl Magn Res* **29**, 1-49 (1996)
59. M Spraul, M Hofmann, P Dvortsak, J K Nicholson, I D Wilson, *Anal Chem* **65**, 327-330 (1993)
60. M Spraul, M Hofmann, J C Lindon, J K Nicholson, I D Wilson in Biofluid and Tissue Analysis for Drugs, Including Hypolipidaemics; eds E Reid, H M Hill, I D Wilson, Methodology Surveys in Bioanalysis of Drugs, Vol **23**, Royal Society of Chemistry, 21-32 (1994)
61. I D Wilson, J K Nicholson, M Hofmann, M Spraul, J C Lindon, *J Chromatogr* **617**, 324-328 (1993)
62. M Spraul, M Hofmann, J C Lindon, R D Farrant, M J Seddon, J K Nicholson, I D Wilson, *NMR Biomed* **7**, 295-303 (1994)
63. M Spraul, M Hofmann, J C Lindon, J K Nicholson, I D Wilson, *Anal Proc* **30**, 390-392 (1993)

64. A W Nicholls, J C Lindon, S Caddick, R D Farrant, I D Wilson, J K Nicholson, *Xenobiotica* **27**, 1175-1185 (1997)
65. A W Nicholls, J C Lindon, R D Farrant, J P Shockcor, I D Wilson, J K Nicholson, *J Pharm Biomed Anal* (in press)
66. U G Sidelmann, C Gavaghan, H A J Carless, R D Farrant, J C Lindon, I D Wilson, J K Nicholson, *Anal Chem* **67**, 3401-3404 (1995)
67. U G Sidelmann, C Gavaghan, H A J Carless, M Spraul, M Hofmann, J C Lindon, I D Wilson, J K Nicholson, *Anal Chem* **67**, 4441-4445 (1995)
68. U G Sidelmann, E M Lenz, M Spraul, M Hofmann, J Troke, P N Sanderson, J C Lindon, I D Wilson, J K Nicholson, *Anal Chem* **68**, 106-110 (1996)
69. E M Lenz, D Greatbanks, I D Wilson, M Spraul, M Hofmann, J Troke, , J C Lindon, J K Nicholson, *Anal Chem* **68**, 2832-2837 (1996)
70. O Corcoran, M Spraul, M Hofmann, I M Ismail, J C Lindon, J K Nicholson, *J Pharm Biomed Anal* **16**, 481-489 (1997)
71. A E Mutlib, J T Strupczewski, S M Chesson, *Drug Metab Disposition* **23**, 951-964 (1995)
72. J P Shockcor, S E Unger, I D Wilson, P J D Foxall, J K Nicholson, J C Lindon, *Anal Chem* **68**, 4431-4435 (1996)
73. K I Burton, J R Everett, M J Newman, F S Pullen, D S Richards, A G Swanson, *J Pharm Biomed Anal* **15**, 1903-1912 (1997)

Chapter 9

G. Gemmecker

9 NMR as a Tool in Drug Research

9.1 Introduction

For many years now, multidimensional NMR spectroscopy has been routinely used for the structure determination of small organic molecules, solving questions of molecular constitution as well as conformation. In recent years, the applicability of NMR studies has been continuously extended to larger biomacromolecules, mainly proteins and nucleic acids. Recent progress in protein expression, isotopic labeling and NMR methodology has made possible the determination of high-resolution protein structures up to ca. 20–30 kDa in an almost routine manner and with a resolution generally comparable to that of X-ray structures. As a result of this development, nowadays every fourth structure newly deposited in the Brookhaven Protein Data Bank is based on NMR data [1].

In addition to being an alternative to X-ray diffraction for the structure determination of small proteins difficult to crystallize, NMR also offers many possibilities to study intermolecular interactions and reactions under conditions that are not suited to crystallization. With the help of NMR spectroscopy, information can also be gained on the dynamic behavior of the systems under study. These aspects are important in leading us from the often merely static definition of "structure" towards a much more realistic view of biologically active molecules and their interactions.

Biological mechanisms are almost always based — at the molecular level — on more or less specific intermolecular interactions involving biomacromolecules (e.g., receptor-hormone; transporter-substrate; enzyme-inhibitor; antibody-antigen; nucleic acids-regulatory proteins). Understanding these interactions is a prerequisite for a modern approach to drug design, since the majority of newly developed drugs specifically target one or more of these systems in order to suppress, enhance or modify the intermolecular interactions involved.

NMR is uniquely able to provide information about the underlying interactions on various levels, from a merely qualitative indication of complex formation up to high-resolution structures and kinetic data. Therefore, NMR can make important contributions to drug design directed at molecular targets and to helping to understand the affinities and specificity of intermolecular interactions on a structural basis.

In the following, the different approaches to the study of intermolecular interactions by NMR will be presented in more detail. Since most of the systems studied so far are protein complexes (with organic ligands, nucleic acids, other proteins etc.), we will limit this discussion to protein-ligand interactions. However, all the approaches discussed in the following can also be employed for studies of ligand interactions with DNA, RNA or oligosaccharides, as numerous applications show.

9.2 Protein Structures from NMR

The most obvious contribution of NMR to drug design has been the part it has played in the structures determination of increasing numbers of protein. Since the first high-resolution protein NMR structure was published in 1985 [2], [3], the quality of NMR protein structures has been continuously improved and their size limit extended. High-resolution structures of biomacromolecules are routinely used for molecular modeling approaches to drug development and improvement, which are, however, beyond the scope of this contribution [4]. Today, most NMR protein structures (except for very small proteins < 10 kDa) are derived from isotope-labeled protein samples that have been overexpressed in genetically modified bacterial cultures. The development of the tools for genetic engineering and protein overexpression in isotopically enriched media during the last decade has certainly made a crucial contribution to NMR protein structure determination (for a recent review, see [5]).

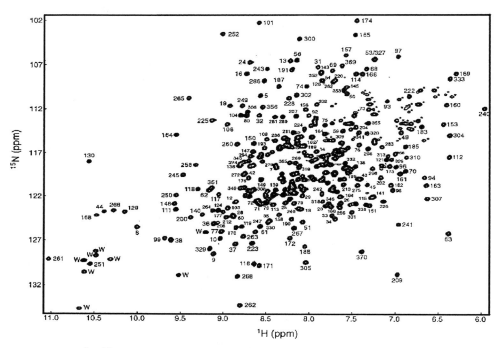

Figure 9-1: $^1H,^{15}N$-HSQC spectrum of the amide signals of a 45-kDa maltose binding protein / β-cyclodextrin complex. With the help of a deuterated $^{15}N,^{13}C$-labeled sample, all backbone signals could be sequentially assigned (reproduced with permission from [6]; Copyright 1997 American Chemical Society).

While the cheapest labeling scheme, [U-^{15}N] (uniformly ^{15}N) enrichment can be used for proteins up to ca. 10-12 kDa; larger proteins are usually [U-^{15}N, ^{13}C]-labeled to facilitate signal assignment as well as the measurement of NMR parameters needed for the determination of the three-dimensional structure. For most proteins of ca. 20 kDa and beyond, additional partial deuteration is also required to reduce relaxation and signal linewidth which otherwise increase steadily with molecular weight [7], [8]. Today NMR spectroscopy can solve protein structures up to the 30–40 kDa range (Figure 9-1), and new developments promise to extend the size range even further [9], [10].

The resolution of X-ray structures is difficult to compare with that of NMR structures because of the very different methods of structure calculation from the experimental data. However, it is generally accepted that well-resolved NMR structures are comparable to X-ray structures with a ≤ 2 Å resolution. In addition to the coordinates of a three-dimensional protein structure, another aspect of many NMR studies important for drug development is the information about internal dynamics [11]. Many protein-ligand complexes do not merely resemble rigid lock-and-key systems, but involve conformational rearrangement of one or both molecules. Binding-site flexibility in the protein, for example, will have implications on ligand specificity [12], and flexibility of protein and/or ligand will affect the thermodynamics and kinetics of complex formation. NMR has proven a valuable source of information about local molecular dynamics on many different time scales [13], [14].

9.3 Protein-Ligand Interactions

9.3.1 NMR of Molecular Complexes

While NMR spectroscopy has become an accepted alternative to X-ray diffraction for elucidating high-resolution structures of biomacromolecules such as proteins, DNA, RNA and oligosaccharides, it can be applied to drug design in many more ways. Besides delivering the three-dimensional structures of the free proteins as raw material for modeling studies on ligand binding, NMR can directly yield valuable experimental data on biologically important protein-ligand complexes themselves:

1. *Qualitative and quantitative binding assay*: changes in NMR parameters allow us to detect and quantitatively determine binding affinities of potential ligands.
2. *Determination of binding site*: based on changes in the assigned NMR signals of a protein upon ligand addition, a ligand's binding site on the protein can be located.
3. *Conformational information*: the conformations of the protein and/or ligand in the complex can be determined and compared to their free states.
4. *Dynamic information*: in addition to the static structure, local flexibilities can be measured for the complex and the free components to yield a more accurate picture of the intermolecular interactions.

9.3.2 Aspects of Binding Affinity

The behavior of protein-ligand complexes in NMR measurements depends largely on their thermodynamic and kinetic properties (i.e., dissociation constants and on/off rates). The characteristics of the system under study and the kind of information desired determine the adequate approach for an NMR investigation.

Protein concentrations in solution are often limited by solubility and aggregation (and, of course, protein availability) to a maximum of 0.5–5 mM. Today most NMR structural studies are performed at 1–2 mM protein concentrations, although it can be expected that new technologies leading to increased sensitivity in NMR experiments (increased field strength, improved probe design, cryo-probes etc.) will lower the concentration requirements to the 100-mM range within the next few years.

For a typical millimolar protein concentration, three different regimes can be distinguished (Figure 9-2):

Thus we have:

1. strong binding, with dissociation constants K_D in the sub-micromolar range,
2. moderate binding, with micromolar K_D values and
3. weak interactions with millimolar dissociation constants.

(for $K_D > 1$ M, essentially no complex formation will occur at all, and no observable effects be detected in NMR measurements). For strongly binding ligands, under normal protein NMR conditions, dissociation constants in the nanomolar range lead to negligible concentrations of free protein and ligand under stoichiometric conditions (solid line in Figure 9-2), and we can assume that both the protein and ligand exist almost exclusively (>95%) in the bound state. The complex behaves like a single stable molecule, and its spectra will not show any signs of the free species or of exchange between the free and bound species.

Protein complexes with high-affinity ligands, e.g., enzyme inhibitors, can be treated as single molecules in NMR and X-ray studies and thus pose no special problems, except for their larger size and possible changes in solubility. Because of their high stability, such complexes usually crystallize readily and are therefore easily accessible to X-ray diffraction studies.

In the moderately binding case, under stoichiometric conditions both protein *and* ligand will exist as bound *and* free species in significant percentages (solid line in Figure 9-2). Depending on the kinetic stability of the complex, exchange between the free and bound species might also affect the NMR spectra (see next section). For a 1 mM dissociation constant, in a stoichiometric mixture of protein and ligand (1 mM each) only half of the molecules exist as protein-ligand complex, the other half being free ligand and free protein (in slow or fast exchange with the complex). It is obvious that under such conditions the NMR spectrum of the system will be far more complex and difficult to analyze than for the isolated complex alone. However, for micromolar K_D values the protein can be driven into a completely complexed state by adding a moderate (10- to 100-fold) excess of the ligand (dotted curve in Figure 9-2). Signals of the excess component can be eliminated from the NMR spectra by isotope filtering techniques (see below). Of course, ligand solubility has to be sufficiently high, but this can often be accomplished by adding small amounts of organic solvents (e.g., DMSO) to the aqueous protein solution. Chemical and biological processes often include such moderately strong intermolecular interactions leading to a dynamic equilibrium between free and bound states of the components involved. Since the resulting mixtures of free and bound species in most cases cannot be crystallized, NMR is the method of choice for gaining structural information on these very interesting systems.

In the weakly bound case, it will not be possible to force a complex component into a completely complexed state, even with a large excess of the other compound. Under NMR conditions, only a small fraction of the molecules will exist in the complexed state, so that a conventional three-dimensional structure determination becomes impossible. Nevertheless, a lot of information about the complex can be gained from NMR measurements (usually involving isotope filtering), e.g., indications of conformational changes upon complexation and the molecular regions involved in binding.

When screening for lead compounds from a library weakly binding ligands are the routine case that will still have to undergo extensive optimization to yield a strongly binding ligand. Again, X-ray diffraction methods are generally not applicable to these cases and NMR becomes the method of choice for structural studies.

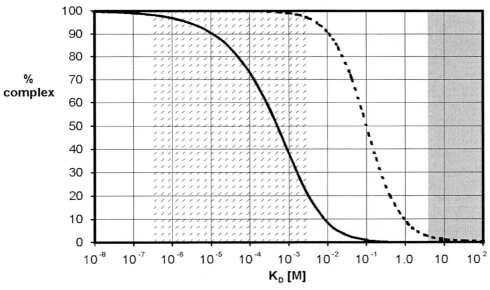

Figure 9-2 Percentage of 1:1 protein-ligand complex formation for a 1 mM protein solution with a ligand concentration of 1 mM (*solid line*) and 100 mM (*dotted line*), as a function of the dissociation constant K_D. In the intermediate binding regime (*dotted region*), proteins can be completely complexed only by an excess of ligand. For even lower affinities only partial complexation is possible, and for molar K_D values (*area shaded in gray*) no significant effects can be observed under NMR conditions.

9.3.3 Exchange Time Scales for NMR Parameters

Except for high-affinity ligands, the protein-ligand complex usually coexists in solution in an equilibrium with significant amounts of free protein and/or free ligand, depending on their relative molar ratios. The dissociation constant K_D can be expressed as the ratio between off- and on-rate (governing dissociation and formation of the complex, respectively):

$$K_D = \frac{k_{off}}{k_{on}}$$

with the lifetime of the complexed state being the inverse of the off-rate k_{off}.
Assuming that the on-rate k_{on} is diffusion limited (i.e., ca. 10^8 M^{-1} s^{-1}), one can estimate the range of complex lifetimes for different affinities (the lifetimes of the *free* species depend also on their molar ratio!):

Table 9-1: Affinities and Lifetimes.

K_D	k_{off}	lifetime of complex
1 nM	0.1 s^{-1}	10 s
1 mM	100 s^{-1}	10 ms
1 mM	10^5 s^{-1}	10 ms

Table 9-2: Time Scales for Exchange Effects of NMR Parameters

Parameter	Typical range [Hz]	Typical differences between free and bound state [Hz]	lifetime at intermediate exchange
Chemical shifts	0–10000	100–1000	1–10 ms
Scalar couplings	0–200	1–10	0.1–1 s
NOE (build-up rates)	0–5	0.1–5	0.2–10 s
Relaxation rates: T_1	0.1–10	0.1–10	0.1–10 s
T_2	1–100	1–50	0.02–1 s

What are the implications of this for NMR measurements? In order to measure NMR parameters (represented by a frequency or rate) separately for each state (i.e., free and bound state), the state's lifetime has to be significantly longer than the inverse of the differences of this parameter in the two states (corresponding to the intermediate exchange regime). For the most important NMR parameters, Table 2 gives an estimate of the time-scales involved.

Thus, to observe the chemical shifts of the free and bound species *separately* (i.e., as two sets of signals), their lifetimes will have to be in the range of hundreds of milliseconds to seconds ("slow exchange"). On the other hand, if the individual lifetimes are in the sub-millisecond range — much shorter than in the intermediate exchange regime — , then only a single set of signals will be observed, with the chemical shifts reflecting an average over both the free and the complexed species, weighted by their relative concentration ("fast exchange").

As one can see from Table 2, the fast and slow exchange regimes are not fixed, but depend on the NMR parameters under consideration and also on their specific differences in the two species. A large chemical shift difference between free and bound state might still lead to two separate signals for this resonance at a certain complex lifetime, while another signal with only a small chemical shift difference can already be in fast exchange, i.e., displaying just a single signal with an averaged chemical shift, under the same conditions.

For the intermediate exchange regime itself, when the lifetimes of the free and bound species are close to the inverse of the differences of their NMR parameters, a very complex behavior occurs. With small organic molecules undergoing conformational changes, the chemical shift "exchange broadening" can yield a wealth of kinetic data. However, in the case of macromolecular complexes individual lineshapes are not easily accessible for a quantitative evaluation (due to limited spectral resolution, low signal-to-noise and spectral overlap), so the most prominent effect of intermediate exchange is merely the loss of signal intensity due to the line broadening.

Except for high-affinity ligands with nanomolar K_D values, lifetimes usually range in the millisecond range for most complexes. From Table 2 it can be seen that the parameters least sensitive to averaging are the chemical shifts and, to a lesser degree, T_2 relaxation rates. The transferred-NOE measurements actually exploit the fast averaging of NOEs (see next section).

9.3.4   Transfer NOE

The transfer nuclear Overhauser enhancement (tr-NOE) can be used to gain information about the bound conformation of a weakly binding ligand. The method requires a relatively low affinity ($K_D \geq 100$ mM) and sub-millisecond lifetimes of the ligand's free and bound state. In this case, only one averaged set of signals will be observed for the ligand, and the intramolecular NOEs of the ligand will also be averaged over its free and bound conformation.

In order to interpret these NOEs in terms of the bound conformation, the contributions from the free state have to be negligible. The NOE effect vanishes for molecular orientation correlation times t_c in the order of the inverse of the resonance frequency (i.e., a few nanoseconds for 300-600 MHz ^{1}H frequency) [15] . This is the case for molecules with molecular masses of ca. 100–1000 Da (depending on temperature, solvent viscosity and spectrometer frequency).

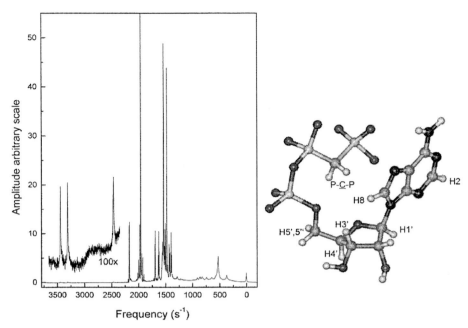

Figure 9-3: Application of tr-NOE. *Left*: ^{1}H spectrum of 1 mM ATP with bovine heart mitochondria (20 mg/ml); the *insert* shows the low-field ATP signals with 100-fold vertical expansion. *Right*: conformation of a nucleotide bound to the mitochondrial ADP/ATP carrier, derived from tr-NOE studies (reproduced with permission from [16], Copyright 1999 American Chemical Society.)

If such a small ligand binds to a macromolecular receptor, then the correlation time t_c of the *bound* state will be determined by the much larger molecular mass of the complex, leading to intense NOE effects. Therefore, even a small fraction of bound ligand will dominate the averaged NOE patterns observed for the ligand signals. *In praxi*, one generally uses a large (ca. 10- to 100-fold) excess of ligand, so that the ligand signals are easily observable and no isotopic labeling of either protein or ligand is required. Since neither the signals of the bound state of the ligand nor the protein itself have to be observable in the spectrum, this method can be applied to complexes with very large biomacromolecules, e.g., the complete bacterial ribosome [17] . As an example, the conformation of a nucleotide bound to the mitochondrial ADP/ATP carrier (a hexameric structure of ca. 200 kDa) is shown in Figure 9-3.

The main limitation of the tr-NOE method is its restriction to small weakly binding ligands. Also, from the obtained averaged NOE effects — although dominated by the bound conformation — no precise interatomic distances can be derived, but only rather rough estimates, so that structural models for the protein-ligand interaction have to rely on additional information and/or extensive modeling studies. Other NMR parameters like J couplings cannot be evaluated in these systems, since they are dominated by contributions from the prevailing free state of the ligand.

9.3.5 Isotope Filters

In NMR studies of protein-ligand interactions, isotope filters are an important concept for simplifying spectra of systems otherwise too complex for an analysis. The use of isotope filters allows us to observe only one molecular component of a complex at a time, which greatly facilitates the tasks of signal assignment and solving the three-dimensional structure especially for large complexes (e.g., protein-protein or protein-nucleic acid complexes, or complexes with more than two components). Isotope filters are also important when binding affinities are only moderate, so that an excess of the other component(s) is required to bring one component of interest into a uniformly complexed state. Observable NMR signals can then be restricted to the completely complexed component without being obscured by the mixture of free and complexed states of the excess components.

The use of isotope-filtering NMR techniques requires that different parts of the system under study differ in their isotopic enrichment pattern. Usually, these are the different molecular components of a complex, although recently the technique has been extended even to different parts of a single protein [18]. In the standard procedure, only *one* component is labeled with stable NMR-active isotopes (usually ^{15}N and/or ^{13}C), and the other one used in natural isotopic abundance. The NMR-active isotopes can then be used for signal assignment via 2D or 3D heteronuclear correlation spectra, and for improved resolution by adding a third heteronuclear dimension to 2D ^1H TOCSY or NOESY experiments. Since these pulse sequences require ^{15}N or ^{13}C spins, only the labeled part of the complex will show signals, and the other unlabeled components not containing ^{15}N or ^{13}C (except for the negligible low natural abundance of 0.3 % and 1.1 %, respectively) are not observed in these experiments.

Isotope filters (so-called *half filters*) can be applied separately in each spectral dimension [19], [20], so that 2D and 3D NMR experiments can be set up showing only signals *within* the labeled component. Alternatively signals between the labeled and unlabeled components can be selected, which is especially useful for observing intermolecular NOE contacts. In addition, selective deuteration of single complex components can also be used (alone or in combination with $^{15}N/^{13}C$ enrichment) for suppressing ^1H signals of single complex components or even specific signal subsets (e.g., sidechain signals) [9], [6].

Ideally, in the simple case of a 1:1 protein-ligand complex, one would need one sample with labeled protein and unlabeled ligand, and another sample containing unlabeled protein and labeled ligand (with the unlabeled components in excess in cases of only moderate binding). Bacterial overexpression techniques offer an easy and reasonably priced access to labeled proteins, so that, e.g., for protein-protein interactions (where the "ligand" is another protein) both components can be easily accessed (Figure 9-4). Also in the classic studies of the immunosuppressants cyclosporin A [21] and FK506/ascomycin in their complexes with the corresponding binding proteins cyclophilin [22] and FKBP [23], respectively, not only the proteins, but also the ligands were available in labeled and unlabeled form from fermentation.

144 NMR as a Tool in Drug Research

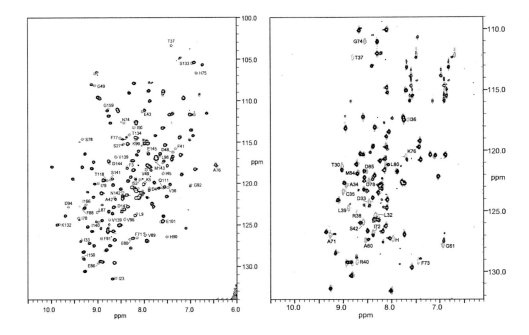

Figure 9-4 Application of differential labeling and isotope editing to the 25-kDa complex between the IIA and IIB domains of the *E. coli* glucose transporter [24]. *Left*: ^{1}H,^{15}N-HSQC spectra of [U-^{15}N]-labeled domain IIA in the free state (*gray*) and complexed with unlabeled IIB (*black*). *Right*: ^{1}H,^{15}N-HSQCs of [U-^{15}N]-IIB free (*gray*) and complexed with unlabeled domain IIA. In both cases a ca. 2-fold excess of the unlabeled protein was necessary to result in complete complexation of the ^{15}N-labeled domain ($K_D \approx$ 30 mM). Assignment labels are shown only for signals shifting significantly upon complex formation (cf. Section 3.7).

However, most ligands are usually not available with isotopic enrichment, especially when they are the product of a multi-step chemical synthesis. In this case, only the *protein* part of the complex can be accessed by modern isotope-edited NMR techniques, while the *ligand* signals will only be visible in conventional 1D or 2D homonuclear ^{1}H-NMR techniques. In a normal 1D or 2D ^{1}H spectrum, however, the ligand signals will inevitably be obscured by the hundreds to thousands of protein ^{1}H signals (Figure 9-5, left). Several techniques have been proposed to selectively suppress the ^{1}H signals of an isotopically labeled protein, based on their coupling to the NMR active isotopes (^{15}N and ^{13}C) [25]-[29] Figure 9-5 (right) shows the effects of suppressing the protein signals for a complex between [U-^{13}C,^{15}N]-labeled FKBP and the unlabeled ligand rapamycin. Similar studies have also been applied to *synthetic derivatives* of ascomycin that were also not accessible in isotopically labeled form [30].

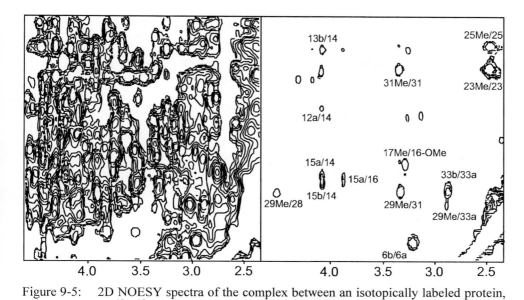

Figure 9-5: 2D NOESY spectra of the complex between an isotopically labeled protein, [U-^{13}C,^{15}N]-FKBP, and the unlabeled ligand rapamycin.
Left: in the conventional spectrum ligand signals are totally obscured by the large number of protein signals;
right: in the isotope-filtered experiment selecting for ^{12}C, ^{1}H signals of the protein are largely suppressed, and ligand signals (shown with assignment labels) are readily visible [26].

9.3.6 Measuring Binding Affinities by NMR

Binding affinities can be quantitatively measured from NMR spectra of a titration series. In most cases, it will be easier to observe the signals of the (isotope-labeled) protein while adding the ligand, e.g. in a 2D heteronuclear shift correlation experiment. In the slow exchange case, upon addition of the ligand the signals of the free protein will gradually grow weaker, while a new set of signals arises for the complexed protein. For a quantitative measurement, the intensities of the signals from free and bound protein have to be measured during the titration (which requires a reasonable signal-to-noise ratio in the spectrum).

From Figure 9-6 one can see that only K_D values in the micro- and millimolar range can be quantified, while high affinity ligands with $K_D \leq 1$ mM lead to essentially identical titration curves. A lower limit for affinity measurements is only given by the fact that at least a detectable amount of complex must be generated by the highest possible concentration of ligand, so that K_D values up to the molar range are accessible (Figure 9-6). In addition, the appearance of a second set of signals naturally raises the possibilities of signal overlap.

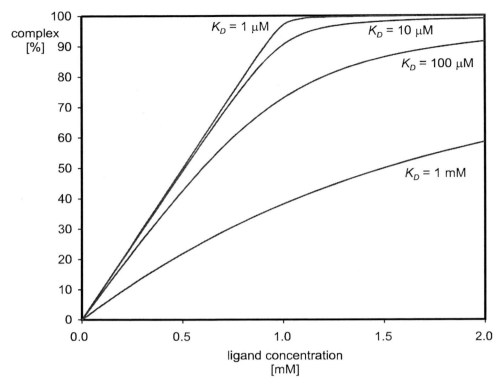

Figure 9-6: Percentage of complexed protein for a titration with a ligand (increasing ligand concentration), for a protein concentration of 1 mM (neglecting dilution) and different K_D values from 1 µM to 1 mM.

In the fast exchange regime, there is always only one set of protein signals visible, but their chemical shifts will move from that of the free protein gradually to the positions of the ligand-bound state (Figure 9-7). Since this case does not require a reliable measurement of signal intensities, only of exact chemical shifts, even NMR spectra with low signal-to-noise can be used. In addition, the assignment of the protein signals in the complexed state can be derived from the free protein's signals by following the shifting signals during the titration.

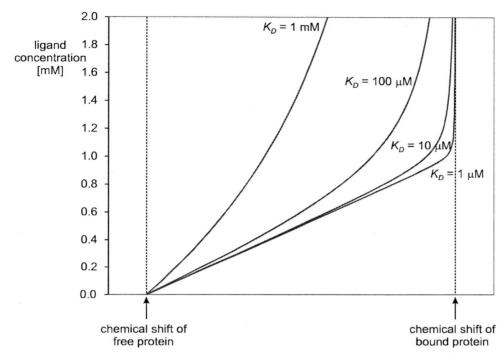

Figure 9-7: Titration curves for the chemical shifts of protein signals upon addition of ligand, in the fast exchange regime (only one signal set with averaged chemical shift). The curves are calculated for a 1 mM protein concentration and different ligand affinities from 1 μM to 1 mM.

Of course, the effects described here (appearance of a second signal set or chemical shift changes upon addition of a ligand) can also be used for a mere *qualitative* detection of binding affinity. Recently, NMR has even been employed for the screening of whole compound libraries for lead compounds in the SAR-by-NMR approach [31], [32] (cf. Section 3.9). Since these compounds have not yet been optimized for the specific target, the affinities are usually in the upper micro- to millimolar range, with complex lifetimes below 1 ms, i.e., the complexes are in the fast exchange regime. This is an advantage, because it is much easier to scan protein spectra for signal shifts — even small ones — than having to look for additional (low intensity!) peaks of the bound conformation in the slow exchange case.

The use of isotope-labeled protein for ligand titrations is advisable, since this allows one to effectively suppress the ligand signals, which becomes especially important when a large excess of the ligand is needed.

9.3.7 Binding Site Localization

The application of NMR is not limited to the mere detection and quantization of ligand affinities. Due to the correlation between individual NMR signals and specific atom positions in the molecular structure, a rather straightforward localization of the effects observed upon ligand binding is possible, yielding information about molecular regions involved in binding.

Of course, a binding site can be defined by finding NOEs between ligand and protein. A full NOE analysis, however, requires the complete assignment of both the protein and ligand signals in the complexed state. While this might be possible with reasonable effort for high-affinity complexes, matters are more complicated in the moderate binding regime. First, a complete resonance assignment of the complex is complicated by the appearance of signals from the free species, and differentially isotope-labeled samples with one or the other compound in excess have to be prepared and investigated.

In addition, NOEs build up only slowly, usually requiring mixing times in the order of 100 ms (see Table 9-2), and the required long complex lifetimes will only occur in high-affinity complexes with nanomolar K_D values (Table 9-1). For all other complexes, the observed NOEs will inevitably be averaged over the free and bound states (cf. tr-NOE, Section 3.4). While this is critical for the evaluation of intramolecular NOEs, *intermolecular* NOEs between protein and ligand obviously can only stem from the complexed states, so that at least a qualitative interpretation is possible.

However, a much faster access is possible to determine only the protein's binding site for a certain ligand. The binding of another molecule disturbs the shielding of spins in the binding site, e.g., through electrostatic or anisotropic interactions from ligand groups, or through small rearrangements in the local conformation of the binding site residues (e.g., changes of backbone and sidechain dihedral angles) — even when the global structure of the protein is not significantly changed upon ligand binding. As a result, the proton and heteronuclear chemical shifts of atoms in the binding interface will generally show significant changes, and from the assignment of the shifted resonances the binding site can often be easily mapped. Of course, in the few cases of a complete structural rearrangement of the protein-upon-ligand binding, all protein signals will be greatly affected, and no localization of the binding site is possible with this method.

Since 2D $^1H, ^{15}N$-HSQC protein spectra can be acquired in a short time (ca. 10–20 min.), usually show very good signal resolution and require only relatively inexpensive [U-^{15}N]-labeling of a protein, this type of spectra has been most widely applied to binding site mapping. Once the tedious work of assigning all resonances and solving the three-dimensional structure of the protein has been done (usually requiring also $^{15}N/^{13}C$-labeled and sometimes even additionally deuterated samples), the unlabeled ligand is simply added to a ^{15}N-labeled protein sample and the effects on the individual resonances observed (Figure 9-4).

For a detailed analysis of the signal shifts a complete assignment of all signals in the complexed state is required. Additional NMR data (e.g., from intermolecular NOEs or structural information about the bound conformations of protein and ligand) and modeling studies are needed to determine the relative orientation of the two molecules. As an example of the versatility of this method, Figure 9-8 displays the mutual binding sites of the two protein domains from the HSQC spectra in Figure 9-4, and the structural model of the complex derived from additional docking studies is shown in Figure 9-9.

However, often a mere qualitative analysis (i.e., noting all significant changes in the spectrum) suffices to map the ligand's binding site on the protein structure, as described in the next Section.

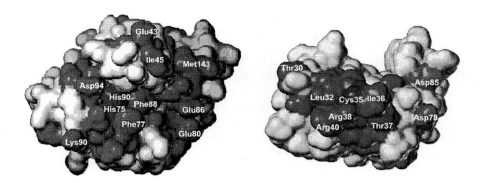

Figure 9-8: Binding interfaces of the two *E. coli* glucose PTS transporter domains IIA (*left*) and IIB (*right*), determined from chemical shift changes upon mutual complexation (cf. Figure 9-4). Residues affected by complex formation are shaded in dark gray (reproduced with permission from [24]; copyright 1997 American Chemical Society).

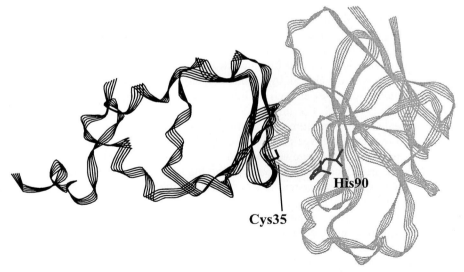

Figure 9-9: Model of the 25-kDa protein-protein complex between domains IIA (*right, in gray*) and IIB (*left, in black*) derived from the binding interfaces (Figure 9-8) and docking studies. For clarity, only the protein backbones are depicted as ribbons. The two residues His90 (domain IIA) and Cys35 (domain IIB) participate in the transphosphorylation reaction occurring in the complexed state [24].

9.3.8 Solvent-Accessible Surfaces

Complementary to mapping the contact surfaces between two molecules (e.g., from chemical shift changes), the solvent-accessible surface of a bound molecule consists of all surface regions not involved in binding or otherwise blocked by the complex partner. This information not only allows us to indirectly deduce the binding site, but is also important for knowing where modifications can be made to a ligand without interfering with its intermolecular interactions.

The usefulness of this approach has been demonstrated in the two complexes cyclosporin A / cyclophilin and ascomycin / FKBP [33], [34], each with ^{13}C-labeled ligand and unlabeled protein. The ^{1}H T_1 relaxation rates of the ligand signals (in a ^{13}C-edited experiment suppressing the protein signals) were compared with and without ca. 4-5 mM HyTEMPO (4-hydroxy-2,2,6,6-tetramethyl-piperidinyl-1-oxy) in the solution of the complex: the unpaired electron of the N-oxide group of HyTEMPO greatly accelerates T_1 relaxation for all ligand regions that are solvent accessible.

Similar measurements can also be conducted using the photo-CIDNP effect (photochemically induced nuclear polarization). Here radicals are transiently generated by irradiating solutions of appropriate reagents, and the electron polarization of the recombination products induces large intensity changes for the NMR signals of all spins that come into close contact with them (for an application to glycoside binding to lectins, see [35]).

9.3.9 Screening Protein-Ligand Interactions

Due to the paradigm shift in the pharmaceutical industry towards combinatorial methods and the screening of compound libraries, several techniques have been developed recently to combine the possibilities of NMR with these drug development approaches.

Based on the previously described binding site mapping from chemical shift changes, a screening scheme was developed by Fesik and co-workers delivering qualitative and quantitative binding affinities as well as binding site information useful for developing improved ligands [31], [36]. For this "SAR-by-NMR" approach, [U-^{15}N]-labeled protein samples are treated with a mixture of potential ligands and the ^{1}H,^{15}N-HSQC spectrum is then compared to that of the free protein. Isotopically labeled protein is required because of the large number and unknown positions of the ligand signals (typically, 10–20 potential ligands are added simultaneously). The high sensitivity and good resolution of 2D ^{1}H,^{15}N-HSQC spectra allow one to run ca. 100 of them in 24 h, screening well over 1000 compounds per day.

If no chemical-shift changes occur upon adding the ligand mixture, then any significant affinity (with K_D values up to ca. 100 mM) can be excluded for all test compounds (typical concentrations are ca. 0.5 mM for the protein and 10 mM for each ligand, see Figure 9-2). If significant signal shifts are detected, then the compound(s) responsible have to be identified by further testing subsets of the original combination.

The main advantages of this screening procedure are:

1. high throughput due to high sensitivity of the $^1H,^{15}N$-HSQC experiment and easy evaluation of signal shifts (> 1000 samples per day);
2. high sensitivity even for low affinity ligands (up to millimolar K_D values);
3. easy access to quantitative K_D data from following the shifting protein signals in a titration series (for these weakly binding ligands the system is usually in the fast exchange regime, see Table 9-1 and Figure 9-7);
4. easy access to binding site information from the sequential assignment of the shifted protein resonances.

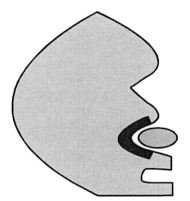 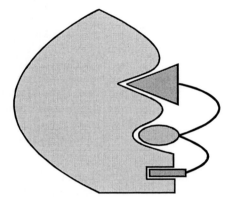

Figure 9-10: Principle of SAR-by-NMR. Protein signals shifts upon ligand addition not only allow for fast screening of potential ligands, they also reveal the specific binding sites (region shown in *black*, *left* panel). Weakly binding ligands to adjacent sites can then be systematically connected (based on this structural information) to yield high-affinity ligands (*right* panel).

Especially the last point makes it easy to do the next step from *identifying* some weakly binding compounds to *developing* improved lead compounds of the next generation (Figure 9-10). One first has to find at least two compounds binding to different, yet spatially close protein regions, and identify their binding sites and binding orientation in more detail (e.g., from NOE data). An improved ligand can then be constructed by connecting the two initial ligands, in their proper relative orientation, with an appropriate linker (the ideal length of which can also be guessed from the distance of the two binding sites).
Several successful applications of this SAR-by-NMR approach have already been published [32], [37], [38], one of which is shown in Figure 9-11.
Besides chemical shift changes as in SAR-by-NMR, other NMR parameters might also be used to detect ligand binding in screening procedures. As described in the tr-NOE section, the intensity of NOE effects changes dramatically when a small ligand binds (even transiently) to a biomacromolecule. Also, the narrow lines of a small ligand broaden considerably upon binding to a protein due to changes in T_2, and its diffusion rate in

solution drops sharply. All these NMR parameters can therefore be employed to detect and identify binding ligands from mixtures of varying complexity.

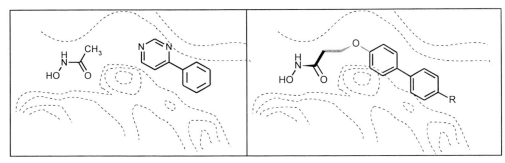

Figure 9-11: Application of SAR-by-NMR to the development of a metalloprotease inhibitor. Screening of a library of small organic compounds revealed acetohydroxamic acid and 4-phenylpyrimidine as weakly binding to stromelysin (K_D 17 mM and 20 mM, respectively) at adjacent sites (*left*, the contour lines symbolize the protein surface). After introduction of an appropriate linker (shown in grey, right panel) and systematic structural variations, a highly potent inhibitor of stromelysin and gelatinase A (IC_{50} = 25 nM, *right*) could be developed within a few months. [37], [38].

A tr-NOE approach has been used, e.g., to find oligosaccharides binding to agglutinin [39] and antagonists for E-selectin [40]. Selecting binding ligands based on their lower diffusion rate can be accomplished using diffusion-edited NMR spectroscopy and has also been used for detecting protein-ligand interactions, mostly in combination with isotope filtering [41], [42], [43].

9.3.10 Conclusion

Based on long-established techniques as well as on very recent developments, NMR has a lot to offer in the field of drug discovery and development. Its applicability to ever larger biomacromolecules has made it one of the premier methods for solving three-dimensional structures of biologically active molecules and complexes in general. Structures of proteins and protein-ligand complexes (and increasingly also information about their dynamics) are needed for modeling the interactions between drugs and their macromolecular targets, and NMR spectroscopy can deliver valuable information in these areas. In addition, specific NMR techniques have been developed to directly yield useful information for drug development purposes, such as NMR-based screening and lead compound improvement.

In the fascinating field of intermolecular interactions, NMR will continue to make important contributions to both our theoretical understanding of biomolecular mechanisms and its application to a modern, structure-based drug design.

References

1. Protein Data Bank, Brookhaven National Laboratory (http://www pdb bnl gov); currently being transferred to the Research Collaboratory for Structural Bioinformatics, RCSB (http://www rcsb org)
2. M P Williamson, T F D Havel, K Wüthrich, *J Mol Biol* **182**, 295-315 (1985)
3. K Wüthrich, NMR of Proteins and Nucleic Acids, 1986, Wiley, New York
4. H -J Böhm, G Klebe, H Kubinyi, Wirkstoffdesign, 1996, Spektrum Akad Verlag, Weinheim, ISBN 3-8274-0012-0
5. For a recent review: V Dötsch, G Wagner, *Curr Opin Struct Biol* **8**, 619 -623, (1998)
6. K H Gardner, X Zhang, K Gehring, L E Kay, *J Am Chem Soc* **120**, 11738-11748 (1998)
7. R A Venters, C C Huang, B T Farmer II, R Trolard, L D Spicer, C A Fierke, *J Biomol NMR* **5**, 339-44 (1995)
8. K H Gardner, L E Kay, *Ann Rev Biophys Biomol Structure* **27**, 357-406 (1998)
9. C H Arrowsmith, Y S Wu, *Progr NMR Spectrosc* **32**, 277-286 (1998)
10. V Dötsch, G Wagner, *Curr Opin Struct Biol* **8**, 619-623 (1998)
11. G C K Roberts, *Curr Opin Biotechnol* **10**, 42-47 (1999)
12. L E Kay, D R Muhandiram, G Wolf, S E Shoelson, J D Forman-Kay, *Nature Struct Biol* **5**, 156-163 (1998)
13. J W Peng, G Wagner, *Biochemistry* **31**, 8571-8586 (1992)
14. A G Palmer III, *Curr Opin Struct Biol* **7**, 732-737 (1997)
15. D Neuhaus, M P Williamson, The nuclear Overhauser effect in structural and conformational analysis, 1989, VCH, New York
16. T Huber, M Klingenberg, K Beyer, *Biochemistry* **38**, 762 -769 (1999)
17. G Bertho, G J Gharbi-Benarous, M Delaforge, J P Girault, *Bioorg Med Chem* **6**, 209-21 (1998)
18. T Yamazaki, T Otomo, N Oda, Y Kyogoku, K Uegaki, N Ito, Y Ishino, H Nakamura, *J Am Chem Soc* **120**, 5591-5592 (1998)
19. G Otting, H Senn, G Wagner, *J Magn Reson* **70**, 500-505 (1986)
20. G Otting, K Wüthrich, *Quart Rev Biophys* **23**, 39-96 (1990)
21. G Gemmecker, M Eberstadt, S Golic Grdadolnik, H Kessler, A Buhr, B Erni, *Biochemistry* **36** (24), 7408–7417 (1997)
22. a) C Weber, G Wider, B von Freyberg, R Traber, W Braun, H Widmer, K Wüthrich, *Biochemistry* **30**, 6563-74 (1991)
    b) S W Fesik, R T Gampe Jr , H L Eaton, G Gemmecker, E T Olejniczak, P Neri, T F Holzman, D A Egan, R Edalji, R Simmer et al, *Biochemistry* **30**, 6574-83 (1991)
23. Y Theriault, T M Logan, R Meadows, L Yu, E T Olejniczak, T F Holzman, R L Simmer, S W Fesik, *Nature* **361**, 88-91 (1993)
24. R P Meadows, D G Nettesheim, R X Xu, E T Olejniczak, A M Petros, T F Holzman, J Severin, E Gubbins, H Smith, *Biochemistry* **32**, 754-65 (1993)
25. G Otting, K Wüthrich, *J Magn Reson* **85**, 586-594 (1989)
26. G Gemmecker, E T Olejniczak, S W Fesik, *J Magn Reson* **96**, 199-204 (1992)
27. M Ikura, A Bax, *J Am Chem Soc* **114**, 2433-2440 1992

28. K Ogura, H Terasawa, F Inagaki, *J Biomol NMR* **8**, 492-498 (1996)
29. C Zwahlen, P Legault, S J F Vincent, J Greenblatt, R Konrat, L E Kay, *J Am Chem Soc* **119**, 6711-6721 (1997)
30. A M Petros, M Kawai, J R Luly, S W Fesik, *FEBS Lett* **308**, 309-314 (1992)
31. S B Shuker, P J Hajduk, R P Meadows, S W Fesik, *Science* **274**, 1531-1534 (1996)
32. P J Hajduk, J Dinges, G F Miknis, M Merlock, T Middleton, D J Kempf, D A Egan, K A Walter, T S Robins, S B Shuker, T F Holzman, S W Fesik, *J Med Chem* **40**, 3144-3150 (1997)
33. S W Fesik, G Gemmecker, E T Olejniczak, A M Petros, *J Am Chem Soc* **113**, 7080-7081 (1991)
34. A M Petros, P Neri, S W Fesik, *J Biomol NMR* **2**, 11-8 (1992)
35. H C Siebert, R Adar, R Arango, M Burchert, H Kaltner, G Kayser, E Tajkhorshid, C W von der Lieth, R Kaptein, N Sharon, J F Vliegenthart, H J Gabius, *Eur J Biochem* **249**, 27-38 (1997)
36. H Kessler, *Angew Chemie Int Ed Engl* **36** (8), 829–831 (1997)
37. P J Hajduk, G Sheppard, D G Nettesheim, E T Olejniczak, S B Shuker, R P Meadows, D H Steinman, G M Carrera Jr., P A Marcotte, J Severin, K Walter, H Smith, E Gubbins, R Simmer, T F Holzman, D W Morgan, S K Davidsen, J B Summers, S W Fesik, *J Am Chem Soc* **119**, 5818-5827 (1997)
38. E T Olejniczak, P J Hajduk, P A Marcotte, D G Nettesheim, R P Meadows, R Edalji, T F Holzman, and S W Fesik, *J Am Chem Soc* **119**, 5828-5832 (1997)
39. B Meyer, T Weimar, T Peters, *Eur J Biochem* **246**, 705-709 (1997)
40. D Henrichsen, B Ernst, J L Magnani, W -T Wang, B Meyer, T Peters, *Angew Chem Int Ed* **38**, 98-102 (1999)
41. M J Shapiro, J R Wareing, *Curr Opin Chem Biol* **2**, 372-375 (1998)
42. P J Hajduk, E T Olejniczak, S W Fesik, *J Am Chem Soc* **119**, 12257-12261 (1997)
43. N Gonnella, M Lin, M J Shapiro, J R Wareing, X Zhang, *J Magn Reson* **131**, 336-338 (1998)

Chapter 10

B. Chankvetadze, G. Blaschke, G. Pintore

10 Ligand-Cyclodextrin Complexes

10.1 Introduction

Historically, nuclear magnetic resonance (NMR) spectroscopy was the instrumental technique that provided the first experimental evidence for inclusion complex formation of cyclodextrins (CD) and guest molecules in the liquid phase [1]. Since that time NMR spectroscopy remains one of the key techniques for studying CD complexes with their ligands [2].

What are the unique features of NMR spectroscopy which other techniques do not offer?

1. NMR spectroscopy allows a clear differentiation between inclusion and other possible (external) interactions. This is an important advantage over other techniques that are more global and do not provide convincing proof of an inclusion.
2. NMR spectroscopy provides considerable information about the environment of individual atoms and their involvement in intermolecular and intramolecular interactions.
3. Enantiotopic signals may be distinguished in a chiral environment in the NMR spectrum. This allows a racemic sample to be used for a study of enantioselective binding.
4. NMR spectroscopy may provide detailed structural information on the inclusion complexes via the Nuclear Overhauser Effect (NOE) as well as on the dynamics of the complexes by classical line-shape analysis and by studying different spin-matrix effects.

Thus, NMR spectroscopy has the potential to provide almost complete information on ligand-CD interactions (stoichiometry, binding constants, free energy, enthalpy and entropy of complex formation, dynamics and structure of the complexes) in a solution. The kinetics of complex formation is usually too fast on the NMR time scale and therefore difficult to follow using this technique. However, an information on the kinetics of complex formation may be also obtained from NMR data.

The object of this chapter is to provide the reader with a methodology where and how NMR spectroscopy may be used for obtaining information on ligand-CD interactions in solution.

10.2 Cyclodextrins and their Properties

CDs (Figure 10-1) were discovered by the French scientist Villiers, who obtained this material from the potato starch digest of *Bacillus amylobacter* and named it "cellulosine" because of its similarity in some respects to cellulose [3].
Later, Schardinger found that one of the thermophilic bacteria, which he called strain II was able to dissolve starch and form crystalline "dextrins" [4]. Although Schardinger did not propose a structure for his crystalline dextrins, he made several observations regarding their cyclic structure. One of his important findings was that: "With various compounds, the crystalline dextrins form loose complexes" [4].

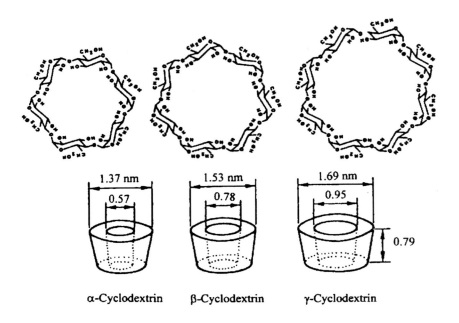

Figure 10-1: Structure and approximate geometric dimensions of α-, β-, and γ-CD molecules

The next major contribution to CD chemistry was the method of Freudenberg and Jacobi [5] for the isolation of pure α- and β-dextrins. They also isolated an additional crystalline dextrin and named it γ-dextrin. In 1936, the ring structures of α-, β- and γ-dextrins were tentatively proposed by the same group [6]. This hypothesis was verified experimentally by Freudenberg and Meyer-Delius, who found that Schardinger's dextrins are cyclic oligosaccharides composed solely of D-glucosyl residues bonded by α-(1,4)-glycosidic linkages [7].
Freudenberg and co-workers correctly proposed a cyclic structure for the CDs. However, the numbers of D-glucosyl moieties which they assigned to the α- and β-dextrin rings (five and six, respectively) were incorrect. The correct values of six and seven were determined

by French and Rundle [8], who also proposed the names "cyclohexaamylose" and "cycloheptaamylose" for α- and β-dextrins, respectively.
According to current knowledge, CDs are cyclic, α-(1,4)-linked oligomers of D-glucopyranose, each glucopyranosyl residue being in the 4C conformation. The properties of α-, β- and γ- CDs (Figure 10-1) are summarized in Table 10-1 [9].
Freudenberg and co-workers [10] refined the finding of Schardinger [4] and proposed that CD complexes with most of their ligands are of the inclusion type. The first experimental evidence for molecular inclusion by CDs in the solid state came from X-ray crystallography. Hybl et al. [11] determined the structure of the complex of α-CD with potassium acetate by using three-dimensional X-ray diffraction data. In the solid state, the acetate anions are included in the α-CD cavity.

Table 10-1: Characteristics of α-, β- and γ-CDs (reproduced by permission from [9]).

Characteristics	α	β	γ
Number of glucose units	6	7	8
Molecular mass	972	1135	1297
Solubility in water, at room temp, mg/ml	145	18.5	232
$[\alpha]_D$ 25°C	150±0.5	162.5±0.5	177.4±5
Cavity diameter, Å	4.7-5.3	6.0-6.5	7.5-8.3
Height of torus, Å	7.9±0.1	7.9±0.1	7.9±0.1
Diameter of outer periphery, Å	14.6±0.4	15.4±0.4	17.5±0.4
Approx. volume of cavity; Å	174	262	427
Approx. cavity volume in 1 mol CD (ml)	104	157	256
In 1 g CD (ml)	0.10	0.14	0.20
Crystal forms (from water)	Hexagonal plates	Monoclinic parallelograms	Quadratic prisms
Crystal water, wt %	10.2	13.2-14.5	8.13-17.7
Diffusion constant at 40°C	3.443	3.224	3.000
Hydrolysis by *A. oryzae* α-amylase	Negligible	Slow	Rapid
V_{max} value, min^{-1}	5.8	166	2300
Relative permittivity (on incorporating the toluidinyl group of 6-*p*-toulidynil-naphthalene 2-sulfonate) at pH=5.3, 25°C	47.5	52.0	70.0
(on incorporating the naphthalene group)	too small	29.5	39.5
pK (by potentiometry) at 25°C	12.332	12.202	12.081
Partial molar volumes in solution mL mol^{-1}	611.4	703.8	801.2
Adiabatic compressibility in aqueous solutions, mL (mol^{-1} bar^{-1})×10^4	7.2	0.4	–5.0

Some compounds owe their inclusion complex formation ability to the crystal structure in the solid state. One example is urea, for which an inclusion complex formation ability is not a microscopic property of a single molecule but a property of macroscopic solid matter. Therefore, urea is not able to form inclusion complexes in solution. In contrast, the inclusion complex formation ability of CDs is a property of the molecule allowing the inclusion of guest molecules in the solid state as well as in solution. This property of CDs in a solution was first experimentally confirmed by NMR spectroscopy, as mentioned above [1]. Cramer discovered another important property of CDs, i.e. their ability "to distinguish not only molecules with different shape but optical antipodes too" [12]. At present most of applications of CDs are based on these two properties: formation of (inclusion) complexes and the stereoselectivity of this process [13].

10.3 Application of Cyclodextrins in Drug Development and Drug Analysis

Together with the inclusion complex formation ability and the stereoselectivity of this process, CDs possess a number of properties that make them useful candidates for modifying of the properties of drugs. It is important that CDs are nontoxic, biodegradable and not very expensive compounds that are commercially available from several suppliers worldwide.

In drug development, CDs are used for increasing the water solubility of lipophilic compounds or, in contrast, for decelerating of the dissolution or biotransformation of very hydrophilic compounds by their encapsulation in a hydrophobic CD-cavity, for preserving the odor of odorants or the taste of food additives, or for masking unpleasant odors and tastes. Selected examples of CD applications in drug development and technology are summarized in Table 10-2 [14], [15].

Another important application of CDs is (chiral) drug analysis. In this field, CDs are used in combination with UV-VIS spectrometry, fluorescence and NMR spectroscopy, mass spectrometry, in electrochemical analysis and most widely in various separation techniques such as thin-layer chromatography (TLC), gas chromatograpy (GC), high-performance liquid chromatography (HPLC), super- and subcritical fluid chromatography (SFC) [13] and capillary electrophoresis (CE) [9], [16].

The studies of ligand-CD interactions are basically driven by the following goals: CDs as well-organized and well-characterized objects of medium size represent good models of enzymes and molecular receptors. A knowledge of ligand-CD complexation can significantly contribute to the better understanding of the nature of the noncovalent intermolecular interactions that play an important role in the living body. Secondly, the understanding of intrinsic mechanisms of complexation and chiral recognition by CDs allows us to optimize technological schemes for the application of CDs as carriers for biologically active compounds [14], [15] and in separation science and other analytical techniques [9], [16].

Table 10-2: Selected applications of cyclodextrins in drug development and technology [14], [15].

Effect	Examples
Control of solubility	Itraconazole/β-CD oral delivery form (improved solubility); glibornuride/β-CD; testosterone (progesterone)/branched CD; lotepredrol/α-CD (nasal delivery form); flufenamic acid/triacetyl-β-CD complex (limited solubility); hydrocortisone, digitoxin, diazepam, indomethacin, chlorpromazine.
Stabilization	Nicardipine/β-CD (stability increase to the light); digoxin/β-CD (stability increase to acid-catalyzed hydrolysis), O^6-benzyl-guanine/SBE-β-CD (stability increase to acid-catalyzed hydrolysis); thymopentin/2-HP-β-CD (stability enhancement in aqueous solution).
Absorption enhancement	Hexetil hydrochloride/α-CD (improvement of oral bioavailability); ketoconazole/β-CD (improvement of oral bioavailability); ethyl 4-biphenylyl acetate /2-HP-β-CD (improvement of rectal absorption); estradiol/DM-β-CD (enhancement of nasal absorption); nifedipine/2-HP-ß-CD (improvement of the oral bioavailability).
Alleviation of toxicity	Chlorpromazine/β-CD derivatives (reduced muscular damage caused by chlorpromazine); CD complexes of acidic antiinflammatory drugs exhibit less ulcerogenic potency; gentamicin/CD-sulfate complex is characterized by less renal impairment.
Cyclodextrin-based drug delivery systems	The hydrophilic cyclodextrins have been extensively applied to enhance the oral bioavailability of steroids, cardiac glycosides, nonsteroidal anti-inflammatory drugs, barbiturates, antiepileptics, benzodiazepines, antidiabetics, vasodilators, etc.; Delayed and prolonged release of diltiazem and molsidormine was achieved by their complexation with CD derivatives, modified release of nifedipine can be achieved by its complexation with 2-HP-β-CD; furosemide and piretanide (loop diuretics) release can be modified after complexation with DM-β-CD. Prednisolone dosage forms can be optimized also by complex-formation with 2-HP-β-CD.
Site-specific delivery	Estradiol (testosterone, dexamethasone, benzyl penicillin)/HP-β-CD; N-leucine-enkephalin/6-amino-6-deoxy β-CD (covalent compound);
Cell surface targeting	Methicilline (galactosa, α–glucosylgluconamide)-β-CD conjugate;

10.4 NMR Spectroscopic Studies of Ligand-CD Complexes

The terms ligand and substrate are operational and do not relate to the nature of the interaction or the properties of the complex. In principle, it is arbitrary which interactant is made to serve as the substrate and which as the ligand. In this text CDs are considered as the hosts (receptors) and their guests as ligands.

10.4.1 Stoichiometry of Ligand-CD Complexes

Since maximal fractions of both counterparts are complexed in intermolecular complexes in their stoichiometric mixture, the information about a stoichiometry is useful for analytical as well as for technological purposes. On the other hand, this information can be applied to the calculation of the binding constants.

Several techniques such as the continuous variation method, mole ratio method, solubility phase diagrams, the direct measurement of complex molecular masses, etc. are available for a determination of the stoichiometry of intermolecular complexes. The commonly used method is the continuous variation method, which was originally proposed by Ostromislensky in 1911 [17] and modified by Job [18].

A Job's graph is constructed by plotting an appropriate physico-chemical parameter (absorbance, chemical shift, solubility, etc.) against the molar fraction of a given component in a binary mixture. In a single, stable complex, the plot has a triangular form with a maximum (the apex of the triangle) indicating the stoichiometric ratio of components in a given complex. The formation of weak complexes results in curved plots.

For the construction of Job's plot, a series of solutions are prepared by mixing different volumes of equimolar solutions of the two components and diluting them to a constant volume, resulting in solutions having identical overall molar concentrations but different molar ratios of the components. The appropriate parameter is measured for these solutions and plotted against the molar ratio of the components. The maximum of this plot indicates the stoichiometry of a ligand-CD complex.

The free and complexed ligands as well as the CDs have different NMR characteristics (i.e., chemical shifts). However, ligand-CD interactions are commonly fast on the NMR time scale. Therefore, a set of averaged signals between the signals of free and complexed ligands (or CD) will be observed in the NMR spectrum instead of two sets of signals corresponding to the free and complexed ligand (or CD) (Figure 10-2). This so-called complexation induced-chemical shift (CICS) relates to the concentration of the complexed ligand (or CD). A higher CICS for a given ligand-CD mixture means a higher degree of complexation. Therefore, for a given total amount of ligand and CD, the maximal CICS will be observed at the ratio of the components where the concentration of complex reaches a maximum. According to the continuous variation method, this will take place at a stoichiometric ratio (Figure 10-3) [19]. The continuous variation method based on NMR spectroscopy is widely used for the determination of ligand-CD complexes. Some of these studies are summarized in Table 10-3.

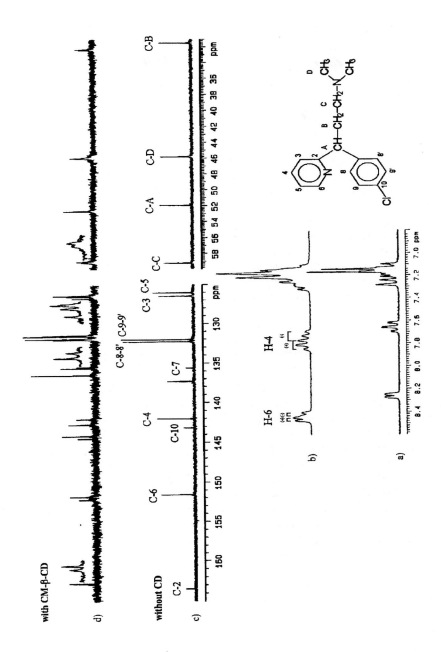

Figure 10-2: The aromatic part of ^{1}H-NMR spectra of (+)/(-)=2/1 mixture of chlorpheniramine in the absence **a** and in the presence **b** of an equimolar amount of CM-β-CD, and ^{13}C-NMR spectra of the same mixture (**c** and **d** respectively). ^{2}H$_2$O was used as a solvent (reproduced by permission from [19]).

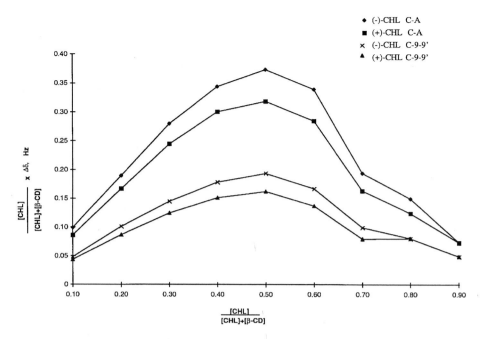

Figure 10-3: Job's plots for (±) chlorpheniramine(CHL)/β-CD complex (reproduced by permission from [19]).

The clear advantage of NMR spectroscopy for the construction of Job's plots is that this technique provides a multiple set of data that makes the results statistically more reliable. A disadvantage of NMR spectroscopy is the necessity for a set of measurements and large quantities of the substances. As a result NMR spectroscopy is a relatively expensive and time-consuming technique. However, most spectrometric techniques suffer from the last disadvantage. From this viewpoint, soft ionization mass-spectrometric (MS) techniques, such as fast-atom bombardment (FAB-MS) and electrospray ionization (ESI-MS), possess some advantages over NMR spectroscopy. MS techniques allow one to determine an M/Z ratio, and in this way a stoichiometry of the complex from a single experiment. MS is a useful technique for identification of complexes possessing different stoichiometry. However, this is not a powerful enough technique to differentiate between various complexes of the same stoichiometry.

Table 10-3 Selected applications of NMR spectroscopy to determination the stoichiometry and binding constants of ligand-CD complexes.

Ligand	CD	Stoichiomety	Binding constant, M^{-1}	Ref.
Benzhydrylamine derivatives	β-CD	1:2	Not determined	[20]
Benzoic acid	β-CD	1:1	48.5	[21]
(±)-1,1′-binaphthyl-2,2′-diyl-hydrogen phosphate	β-CD	1:1	261 R-(-) 338 S-(+)	[22]
	CM-β-CD	1:1	87 R-(-) 147 S-(+)	
Bromoadamantane	α-CD	1:2	Not determined	[23]
	β-CD	1:1	Not determined	[24]
	γ-CD	1:1	Not determined	
Chalcone derivative [2′-carboxymethoxy-4,4′-bis (3-methyl-2-butenyloxy] chalcone	α-CD	1:1, 1:2	K_{11}=80±15 K_{12}=1100±50	[25]
	β-CD	1:1, 1:2	K_{11}=2000±50 K_{12}=270±20	
	γ-CD	1:1, 1:2	K_{11}=1500±300 K_{12}=40±10	
Chlorogenic acid	β-CD	1:1	758±34/23	[26]
(±)-Chlorpheniramine	β-CD	1:1	757 (+)-CHL 446 (-)-CHL	[19]
	CM-β-CD	1:1	827 (+)-CHL 700 (-)-CHL	
	TM-β-CD	1:1	53 (+)-CHL 58 (-)-CHL	
(±)-Dimethindene	β-CD	1:1	457 S(+)-DM 504 R(-)-DM	[27]
Dimethindene	β-CD	1:1	Not determined	[28]
1,4-Dimethyl-bicyclo [2.2.2] octane	α-CD	1:2	Not determined	[29]
	2-mono-tosyl-α–CD	1:1		
Etilefrine	β-CD	1:1	Not determined	[30]
	heptakis (2,3-di-O-acetyl) β-CD	1:1		
Fencamfamine	Diacetyl-β-CD γ-CD	mixed	Not determined	[31]
(S)/(+)-Fenoprofen	β-CD	1:1	2010±20	[32]
(±)-Flurbiprofen	β-CD	1:1 (2:2)	5700 (-)-FP 3500 (+)-FP	[33]
Gliclazide	β-CD	1:2	1094	[34]
Indomethacin	β-CD	1:1	760	[35]

Ligand	CD	Stoichiometry	K	Ref
(±)-Metomidate	β-CD	1:1	483 and 655	[36]
	CM-β-CD	1:1	350 and 423	
	SBE-β-CD	1:1	397 and 564	
(±)-Mianserine	β-CD	1:1	Not determined	[37]
	CM-β-CD	1:1	Not determined	
	SBE-β-CD	1:1	Not determined	
Monohydroxy-pyridines	β-CD	1:1	2-hydroxypyridine 41 ± 9 3-hydroxypyridine 95 ± 5 (lactim isomer) 3-hydroxypyridine 46 ± 3 (lactam isomer) 4-hydroxypyridine 79 ± 25	[38]
1-Naphtalenesulfonate	β-CD	1:1	209	[39]
Naproxen	β-CD	1:1	420 (alkaline medium)	[40]
Naproxen	α-CD	1:1	Not determined	[41]
	β-CD	1:1	1702	
	γ-CD	1:1 or 2:1	Not determined	
	Me-β-CD	1:1	6892	
	HE-β-CD	1:1	2145	
Naproxen	HP-β-CD	1:1	2581	
(±)-α-Pinene	α-CD	1:2	$4 \cdot 10^6$ (+)-enantiomer $1.6 \cdot 10^7$ (−)-enantiomer	42
	β-CD	1:1	Not determined	
	γ-CD	1:1	Not determined	
	TM-α-CD	1:2	$1.2 \cdot 10^5$ (+)-enantiomer $2.1 \cdot 10^5$ (−)-enantiomer	
	TM-β-CD	1:1	$1 \cdot 10^3$ (+)-enantiomer $1.3 \cdot 10^3$ (−)-enantiomer	
	TM-γ-CD	not detected		
Piroxicam	β-CD	1:1 (two)	113 (averaged for two complexes)	43
Propranolol	β-CD	1:1	Not determined	44
Steroids	β-CD	1:1 or 1:2 depends on the structure	Not determined	45
(±)-Verapamil	β-CD	1:1	272 ± 34 (+)-VP 207 ± 59 (−)-VP	46
	TM-β-CD	1:1	6 ± 1 (+)-VP 30 ± 7 (−)-VP	

Another critical point that needs to be considered when using NMR spectroscopy for constructing a Job's plot is that the property to be measured should be additive. In the ligand-CD pairs where multiple complexation takes place, one and the same group may be involved in several complexes. Consequently, it will be affected in different ways. As a result, the requirement for additivity will be not fulfilled, and it will be difficult to observe a clear relationship between the CICS and the degree of complexation. This can make NMR spectroscopy impracticable for a determination of the stoichiometry of ligand-CD complexes. In addition, the continuous variation method provides information on the ratio of the components and not on the absolute stoichiometry. This means that it is impossible to distinguish, for example, between 1:1 and 2:2 complexes using this method. The above-mentioned MS techniques do not suffer from this disadvantage and may be used as complementary tools to NMR spectroscopy. However, a risk of a false peak formation make MS techniques not absolutely reliable and an application of a complementary technique is advised together with MS.

10.4.2 Binding Constants of Ligand-CD Complexes

The binding or equilibrium constant (K_e) characterizes the equilibrium between the free ligand ([A]$_e$), the free receptor ([B]$_e$) and the ligand-receptor complex(es) ([AB]$_e$). A higher value of the binding constant

$$K_e = \frac{[AB]_e}{[A]_e[B]_e} \qquad (1)$$

means a higher degree of complexation. In principle, any technique that provides the information about the concentration of a ligand-CD complex can be used for the determination of the equilibrium constants.

Benesi and Hildebrand derived the equation for the calculation of association constants of iodine with aromatic hydrocarbons by spectrophotometry in 1949 [47]. At present, many techniques used for the determination of binding constants represent the various modifications of the Benesi-Hildebrand equation [47]. That commonly used in spectrometric techniques is Scott's modification of this technique [48].

As already mentioned in the part a), a statistically averaged CICS will be most useful for the characterization of the degree of ligand-CD complexation.

Let us assume that the initial molar concentration of the solute is [A]$_i$ and a complex between A and B is formed in of 1:1 stoichiometry. Then in the case of complete complexation of the analyte the concentration of the complex will be [A]$_i$. The equilibrium concentration of the complex [AB]$_e$ can be expressed as follows:

$$[AB]_e = \frac{\Delta\delta_{obs}}{\Delta\delta_s}[A]_i \qquad (2)$$

Where $\Delta\delta_{obs}$ is the observed chemical shift and $\Delta\delta_s$ is the chemical shift at the saturation (that is when the analyte is completely complexed). Combination of eq. 1 and 2 gives:

$$K_e = \frac{\frac{\Delta\delta_{obs}}{\Delta\delta_s}[A]_i}{([A]_i - \frac{\Delta\delta_{obs}}{\Delta\delta_s}[A]_i)([B]_i - \frac{\Delta\delta_{obs}}{\Delta\delta_s}[A]_i)} \qquad (3)$$

After simple mathematical transformations [16] and assuming that one of the components (in our case B) is taken in excess equation 3 can be rewritten as follows:

$$\frac{1}{\Delta\delta_{obs}} = \frac{1}{K_e \Delta\delta_S [B]_i} + \frac{1}{\Delta\delta_S} \qquad (4)$$

The plot $1/\Delta\delta_{obs}$ versus $1/[B]_i$ for a 1:1 complex will be linear. The slope is equal to $1/K_e\Delta\delta_s$, and intercepts on the ordinate is $\frac{1}{\Delta\delta_s}$ (Figure 10-4) [19].

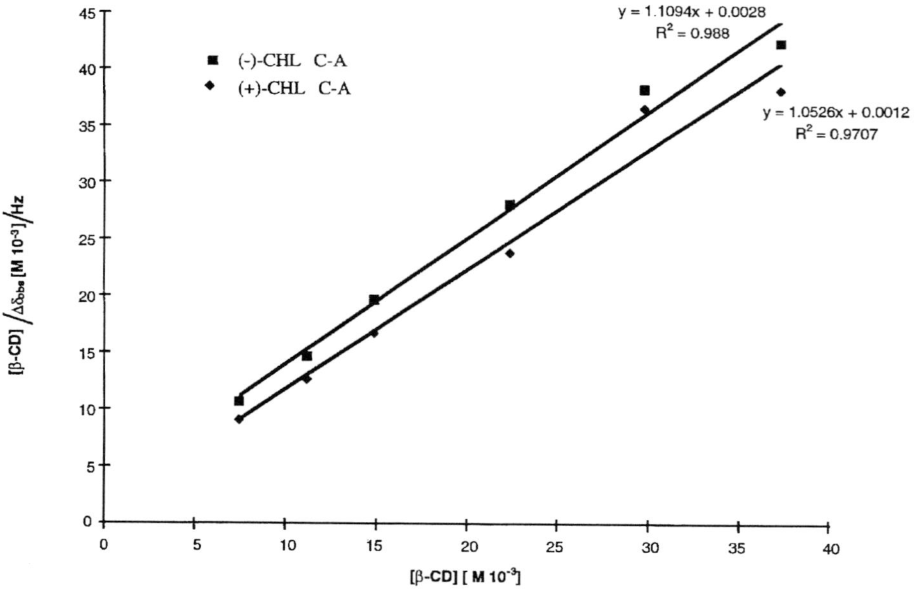

Figure 10-4: Scott's plots for (±)-CHL/β-CD complex (reproduced by permission from [19]).

An alternative way for the graphical solution of equation 3 was proposed by Foster and Fyfe [49]. According to this technique, equation (4) is rewritten in the following form:

$$\frac{\Delta\delta_{obs}}{[B]_i} = K_e(\Delta\delta_S - \Delta\delta_{obs}) \qquad (5)$$

Plotting $\Delta\delta_{obs}/[B]_i$ versus $\Delta\delta_{obs}$ yields $-K_e$. The value of $\Delta\delta_S$ can be obtained by extrapolation to an infinitely dilute solution.

The enthalpy (ΔH) and entropy (ΔS) of the complexation can be determined based on the temperature dependence of the equilibrium constant (K_e) using the classical Gibbs-Helmholtz equation [33], [50], [51]:

$$\ln K_e = -\frac{\Delta H}{RT} + \frac{\Delta S}{R} \qquad (6)$$

Where T is the absolute temperature and R is the ideal gas constant.

The ligand-CD association constants determined using NMR-spectroscopy are summarized in Table 10-3 [20-46]. Several book chapters and review papers describe the methodology, statistical variance and applicability of the measurements of binding constants using NMR spectroscopy [50-52]. These data represent an important value for an understanding the relations between the chemistry and binding properties of CDs. On the other hand, the binding constants determined by NMR spectroscopy may be used in order to explain results observed in other techniques such as HPLC, GC, CE, etc. [16].

10.4.3 Structure and Dynamics of Ligand-CD Complexes

The information on ligand-CD complexes is extremely useful for a better understanding of the nature of forces involved in complex formation, for the identification of the topological areas of a CD that possess the highest affinity and highest (stereo) discriminatory power, etc.

As mentioned above, several techniques may provide information on the stoichiometry and binding constants of ligand-CD complexes. However, just a few of them are suitable for the determination of the structure of intermolecular complexes in solution. This is especially important in the case of CDs because they possess multiple binding sites. The method of choice for obtaining structural information in the solid state is certainly X-ray crystallography. However, NMR spectroscopy provides most complete structural information about noncovalent complexes in solution.

CICS in CD-ligand complexes is the result of the shielding (mainly on CDs) and deshielding effects that are significant only at a certain distance between the interacting groups. Therefore, it seems reasonable to use CICS for the identification of that part of the guest which is included into the cavity of CD [2], [50], [53], [54], [55], [56], [57]. It is possible not only to calculate the chemical shifts due to complexation in a given ligand-CD pair but also to make assumptions about the structure of a complex [2].

More recent studies show, however, that CICS may be caused not only by an inclusion in the cyclodextrin cavity but also by conformational changes even in a part of a molecule

More recent studies show, however, that CICS may be caused not only by an inclusion in the cyclodextrin cavity but also by conformational changes even in a part of a molecule that is not included into the cavity of a cyclodextrin. Alternatively, a change of the microenviroment (for instance, due to changes in the magnetic susceptibility of water as the result of hydrogen bonding with the glucose units of the cyclodextrins) may affect CICS [2]. The effect caused by conformational changes in the non included part of the molecule may be even stereoselective. The reason for this "apparent" stereoselectivity can be a stereoselective effect of the CD on the molecule as a whole. Moreover, in some cases, the CICS as well as the difference of the CICS between enantiomers can be higher for atoms and groups in the nonincluded part of the ligand than for those in the included part. Therefore, the value of the CICS must be used with some care for the determination of the structure of ligand-CD complexes.

More reliable information on the structure of ligand-CD complexes may be obtained using NMR experiments that are based on the Nuclear Overhauser Effect (NOE).

The basics of these techniques have been described in the Chapter on protein-ligand complexes and in the reference. Just two illustrative examples will be discussed below.

NOE enhancement of resonance signals can be observed between noncovalently bonded atoms in a certain proximity (3–5 A) in space. The known dependence (r^{-6}) of NOE

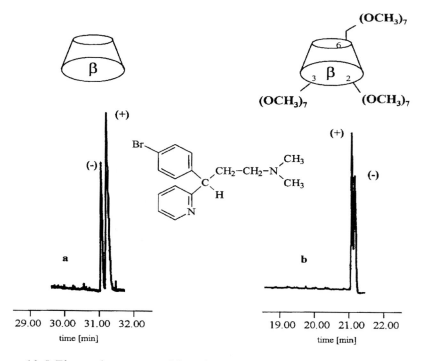

Figure 10-5: Electropherograms of the mixture of BPh enantiomers [(+)/(-)=2/1] in the presence of (a) 18 mg/ml β-CD and (b) 80 mg/ml TM-β-CD.

enhancements from the distance and the possibility to perform a volume integration of the cross-peaks allows us to quantify NOE data. Considering this information, a reliable statistically averaged structure of ligand-CD complex can be determined in solution.

Several instrumental techniques that were developed during the last few decades allow us to detect very fine differences in the (chiral) recognition properties of CDs. Capillary electrophoresis (CE) is one of these techniques [16]. The opposite binding pattern of enantiomers of several chiral analytes has been observed in CE for native and modified CDs [19, 59]. One example is the chiral antihistaminic drug brompheniramine (BPh). The affinity of BPh enantiomers with native β-CD and heptakis-(2,3,6-tri-O-methyl)-β-CD (TM-β-CD) is opposite (Figure 10-5).

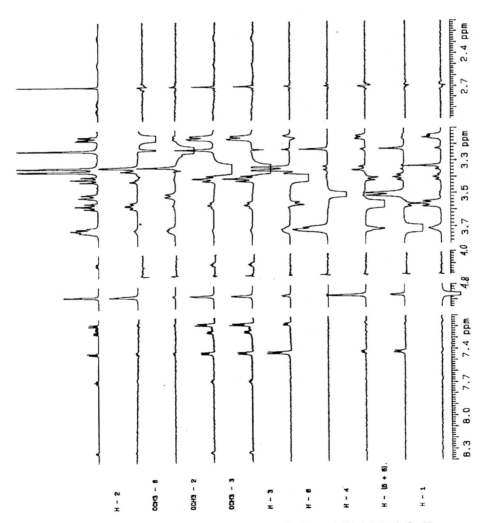

Figure 10-6: 1D-ROESY spectra of (a) (+)-BPh/TM-β-CD and (b) (+)-BPh/β-CD complexes.

In order to understand the origin of this phenomenon, an attempt was made to study the structure of the CD complexes in solution using rotating-frame NOE (one-dimensional CAMELSPIN sequence, 1D-ROESY). The data of the complex of TM-β-CD with (+)-BPh (Figure 10-6) indicate its formation by insertion of the *para*-Br-phenyl moiety of BPh into the cavity of the CD from the secondary side. The pyridine moiety is located in close proximity to the secondary rim outside the TM-β-CD cavity.

In contrast to this, 1D-ROESY cross-peaks were observed between phenyl protons of (+)-BPh, not only with the internal H-3 and H-5 protons of β-CD but also with external protons (Figure 10-7). Further studies are required in order to examine a possible external complexation together with an inclusion of BPh in the case of β-CD.

Another interesting point is the positive NOE-effect on the maleate protons observed by irradiation of internal protons in the β-CD. Although a simultaneous inclusion of both p-bromphenyl and maleate residues is impossible in the cavity of the same CD the formation of sandwich-like complexes with the 2:1 stoichiometry of β-CD to brompheniramine is possible. This hipotetic structure is somewhat supported also with x-ray experiments performed on the single monocrystals obtained from the solution of β-CD and brompheniramine.

1D-ROESY studies on the complexes of the chiral antihistaminic drug dimethindene (DM) with native β-CD and TM-β-CD do not detect the significant differences in the structure of the ligand-CD complexes. Benzimidazole moiety of the molecule is included in the cavity of both CDs entering from the secondary side. In spite of the structural similarity of these two complexes the opposite chiral recognition pattern was observed in CE. In addition, the 1D-ROESY data for the dimethindene-β-CD complex do not confirm a partial inclusion of the pyridine moiety of DM into the cavity of β-CD which has been suggested based on the CICS data [28].

Information about the dynamics of ligand-CD complexes with relatively small association/dissociation rates can be obtained by classical NMR line-shape analysis [2]. In rare cases even different resonance signals can be observed for complexed and free ligands [29], [58].

The rotational correlation times of guest compounds in CD complexes are usually prolonged. Thus, both ^{13}C and deuterium relaxation times provide significant information on the dynamics of ligand-CD complexes and may also be used for the determination of the association constant [59], [60], [61]. This is especially useful in those cases when the data about the dynamics are not accessible from line-shape analysis. Recently available resonance ROESY techniques also allow a semiquantitative estimation of internal motions in ligand-CD complexes [2].

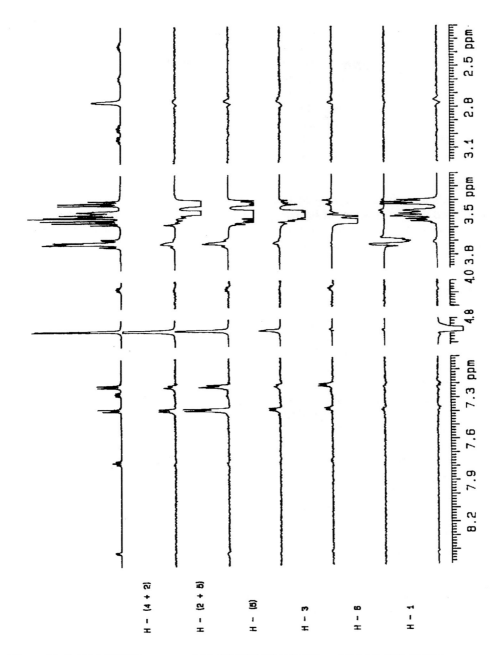

Figure 10-7: 1D-ROESY spectra of (a) (+)-BPh/TM-β-CD and (b) (+)-BPh/β-CD complexes.

Thus, as shown using a few examples in this section, NMR spectroscopy provides significant information on the stoichiometry, binding constants, structure and dynamics of ligand-CD complexes. This information seems to be complete for a description of non-covalent intermolecular complexes. However, the advantages of NMR spectroscopy should not be exaggerated and the data obtained with this technique must be examined by alternative methods. Enormous difficulties arise in the use of NMR spectroscopy when multiple complexes are formed. Most likely chemical shifts will lose the additivity in such cases. As the result, no correlations will be observable between the CICS and the degree of complexation. In this case, alternative techniques such as UV, MS, etc. may be used for the determination of the stoichiometry and the binding constants. Microcalorimetry is sometimes superior for the determination of thermodynamic quantities (free energy, entropy and enthalpy) rather than NMR spectroscopy. Structural data obtained using the NOE technique may be examined using X-ray crystallography. However, it must be considered that the structure of ligand-CD complexes will not necessary be identical in a solution and in the solid state. In certain cases, computed structures (molecular mechanics, molecular dynamics, quantum mechanics) support the data obtained using NMR spectroscopy.

References

1. P V Demarko and A L Thakkar, *Chem Commun* 2-4 (1970)
2. H-J Schneider, F Hacket and V Rüdiger, *Chem Rev* **98**, 1755-1785 (1998)
3. A Villiers, *C R Acad Sci* **112**, 536-538 (1891)
4. F Schardinger, *Centr Bakt Parasitenk* II Abt **29**, 188-197 (1911)
5. K Freudenberg and R Jakobi, *Liebigs Ann Chem* **518**, 102-108 (1935)
6. K Freudenberg, G Blomquist, L Ewald and K Soft, *Chem Ber* **69**, 1258-1266 (1936)
7. K Freudenberg, and M Meyer-Delius, *Chem Ber* **71**, 1596-1600 (1938)
8. D French and R E Rundle, *J Am Chem Soc* **64**, 1651-1653 (1942)
9. J Szejtli, *Chem Rev* **98**, 1743-1753 (1998)
10. K Freudenberg, E Schaaf, G Dumpert and T Ploertz, *Naturwissenschaften* **27**, 850-853 (1939)
11. A Hybl, R E Rundle and D E Williams, *J Am Chem Soc* **87**, 2779-2788 (1965)
12. F Cramer, *Angew Chem* **64**, 136 (1952)
13. S Li and W-C Purdy, *Chem Rev* **92**, 1457-1470 (1992)
14. A R Hedges, *Chem Rev* **98**, 2035-2044 (1998)
15. K Uekama, F Hirayama and I Tetsumi, *Chem Rev* **98**, 2045-2076 (1998)
16. B Chankvetadze (1997) Capillary Electrophoresis in Chiral Analysis, Wiley & Sons, Chichester, UK, 555 pp (1997)
17. I Ostromislensky, *Ber Dtsch Chem Ges* **44**, 268-273 (1911)
18. P Job, *Ann Chim (Paris)* **9**, 113-203 (1928)
19. B Chankvetadze, G Pintore, N Burjanadze, D Bergenthal, D Strickmann, R Cerri and G Blaschke, *Electrophoresis* **19**, 2101-2108 (1998)
20. J Redondo, J Frigola, A Torrens and P Lupon, *Magn Res Chem* **33**, 104-109 (1995)
21. D Salvatierra, C Jaime, A Virgili and F Sanchez-Ferrando, *J Org Chem* **61**, 9578-9581 (1996)

22. B Chankvetadze, G Endresz, G Schulte, D Bergenthal and G Blaschke, *J Chromatogr A* **732**, 143-150 (1996)
23. P M Ivanov, D Salvatierra and C Jaime, *J Org Chem* **61**, 7012-7017 (1996)
24. C Jaime, J Redondo, F Sanchez-Ferrando and A Virgili, *J Org Chem* **55**, 4772-4776 (1990)
25. T Utsuki, K Imamura, F Hirayama and Kaneto, *Eur J Pharm Sci* **1**, 81-87 (1993)
26. P L Irwin, P E Pfeffer, L W Doner, G M Sarers, J D Brewster, G Nagahashi and K B Hicks, *Carbohydr Res* **256**, 13-27 (1994)
27. B Chankvetadze, G Schulte, D Bergenthal and G Blaschke *J Chromatgr A* **798**, 315-323 (1998)
28. A F Casy and A D Mercer, *Magn Reson Chem* **26**, 765-774 (1988)
29. U Berg, M Gustavsson and Astrom, *J Am Chem Soc* **117**, 2114-2115 (1995)
30. S K Branch, U Holzgrabe, T M Jefferis, Malwitz H and M Matchett, *J Pharm Biomed Anal* **12**, 1507-1517 (1994)
31. M Thunhorst, Y Otte, T M Jefferies, S K Branch and U Holzgrabe, *J Chromatgr A* **818**, 239-249 (1998)
32. G Uccello-Barretta, C Chivacci, C Bertucci and P Salvadori, *Carbohydr Res* **243**, 1-10 (1993)
33. P Salvadori, G Uccello-Barretta, F Balzano, C Bertucci and C Chivacci, *Chirality* **8**, 423-499 (1996)
34. J R Moyano, M J Arias-Blanco, J M Gines, A M Rabasko, J I Perez-Martinez, M Mor and F Giordano, *J Pharm Sci* **72**, 72-75 (1997)
35. F Djedaini, S Z Lin, B Perly and D Wouessidjewe, *J Pharm Sci* **79**, 643-646 (1990)
36. G Endresz, B Chankvetadze, Bergenthal D and Blaschke G, *J Chromatgr A* **732**, 143-150 (1996)
37. Chankvetadze B, Endresz G, Bergenthal D and Blaschke G, *J Chromatogr A* **717**, 245-253 (1996)
38. M Cotta Ramusino and S Pichin, *Carbohydr Res* **259**, 13-19 (1994)
39. J Nishijo, Y Ushirode, H Ohbori, M Sugiura and N Fujii, *Chem Pharm Bull* **45**, 899-903 (1997)
40. A Ganza-Gonzalez, J L Vila-Jato, S Anguiano-Igea, F J Otero-Espinar and J Blanco-Mendez, *Int J Pharmac* **106**, 179-185 (1994)
41. G Bettinetti, F Melani, R Mura, R Mannanni and F Giordano, *J Pharm Sci* **80**, 1162-1170 (1991)
42. A Botsi, Perly B and E Hadjoudis, *J Chem Soc Perkm Trans* **2**, 89-94 (1997)
43. G Fronza, A Mele, Redenti E and P Ventura, *J Pharm Sci* **81**, 1162-1165 (1992)
44. D Greatbanks and R Pickford, *Magn Res Chem* **25**, 208-215 (1987)
45. F Djedaini and B Perly, *J Pharm Sci* **80**, 1157-1162 (1991)
46. B Chankvetadze, N Burjanadze, G Pintore, D Strickmann, D Bergenthal and G Blaschke, *Chirality*, accepted (1999)
47. H A Benesi and J H Hildebrand, *J Am Chem Soc* **71**, 2703-2707 (1949)
48. R L Scott, *Recueil* **75**, 787-789 (1956)
49. R Foster and C A Fyfe, *Trans Faraday Soc* **61**(512), 1626-1631 (1965),
50. M V Rekharsky and Y Inoue, *Chem Rev* **98**, 1875-1917 (1998)
51. K A Connors, Binding Constants, Wiley, New York, 441 pp (1987)

52. K A Connors, *Chem Rev* **97**, 1325-1357 (1997)
53. Y Inoue, T Okuda, F H Kuan and R Chujo, *Carbohydr Res* **129**, 9-20 (1984)
54. M Kitagawa, H Hoshi, M Sakurai, Y Inoue and R Chujo, *Carbohydr Res* **163** (1), C1-C3 (1987)
55. M V Rekharsky, R N Goldberg, F P Schwarz, Y B Tewari, P D Ross, Y Yamashoji and Y Inoue, *J Am Chem Soc* **117**, 8830-8840 (1995)
56. Y Yamashoji, T Ariga, S Asano and M Tanaka, *Anal Chim Acta* **268**, 39-47 (1992)
57. C J Hartzell, S R Mente, N L Eastman and J L Beckett, *J Phys Chem* **97** (19), 4887-4890 (1993)
58. H Dodziuk, J Sitkowski, L Stefaniak, J Jurczak and D Sybilska, *Supramol Chem* **3**, 79-81 (1993)
59. B Chankvetadze, G Schulte and G Blaschke, *Enantiomer* **2**, 157-179 (1997)
60. J P Behr and J M Lehn, *J Am Chem Soc* **98**, 1743-1747 (1976)
61. Y Kuroda, M Yamada and I Tabushi, *J Chem Soc, Perkin Trans II*, 1409-1415 (1989)

Chapter 11

J.K. Seydel

11 Ligand-Membrane Interaction

11.1 Introduction

The interest in drug design has been focused mainly on the interaction of ligand molecules with the proteins of the specific receptors and enzymes. Most of the target proteins are embedded in membranes, and it is assumed that biological activity arises as a result of binding to the membrane-bound proteins; the lipid background is considered to play a more passive role. There is, however, increasing evidence that the influence of ligand-membrane interaction on drug activity and selectivity has been underestimated. The so called "nonspecific" interaction of drugs with membrane constituents involves in fact an interaction with specific phospholipid structures. Although the lipid layer is a dynamic fluid, the membranes do not consist of lipids only but possess polarized phosphate groups and neutral, positively or negatively charged head groups and are highly structured and chiral. The interaction with such structures can have a decisive influence on drug partitioning, orientation and conformation and on the physicochemical properties and functioning of the membrane as well. Therefore drug-membrane interactions play an important role in drug transport, distribution, accumulation, efficacy and resistance.

The perturbation of biological membranes by various classes of drugs can lead to changes in membrane curvature and thus to changes in protein conformation and can become an important factor in drug action. The ligand-membrane interactions manifest themselves macroscopically in changes in the physical and thermodynamic properties of "pure" membranes or bilayers. Depending on the composition of the membrane and the structure of the ligand molecules the interaction can favor or prevent drug activity or toxicity.

Luckily most of the perturbations which can occur in biological membranes upon interaction with drug molecules can be studied *in vitro* and quantified by available physicochemical techniques using artificial bilayers which are readily formed or model membranes. One of the most efficient ways of studying ligand-membrane interaction is by use of the various techniques of nuclear magnetic resonance (NMR) spectroscopy.

11.2 Functioning, Composition and Organization of Membranes

11.2.1 The Physiology of Cells and the Importance of Membranes for their Functioning

All living cells are surrounded by one or several membranes. The membrane defines the cell as a living unit and separates the cell more or less from the surrounding; it separates intracellular from extracellular domains. Developed organisms are comparable to cell states in which groups of cells are specialized for special functions and are connected through complex communication networks. Any disorder in the communication of such complex systems influences the functioning of the organisms. It reduces the readiness for reactions, decreases the capability to adapt to changes in the environmental conditions and can lead finally to reduction in efficiency or to death.

The outer membrane, the plasmalemm, efficiently protects the cell from the environment. At the same time, the membrane can carry out important functions for the metabolism of the cell, the uptake of substrates and the elimination of toxic compounds. The substrate exchange with the environment is controlled by transport proteins which are embedded in the membrane (energy-using pumps like Na^+/K^+-ATPase, or other transport units like Na/glucose-cotransport and the ion channels, Na and Ca channels) [1]. It seems a miracle that a membrane of about 8 nm can maintain extreme gradients of intra- and extra-cellular ion and amino acids and protein concentrations. The ratio of intracellular/extracellular ion concentration for example is for Na^+ 10 /140 and for Ca^{2+} 0.0001/2.5. The transmembrane concentration gradient of solvents, ions, pH etc. is necessary for cellular functions, as for example the production of ATP, which cannot function without the transmembrane gradient. Another property of cell membranes besides compartmentalization is the ability to fuse. This concerns intracellular vesicle transport among intracellular organelles as well as for example the fusion of enveloped viruses with target cell membranes.

11.2.2   Composition and Organization of Membranes

11.2.2.1   Mammal Membranes

Membranes consist mainly of proteins and lipids, these frequently having quite different functions. Proteins determine the functional properties, lipids the matrix, i.e. the construction. In principle, cell membranes consist of a phospholipid bilayer into which proteins are integrated. The phospholipid molecules consist of two long-chain fatty acids which are esterified with one of the hydroxy groups of glycerol; the third hydroxy group of glycerin is connected to phosphoric acid, which is substituted by another substructure, for example with choline (phosphatidylcholine), serine or sugar. This constitution confers the amphiphilic character to phospholipids.

The apolar fatty acid chains are lipophilic, while the polar head groups and the polarized phosphate groups are hydrophilic. Because of these properties phospholipids associate readily in water and form the characteristic bilayers with the polar head groups directed to the surrounding water and the fatty acid chains turned toward each other and directed to the inner part of the bilayer. The stability of the bilayer depends on the segregation of the hydrocarbon residues from the watery phase and the polar interaction of the head groups with water, i.e. on the type and charge of the head groups. Also, the phosphate groups bear a partial negative charge. Dissociation occurs at about pH 2–3 (Table 11-1) [2]. Strong polar interactions also arise from carbohydrates attached to certain lipids such as phosphatidylinositol and glycolipids.

Cholesterol with a single hydroxy group possesses weak amphipathic character and leads to a specific orientation in the phospholipid structure.

Table 11-1　Phospholipid charged groups and their electrostatic properties (adapted from [2] with permission from the Macmillan Press Ltd, London UK).

Phospholipid ionizable groups	Electrostatic properties
1. Primary phosphate (phosphatidic acid)	pK_1 3.9 pK_2 8.3
2. Secondary phosphate (Phosphatidylinositol, cardiolipin)	pKa <2.0
3. Secondary phosphate + Quaternary amine (sphingomyelin, phosphatidylcholine)	isoelectric in pH range 3–10
4. Secondary phosphate + amine (phosphatidylethanolamine)	net negative at pH 7.4
5. Secondary phosphate + amine + carboxyl (phospatidylserine)	net negative at pH 7.4

In membranes with charged lipids electrostatic repulsion occurs between similarly charged leaflets on either side of the bilayer and prevents a decrease in thickness of the structure. In addition the electrostatic repulsion supports the lateral cohesion between the hydrocarbon chains and stabilizes the bilayer. The charged surface will attract oppositely charged mobile counter ions from the aqueous phase. At equilibrium they will be distributed according to the electrostatic potential, to form the so-called electrical double layer. Fatty acid composition, phospholipid composition and cholesterol content can be modified in many different ways in intact mammalian cells. These changes alter membrane fluidity [3] and cell surface curvature, and can affect a number of cellular functions [4], including carrier-mediated transport, the activity and properties of certain membrane-bound enzymes, phagocytosis, endocytosis, immunologic and chemotherapeutic cytotoxicity, prostaglandin production and cell growth. The effects of lipid modification on cellular function are very complex. Therefore it is not yet possible yet to make any generalizations or to predict how a given system will respond to a particular type of lipid modification. Cholesterol — an essential component of mammalian cells — is of importance for the fluidity of membranes. Its influence on membrane fluidity has been most extensively studied on erythrocytes, and it was found that increasing cholesterol content restricts molecular motion in the hydrophobic portion of the membrane lipid bilayer. As the cholesterol content in membranes is altered with age, this may have an influence on drug transport and drug regimen. Another important component of membranes is Ca^{2+} ions, which bridge negatively charged head group structures, thus stabilizing the membrane. Displacement of Ca^{2+} ions by cationic drug molecules will necessarily lead to significant changes in membrane organization and properties.

Table 11-2: The fatty acid residues (R) commonly found in membrane lipids (reproduced from [2] with permission from the Macmillan Press Ltd, London, UK).

C-atoms	Fatty Acyl Substituents (R), Chemical structure	Common name
12	$CH_3-(CH_2)_{10}-COO-$	lauric
14	$CH_3-(CH_2)_{12}-COO-$	myristic
16	$CH_3-(CH_2)_{14}-COO-$	palmitic
16	$CH_3-(CH_2)_5-CH=CH-(CH_2)_7-COO-$	palmitoleic
18	$CH_3-(CH_2)_{16}-COO-$	stearic
18	$CH_3-(CH_2)_7-CH=CH-(CH_2)_7-COO-$	oleic
18	$CH_3-CH_2-CH=CH-CH_2-CH=CH-CH_2-CH=CH-(CH_2)_7-COO-$	linolenic
18	$CH_3-(CH_2)_4-CH=CH-CH_2-CH=CH-(CH_2)_7COO-$	linoleic

Table 11-3: Phospholipid composition in Mol% total lipids of some liver cell membranes (adapted from [2] with permission of the Macmillan Press Ltd., London).

Phospholipid	Plasma membranes	Golgi membranes	Lysosomal membranes
Phosphatidylcholine	34.9	45.3	33.5
Phosphatidylethanolamine	18.5	17.0	17.9
Phosphatidylinositol	7.3	8.7	8.9
Phosphatidylserine	9.0	4.2	8.9
Phosphatidic acid	4.4	-	6.8
Sphingomyelin	17.7	12.3	32.9

Eucaryotic cells are generally more complex than procaryotic cells and possess a variety of membrane-bound compartments called organelles. These intracellular membranes allow diverse and more specialized functions. For detailed information the study of special handbooks is recommended [2], [5], [6]. Phospholipid charged groups and their electrostatic properties are given in Table 11-1 [**2**]. Fatty acyl residues (R) commonly found in membrane lipids are summarized in Table 11-2 [2]. Generally, 4 lipid structures are mainly found in eucaryotic cells, phospholipids, sphingolipids, glycolipids and sterols [2]. Various organs have different compositions of phospholipids (Table 11-3).

a. Erythrocyte lipids

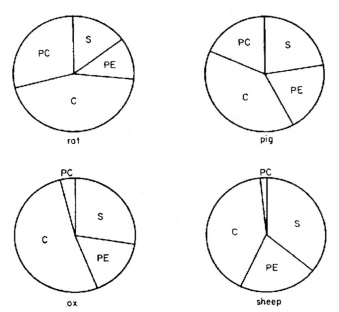

b. Phospholipids of ox endoplasmic reticulum

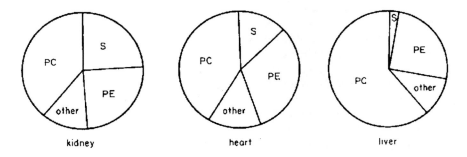

Figure 11-1 a Variation in the ratio of sphingomyelin (S) to phosphatidylcholine (PC) and other constituents such as cholesterol (C) and phosphatidylethanolamine (PE) in the total lipid fraction of Erythrocytes from various species.
b Sphingomyelin relative to total lipids present in the endoplasmic reticulum (reproduced from [2] with permission from the Macmillan Press Ltd, London.

An example is given for the composition of the liver cell membrane [2]. There is also a considerable difference in the proportion of phospholipids in different cell types and different species. Figure 11-1 shows the variation of sphingomyelin (S) to phosphatidylcholine (PC) content in the total lipid fraction of erythrocytes from various species and of ox endoplasmic reticulum in various organs [2]. The lateral mobility of lipids in membranes is another essential aspect [7].

Another important observation is that phospholipids are asymmetrically distributed in the plasma membrane of eucariotic cells. It has been found that phosphatidylserin (PS), phosphatidylethanolamine (PE) and phosphoinositides (PI) are essentially located in the inner monolayer, whereas phosphatidylcholine (PC), sphingomyelin (SH) and glycolipids (GL) are mainly located in the outer monolayer [8]. An example of the asymmetrical distribution of phospholipids in human red cells and in the membrane of the influenza virus is shown in **Table 4** [9], [10].

Table 11-4: Asymmetrical distribution of phospholipids (as mole percent) in membranes of *influenza* viruses [9] and in red blood cells [10]:

	Influenza virus (outside /inside)			
TPL	SH	PC	PE	PS
30/73	6/26	6/7	11/28	4/12
	Red blood cells (outside/inside			
51/51	20/5	22/9	6/25	0/9

TL = total phospholipid; SH = sphingomyelin; PC = phosphatidylcholine;
PE = phosphatidylethanolamine; PS = phosphatidylserine

The stability of the asymmetry is partly due to the activity of the enzymes responsible for lipid synthesis and in part to the activity of specific proteins called "phospholipid flippases". These catalyze the exchange of lipids between both leaflets. The aminophospholipid translocase was first discovered in human erythrocytes and transports aminophospholipids from the outer to the inner monolayer of plasma membranes [11], [12]. The same activity was found in many other cell types. Proteins of the multi drug resistance (MDR) family are also flippases or floppases, respectively. They can transport not only lipids from the inner to the outer monolayer and vice versa but also amphiphilic drugs [13], [14]. These membrane-located transport proteins — as the MDR1 P-glycoprotein — play an essential role in multidrug resistance [14]. The biological function of lipid asymmetry and of proteins involved in the transmembrane traffic of lipids is multiple. A review on the role of translocases in the generation of phosphatidylserine asymmetry has been published recently [15]. For red blood cells it is assumed that the progressive loss of lipid asymmetry, possibly associated with an entry of calcium, corresponds to a signal of ageing which is recognized by macrophages. Drugs which for example compete for calcium bound to PS could interfere with these processes and many other $Ca^2{\pm}$dependent processes such as protein kinase C (PKC) activation.

An important aspect of phospholipid asymmetry in membranes for drug permeation and drug distribution is the generation of surface tension. The increase in surface tension could change the membrane permeability either directly via a change of the bilayer viscosity or via a change of protein conformation.

The curvature of membranes, determined mainly by the volume, size and charge of the head groups of the phospholipids and their distribution in the outer and inner leaflet respectively, is an important factor for the functioning of the embedded proteins. It is an indicator of the internal stress of the lipid layer. The internal stress is the tendency of the lipid system to adopt non-bilayer configuration as for example H_{II} phase. This can be determined by X-ray or neutron diffraction experiments. "The intrinsic radius of curvature is essentially a measure of the average mismatch between the minimum free energy projected areas of the hydrophilic and hydrophobic portions of the lipid molecules" [15]. For many membrane lipids such as unsaturated phosphatidylethanolamines this mismatch is large, corresponding to small values of the intrinsic radius of curvature. For other lipids such as phosphatidylcholine the radius is large and the bilayer therefore relaxed. Results of experiments which correlate the composition of bilayers with the operation or activity of certain intrinsic membrane proteins depend on a limited range of values of the bilayer intrinsic radius of curvature for optimal function. Amphiphilic compounds are known as potent modifiers of the bilayer intrinsic radius of curvature and may act via this property to non-specifically perturb membrane protein function [16]. Catamphiphilic drugs which could interact with the head groups or with the flippases, scramblases or floppases could, therefore, change cell functioning.

Not only does the composition of phospholipids differ in membranes, but also the amount and type of integrated proteins. The ratio of protein/lipid composition of rat tissue membranes is given in Table 11-5 [2]. Membrane proteins can be divided into two classes, intrinsic and extrinsic [17]. Intrinsic membrane proteins are inserted to varying degrees into the hydrophobic core of the lipid bilayer and can be removed only through the use of detergents or denaturants. Extrinsic membrane proteins are associated noncovalently with the surface of the membrane, and can be removed from the bilayer by incubation with alkaline. The intrinsic membrane proteins can be classified into three groups, monotopic, bitopic and polytopic [18]. The conformation of membrane integrated proteins depends not only on the type and sequence of the amino acids but also on the lipid environment, i.e. the composition of the membrane. The effect of lipid environments and external factors such as temperature, pH, ionic strength on the 3D-structure of proteins has been studied in reconstituted membranes [19], and the major driving forces for specific membrane association of proteins are hydrophilic interactions and the peptide amphipathic character [20], [21].

Different amphipathic helical peptides can be membrane-stabilizing or lytic to membranes depending on the structure of the helix, which in turn determines the nature of its association with the membrane. Features of peptides that are responsible for their specific properties have been discussed [22]. The type of peptide is of importance for the functioning of the cells. It has been shown that a mixture of anionic amphiphilic peptides with charge-reversed cationic peptides did induce a rapid and efficient fusion of egg phosphatidylcholine vesicles, but no fusion was observed with each peptide alone, probably because only the mixture has an ordered helical structure [23]. The protein ordering, proceeding in the membrane, seems to depend to a certain degree on the electrostatic surface potential and interface permittivity. They are influenced by electrostatic interaction between protein, polar head groups of the phospholipid and ions in the aqueous media of the membrane surface, and this can be affected by exogenous molecules such as drugs.

Table 11-5: Protein and lipid ratios of rat tissue membranes (reproduced from [2] with permission of the Macmillan Press Ltd, London).

Membrane	Protein/lipid	Cholesterol/polar lipid
Plasma membranes		
Myelin	0.25[1]	0.95
Erythrocytes	1.1	1.0
Rat liver cells	1.5	0.5
Plasma membranes		
Nuclear membranes	2.0	0.11
Endoplasmatic reticulum		
Rough-surface	2.5	0.10
Smooth-surface	2.1	0.11
Mitochondrial membranes		
Inner membrane	3.6	0.02
Outer membrane	1.2	0.04
Golgi membranes	2.4	-

[1] myelin from central nervous system; higher ratios are found in peripheral nerve myelin

The phospholipid-induced conformational change of the intestinal Ca-binding protein in the absence and presence of Ca^{2+} has been described [24].
Perhaps the most important turning point in the development of our current understanding of biomembranes was the fluid mosaic model [17], [25] (Figure 11-2). This model describes the cell membrane as a fluid 2-dimensional lipid bilayer matrix of about 50 Å thickness with its associated proteins, and it allows for the lateral diffusion of both lipids and proteins in the plane of the membrane [26]. This model has been further developed. Concerning the distribution of lipids and proteins, it has been assumed that the membrane consists of solid domains coexisting with areas of fluid disordered membrane lipids which

may also contain proteins [27]. This concept has been integrated into the fluid mosaic model and was called the plate model of biomembranes [28].

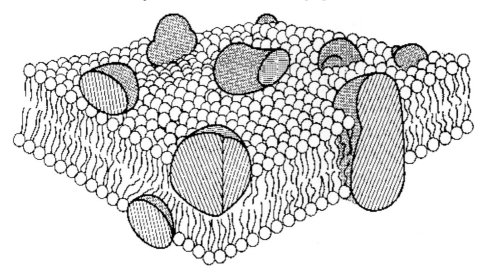

Figure 11-2: The fluid-mosaic model of membrane structure proposed by Singer and Nicolson [17] (reproduced from [25] with permission from Wiley-VCH, D-69451 Weinheim, Germany).

The concept of lamellar anisotropy was first proposed by Siekevitz [29]. Since then many other available data have led to a refined model for the dynamic organization of biomembranes. The most important difference from the fluid mosaic model is that a high degree of spatio-temporal order should also prevail in the liquid crystalline fluid membrane and membrane domains [30]. The interaction forces responsible for the ordering of membrane lipids and proteins are hydrophobicity, coulombic forces, van der Waals dispersion, hydrogen-bonding hydration forces and steric elastic constraints. Specific lipid-lipid and lipid-protein interactions result in a precisely controlled and highly dynamic architecture of the membrane components. On the basis of recent experimental and theoretical progress in the understanding of the physical properties of lipid bilayer membranes, it is assumed that the lipid bilayer component of cell membranes " is an aqueous bimolecular aggregate characterized by a heterogeneous lateral organization of its molecular constituents" and it is assumed that the dynamically heterogeneous membrane states are important for membrane functions such as transport of matter across the membrane and for enzymatic activity [31].

Clearly, the "construction" of membranes represents an important barrier for drug absorption, distribution and efficacy depending on the structure of the membrane and the drug molecules. The hydrophobic inner part of the bilayer becomes an almost complete barrier against the diffusion of polar or charged drug molecules. In contrast apolar drug molecules can easily penetrate the membrane. In this context it should, however, also be pointed out that the composition of mammalian membranes varies in different organs and species. (Table 11-3 to Table 11-5). This could become an important factor for the improvement of selectivity in drug absorption, distribution, accumulation and therefore also for drug efficacy.

11.2.2.2 Artificial Membranes and Liposome Preparation

During the last decades artificial phospholipid membranes have become a subject of intensive research as a model for biological membranes. As discussed before, the biological membranes are composed of complex mixtures of lipids, sterols and proteins. Defined artificial membranes may therefore serve as simple models of membranes which have many striking similarities to biological membranes.

For the study of drug membrane interactions and of the influence of drug structure and membrane composition, artificial membranes simulating especially mammalian membranes can easily be prepared because of the readiness of phospholipids to form automatically lipid bilayers, i.e. their tendency for self-association in water. The macroscopic structure of dispersions of phospholipids depends on the type of lipids and on the water content. The structure and properties of self-assembled phospholipids in excess water have been described [32], and the mechanism of liposome formation has been reviewed [33]. While the individual components, i.e. the membrane proteins and lipids, are composted of atoms and covalent bonds, their association with each other to produce membrane structures is governed largely by the hydrophobic effect. This hydrophobic effect is derived from the structure of water and the interaction of other components with the water structure. Because of their enormous hydrogen-bonding capacity, water molecules adopt a structure in both the liquid and the solid state.

The similarity of the physicochemical properties of artificial and biological membranes allows one to use such preparations (liposomes, vesicles) to study special effects of drugs on defined artificial membranes. A comparison of some important physicochemical properties of biological and artificial membranes is given in Table 11-6 [2]. The structural dependence of phospholipid solutions on the water content is called lysotropic polymorphism. Up to a water content of 30% dipalmitoylphosphatidyl-choline (DPPC) forms lamellar phases consisting of superimposed bilayers. On increasing the water content, heterogenic dispersions are obtained formed by multilamellar structures, the so-called liposomes (see also Section 11.2.3.1. Thermotropic and Lysotropic Mesomorphism). Because of the complex composition of fatty acyl chains a large variety of different molecules is present within each class of phospholipid, but all have in common that they are amphipathic.

The packing of the phospholipids depends on the normalized chain length difference between the *sn*-1 and *sn*-2 acyl chain. Besides the bilayer structure of noninterdigitated acyl chains, mixed interdigitated, partially and totally interdigitated bilayer structures exist. Liposome preparations have been shown to be suitable not only for studying special drug-membrane interaction effects *in vitro* but also for use as drug carriers. Various techniques have been developed and described to prepare homogeneous liposome preparations, unilamellar or multilamellar and of different size. The original preparation of liposomes produces multilamellar vesicles of varying size with diameters in the range 0.2–0.5 mm [34]. Sonication of such preparations produces small unilamellar vesicles which are homogeneous in size distribution and have been extensively used as model membranes [35].

Table 11-6: The physical characteristics of artificial bilayers and biological membranes (reproduced from [2] with permission of the Macmillan Press Ltd, London).

Property	Phospholipid bilayer	Biological membranes
Thickness (nm) [1]	4.5–10	4–12
Interfacial tension (Jm^{-2})	2.0–60	0.3–30
Refractive index	1.56-1.66	approx. 1.6
Electrical resistance (Wm^{-2})	10^7–10^{13}	10^6–10^9
Capacitance ($mF\,m^{-2}$)	3–13	5–13
Breakdown voltage (mV)	100–550	100
Resting potential diffr.(mV)	0.140	10–88

[1]These measurements have been made by electron microscopy, X-ray diffraction and optical methods. The capacitance was determined using an assumed dielectric constant

A method leading to vesicles with high efficacy of encapsulation and a vesicle size distribution of the order of 0.4 mm has also been described [36]. Conditions for obtaining convenient and reproducible preparations of unilamellar liposomes of intermediate size by use of phase evaporation followed by extrusion through polycarbonate membrane have been reported [37]. Large unilamellar vesicles have been obtained using a stainless steel extrusion device operating at elevated temperature (above the phase transition temperature) [38]. A procedure which could be useful in the preparation of asymmetric liposomes has also been reported, e.g. for liposome preparations differing in phosphoinositide content, using chromatography on immobilized neomycin [39]. A new rapid solvent exchange (RSE) method has been developed which avoids the passage of lipid mixtures through an intermediate solvent-free state. This prevents possible demixing of membrane components [40].

11.2.3 Dynamic Molecular Organisation of Membranes

11.2.3.1 Thermotropic and Lysotropic Mesomorphism of Phospholipids

The different thermotropic phase transitions observed when phospholipids in the solid state are heated above the melting point is an important physicochemical property which can very effectively be applied to study drug-membrane interactions. It is the basis for most physicochemical techniques for studying such interactions, especially in calorimetric experiments. The phospholipid does not directly change from the crystalline to the isotropic fluid state, but proceeds at a medium temperature through the liquid-crystalline state. Both transition states from crystalline to liquid-crystalline and from liquid-crystalline to the isotropic fluid state are associated with uptake of heat, i.e. they are endothermic processes (Figure 11-3). The fatty acid chains of the phospholipid in the crystalline state are in the extended conformation and all in *trans* configuration, and form a regular, stiff crystal lattice stabilized by van der Waals forces. Electrostatic interactions of the head groups lead to an additional molecular interaction. During the first transition (crystalline to liquid-crystalline), melting of the hydrocarbon chains occurs, their mobility increases, gauche configuration appears and "kinks" in the chains are observed. Strong head group interaction prevents complete loss of the molecular connection. The second transition from liquid-crystalline to the liquid phase leads to a movement of the phospholipid molecules in any direction. The invertal cylinder H_{II} state is found at high temperatures.

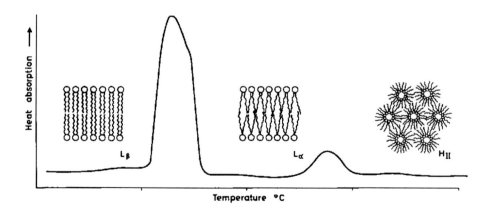

Figure 11-3: Structures adopted by phospholipids in aqueous media. Extended bilayer, L_β and L_α (gel and liquid-crystalline) states exist at low and intermediate temperatures, respectively, and invertal cylinder (hexagonal) H_{II} state is found at elevated temperatures.

Besides this thermotropic mesophormism, a lysotropic mesomorphism is observed [41]. The phase transition temperature, T_t, for the transition from the crystalline to the liquid-crystalline state decreases as a function of water content. The decrease in T_t is due to the destabilization of the crystal lattice in the head group region by the water molecules, which in turn decreases the interaction of the fatty acid chains. Sufficient water leads to the formation of thermodynamically optimal arrangements of the phospholipids in bilayers, so that the fatty acids are directed into the inner part of the layer and the hydrophilic head groups to the outer phase in contact with the watery medium. Analogously to the 3-dimensional crystal lattice of a salt, one can speak of a 2-dimensional "membrane-crystal" which possesses a highly ordered state below the melting point and the phase transition, respectively.

The insertion of water molecules into the membrane increases the space which can be occupied by the fatty acid chains and in consequence increases their mobility. The separation of the head groups and therefore the decrease in T_t is, however, limited because only defined amounts of water can be bound. The maximum amount of water and also the characteristic T_t at maximum hydration are a function of structure and charge of the head groups.

Phospholipids which have taken up the water molecules into the head group region below T_t are in the so-called gel phase. An increase in temperature will lead to a phase transfer from the gel to the liquid-crystalline phase. This is paralleled by an increase in the mobility of the fatty acid chains, where the mobility of the methylene groups increases with increasing distance from the acyl groups, and is also paralleled by the lateral diffusion of molecules within the membrane. The bilayer structure, however, remains intact because of the strong polar interaction in the head group region.

As already mentioned, the phase transition temperature depends on the phospholipid structure, i.e. the number and conformation of double bonds, length of fatty acid chains, charge and volume of phospholipid head group, hydrogen bonds etc. The latter are important for the stabilization of the membrane and an increase in T_t.

11.2.3.2 Phase separation and domain formation

Phase separation is observed when mixtures of phospholipids are incorporated into the membrane where the single components possess certain differences in their structure. In this case the phase separation is indicated in the thermogram by two or more signals. Structural characteristics which could lead to phase separation are differences in chain length and different degrees of saturation. Mixtures of dipalmitoylphosphatidic acid (DPPA) and diphosphatidylcholine (DPPC) or of distearoyllecithin with dimyristoyllecithin, for example, show the phenomena of phase separation. The reason for this is a lateral phase separation caused by domain formation.

Many electron spin resonance (ESR) studies of different systems have shown that phase separation in lipid layers may lead to a domain-like lateral structure. The area of domain formation can be extended over several hundred Å. In this connection the charge-induced domain formation in biomembranes is of special interest for the medicinal chemist. Especially the addition of Ca^{2+} to negatively charged lipids leads to domain formation. Each lipid component is expected to have a characteristic spontaneous curvature. The Ca^{2+}- induced domains lead to protrusions in the membrane plane. The lateral variation in the concentration in the plane of the membrane would then lead to a parallel variation in

11.2.4 Possible Effects of Drugs on Membranes and Effects of Membranes on Drugs

curvature. In biological membranes not only differences in the lipid composition but also membrane-integrated proteins can lead to phase separation caused by local tightening of membrane regions. The resulting microheterogeneity is of the utmost importance for many membrane functions (Figure 11-4) [42].

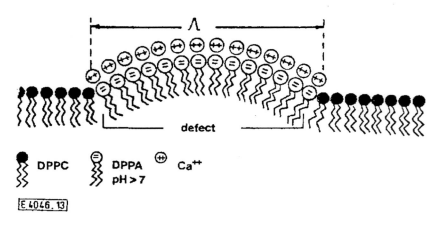

Figure 11-4: Lipid domain of one phospholipid component (e.g. DPPA) forms protrusion within 2-dimensional region of the second component (e.g. DPPC). The protrusion is caused by differences in spontaneous curvature of the two phospholipids (reproduced from [42] with permission from Wiley-VCH, D-69451 Weinheim, Germany).

At the borderline of rigid and fluid phospholipid the properties differ significantly from those of the pure phase, as for example:

1. increased compression of the phospholipids
2. increased permeability for hydrophilic particles
3. decreased order in lipid organization
4. facilitated hydrolysis of lipids by, for example, phospholipase A_2 [43].

The observed phase separation in the surroundings of integrated membrane proteins is of importance for their functioning. According to Carruthers [44], phase separation and domain formation create an area of rigid phospholipids around the proteins in an otherwise fluid surrounding.

For the medicinal chemist it is of interest to note that such phenomena, i.e. phase separation and domain formation, can also be induced in artificial membranes by cationic amphiphilic drugs. An increase in the microheterogeneity in biological membranes must therefore be considered, and by this a decisive change in membrane function in a defined area by indirect physicochemical interaction with amphiphilic drugs.

Artificial membranes have become a subject of intensive research as a model for biological membranes. Even though they are simpler they show striking similarities to biological membranes. Experimental work on such artificial membranes or liposomes has demonstrated that their structural properties may be strongly affected by membrane-associated molecules. Many drugs directly or indirectly influence cell membrane properties. The interaction can lead to disruption of the membrane so that it becomes highly permeable, for example by interaction between protein and phospholipids or by formation of complexes between ligand molecules and head groups or sterols. Several compounds change the fluidity of membranes. This has especially been studied and reported for the action of anesthetics [45], and methods of studying membrane fluidity have been described [46]. As discussed, changes in membrane fluidity can affect receptor and enzyme activity on both the static and the dynamic level. Another aspect is a change in the ability of drugs to pass through the membrane, which can influence their efficacy. Because of the unique structure of a lipid matrix consisting of phospholipids and embedded proteins the interaction of drug molecules with polar head groups, apolar hydrocarbons or both can induce several changes in the membrane and in consequence changes in drug behavior (diffusion, accumulation and conformation) [47].

Events which can arise from ligand-membrane interaction are summarized below, and examples of these events will be presented in this review on NMR techniques useful in the analysis of ligand-membrane interaction.

Possible effects of a drug on membrane properties:

1. Change in the conformation of acyl groups (*trans-gauche*)
2. Increase in membrane surface (curvature)
3. Change in microheterogenicity (phase separation, domain formation)
4. Change in thickness of the membrane
5. Change in surface potential and hydration of head groups
6. Change in the phase transition temperature (fluidity, cooperativity)
7. Change in membrane fusion.

Possible effects of the membrane on drug molecules

1. Diffusion through membrane may become rate limiting
2. Membrane may completely prevent diffusion to the active site (resistance)
3. Membrane may bind drug (accumulation, selectivity, toxicity)
4. Solvation of the drug in the membrane may lead to conformational change of drug molecules or force it into a special orientation (efficacy).

Pioneering work which has shed new light on the importance of drug-membrane interaction for the understanding of drug action and for the drug development process has been contributed by Herbette and coworkers [48], [49], [50], [51], [52], [53] and a general model has been discussed [54]. This is summarized in Figure 11-5 and discusses the various possibilities which could become rate limiting for the interaction of drugs with membrane-bound receptors.

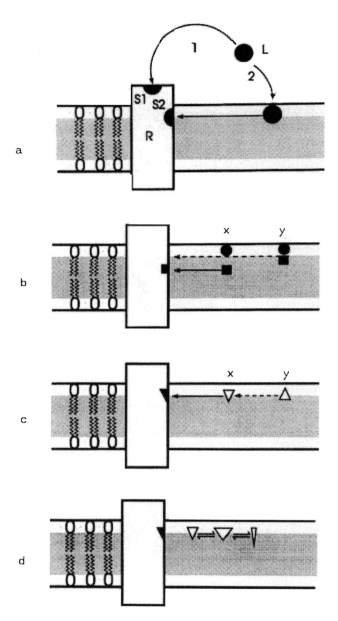

Figure 11-5: Various possibilities in rate limiting steps in the approach to the receptor proteins embedded in the membrane (reproduced from [54] with permission from the American Chemical Society, Washington DC, USA).

The first part of the model describes how a lipophilic substance can localize specific or unspecific binding sites on membrane proteins. The binding can occur directly to the proteins sticking out of the membrane or after distribution into the lipid phase and lateral preferred conformation or the equilibrium between conformers may change in the environment of the bilayer, possessing a different polarity compared to water. The highly structured environment can also reduce the rate of diffusion, depending on the physicochemical properties of the drug molecules. The second part of the model assumes that an influence on the functioning of the membrane-integrated proteins is also possible without a specific drug-protein interaction. This becomes understandable if another parameter is considered which characterizes cells or vesicles, namely the curvature. This is a measure of internal stress. The parameter is based on the tendency of bilayers to take non-bilayer conformation, as for instance a hexagonal phase (see Figure 11-3).

Optimal functioning of certain cell membranes and their integrated proteins affords a defined curvature. Changes in curvature induced by certain membrane-active molecules will lead to an indirect perturbation of the membrane, thus modifying or preventing agonist-receptor interaction and the biological response. For amphiphilic drugs such effects on curvature have been shown.

Because of the defined organization and composition of bilayers we cannot expect that drug transport, distribution, efficacy and the degree of resistance can always be sufficiently described by the octanol/water partition coefficient. The membranes do not consist of the hydrocarbon lipids only but possess polarized phosphate groups and neutral, positively or negatively charged head groups. They are highly structured, can be asymmetric and chiral, so that additional physicochemical properties of the drug molecules are involved in drug-membrane interaction besides overall hydrophobic forces. This is especially true in the case of amphiphilic drugs. The "hydrophobic gravity center" of drug molecules together with hydrogen bonding and charge interaction may become an important factor for the orientation in the membrane.

11.3 NMR Spectrometry; an Analytical Tool for the Study and Quantification of Ligand-Membrane Interactions

11.3.1 Tools for the Analysis and Quantification of Drug-Membrane Interactions

Apart from the biological implications, aqueous dispersions of lipids have attracted more and more the interest of physicists as a system which can be successfully investigated by physical methods despite their great complexity. The development of these methods for the study of drug-membrane interactions is also of great attractiveness and interest for medicinal chemists. The determination and quantification of the various effects of drugs on membranes and vice versa is a precondition for the understanding of drug action. By correlating drug effects on membranes of different composition with structural properties of the drugs under study and by deriving quantitative structure-interaction relations new information can be gained. The physicochemical or spectral parameters describing the strength or type of interaction can be used as "constraints" in the modeling of drug orientation and conformation in the environment of the membrane.

The most frequently used physical methods of investigating events or combinations of events leading to changes in membrane properties or in drug conformation, orientation and localization in the membrane environment are listed below.

Methods of Studying Drug-Membrane Interactions:

1. High pressure liquid chromatography (HPLC)
2. Displacement of $^{45}Ca^{2+}$ from phospholipid monolayers
3. Differential scanning calorimetry (DSC)
4. Fluorimetry
5. **Nuclear magnetic resonance spectrometry (NMR)**
6. Electron spin resonance spectrometry (ESR)
7. Fourier transform infrared spectrometry (FT-IR)
8. X-ray and neutron diffraction
9. Circular dichroism (CD)
10. Others

NMR spectrometry plays a decisive role among the various techniques capable of evaluating the motional characteristics of biological membranes and the influence of ligands on membrane structure and vice versa. The first observation of high-resolution NMR spectra of model membranes goes back to 1966 [55]. Over the past 25 years almost every new NMR technique has been applied to study the dynamics and conformation of membranes, changes in membrane properties by addition of biological components or drugs and to study the effect of membranes on drug orientation and conformation. Several reviews on the application of NMR techniques to phospholipid systems have been published [56], [57], [58]. NMR techniques provide detailed information about molecular conformation and ordering, and relaxation time measurements probe the amplitude and time scale of motions, and allows to study interaction phenomena. NMR spectra can be characterized by the magnetic field at which resonance occurs, depending on the nucleus and the surroundings of the nucleus considered (^1H, ^{13}C, ^{31}P etc.), the degree of spin-spin coupling produced by neighboring effects, the spin-lattice relaxation rate $1/T_1$ and the spin-spin relaxation rate $1/T_2$. The latter expressed as the line width at half maximal height of the resonance signal.

During the interaction between drug and phospholipid molecules one or several of these parameters can change in a manner characteristic of both the drug and the "receptor" phospholipid membrane. Additional information on molecular conformation can be obtained by the nuclear Overhauser effect (NOE), transfer NOE or two-dimensional homonuclear correlated NMR spectrometry (COSY) used to measure the distance between nuclei. For detailed information about these techniques see for example [56], [59], [60].

These methods allow us to detect changes in membrane organization and the localization of drug molecules in the membrane, and provide information on substructures of drug molecules involved in the interaction and on possible changes in membrane and drug conformation. NMR techniques can also be used to follow the rate of drug transfer into the membrane. The following nuclei are of special interest ^1H, ^2H, ^{13}C, ^{19}F, ^{31}P-NMR. NMR techniques in solution as well as solid-state ^{31}P-NMR have been applied. The latter technique has especially been used to study the interaction of peptides and proteins and has been reviewed recently [61].

11.3.2 Study of Membrane Polymorphism by ^{31}P-NMR

For the existence of a phase-separated region, lipids have to move into and out of various phases. The lateral diffusion constant in liquid-crystalline bilayers is about 10^{-8} cm^2 s^{-1}, which corresponds to an exchange frequency between lipid-lipid nearest neighbors of about 10^6 s^{-1}. A necessary precondition for the detection of phases by the NMR technique is that the proportion of observable species in the phase is sufficiently large. Not all NMR-active nuclei may therefore be suitable for the study of phase separation in phospholipids. Limited information can be gained on phospholipid phase transformation from ^{1}H- and ^{13}C-NMR because of problems in resolution. Only certain signals, as for example the choline methyl groups of phospholipids in the outer and inner leaflet of unilamellar bilayers can be identified by ^{1}H- and ^{13}C-NMR when chemical shift reagents are used. However, ^{13}C-NMR can be applied to the study of phospholipid phase transition when the lipid is specifically enriched at the sn-2, carbonyl position [62].

Phosphorus ^{31}P is as well as other nuclei, present in biological membranes and has special advantages. Phospholipid head groups contain an isolated $I = 1/2$ spin system which depends only on chemical-shift anisotropy and dipolar proton-phosphorus interactions. It is therefore a useful probe for structure and motion. The chemical shift of ^{31}P changes with the orientation of the magnetic field with respect to the nucleus. The observed spectrum can therefore be measured over a wide range of about 100 ppm. As the chemical-shift difference for ^{31}P is only ~ 4 ppm the chemical-shift anisotropy, because of orientational effects, controls the spectrum. A typical ^{31}P-NMR spectrum of polymorphic phases of phospholipid bilayers is depicted in Figure 11-6. For details the reader is referred to specific publications [63].

Many different effects and transition states have been studied and reported, as for example the transitions lamellar fluid to hexagonal and lamellar gel to fluid, and have been evaluated as a function of temperature and of membrane-affecting compounds such as acylglycerols, cholesterol or polymyxin. The usefulness of ^{31}P-NMR for the study of lipid polymorphism is also shown in Figure 11-6. In micelles or reversed micelles a single narrow resonance signal is observed. If this lipid is incorporated into a bilayer the motion becomes more restricted, and a broad spectrum with a residual chemical shielding anisotropy of –30 to –50 ppm is observed. The hexagonal phase is characterized by a reduced residual shielding anisotropy, and the sign of it is reversed [56], [64]. The dynamics of phosphate head groups have been experimentally determined and compared to calculated values [65]. Transient nuclear Overhauser effect (NOE) studies have been performed to identify the important proton species contributing to the ^{31}P-^{1}H dipolar interaction [66]. ^{2}H-NMR has been used to probe the membrane surface and to study the gel and liquid crystalline phase and the response of phospholipids in the gel and liquid-crystalline state to membrane surface charges [67], [68].

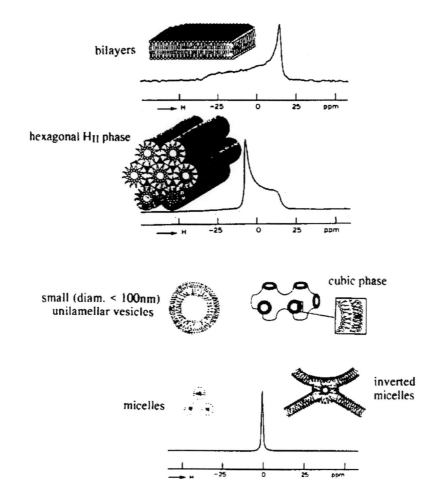

Figure 11-6: NMR-spectra for macroscopic polymorphic phases of phospholipids in bilayers, hexagonal H_{II} and isotropic phases (small vesicles, micelles, inverted micelles and cubic phases). Similar spectra are recorded for any one phase regardless of which phospholipid type forms the phase because the chemical shift differences for different phospholipids is smaller (2–4 ppm) than the anisotropy of chemical shift (~ 30–50 ppm) (reproduced from [63] with permission from Elsevier Science, Amsterdam, The Netherlands).

11.3.3 Effect of Cholesterol and Diacylglycerols

The influence of the amount of cholesterol at various temperatures on the thermotropic phase behavior and organization of saturated phosphatidylethanolamine bilayers has been studied by combining DSC-, FT-IR-, and ^{31}P-NMR. The incorporation of low levels of cholesterol into the bilayer causes a progressive reduction in the temperature, enthalpy and overall cooperativity of the lipid hydrocarbon chain melting transition [69]. The interaction of various cholesterol "ancestors" with lipid membranes on oriented bilayers has also been studied by ^{2}H- and ^{31}P-NMR [70].

The interaction of saturated diacylglycerols (DAGs) with phospholipid bilayers has been examined by various NMR techniques [71], [72], [73] and complementary techniques such as DSC and X-ray diffraction have been used as well [74]. It has been shown that DAGs induce a decrease in the area per phospholipid molecule and cause a parallel increase in lateral surface pressure of the bilayer.

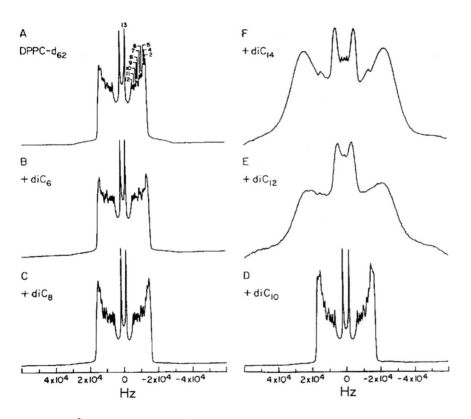

Figure 11-7: ^{2}H-NMR spectra of DPPC-d_{62} in the absence and presence of 25 mol% DAGs at 40 °C. Peaks 1 and 3 are resolved from peak 2 only at higher temperatures (reproduced from [72] with permission from the American Chemical Society, Washington DC, USA).

Since DAGs with diC_8 are the most effective activators of protein kinase C it was concluded that the activation of the enzyme occurs via a transverse perturbation of the lipid bilayer structure [72]. ^{2}H-NMR spectra of dipalmitoyl-phosphatidylcholine (DPPC) in the absence and presence of DAGs of various chain lengths are depicted in Figure 11-7.

The DAGs with chain length longer than C_8 induce lateral phase separation. The modulation of the bilayer to hexagonal transition by DAGs has also been studied. It is found that the effect of additives on the bilayer-to-H_{II} phase transition temperature are generally at least an order of magnitude greater than their effects on the gel-to-liquid crystalline phase transition temperature. DAGs were better modulators than monoacyl-glycerols and triacylglycerols more effective than DAGs [75].

This is of interest because the effect of such additives on the rate of membrane fusion in model systems often correlates with the effect of these additives on the bilayer-to-hexagonal phase transition temperature [75]. The increase in the bilayer-to-H_{II}-phase transition temperature generally inhibits cell fusion, and several compounds which stabilize bilayers show antiviral activity [76].

11.3.4 Effect of Drugs

The various NMR techniques applied to the study of the degree of drug-membrane interactions, the location of drug molecules within the membrane, the effects of drugs on surface charge, drug mobility, transmembrane transport, drug distribution in biological material and the structural dependence of these processes are discussed below and illustrated by special applications. The selection is admittedly subjective, but hopefully serves the purpose of covering those applications which could be of interest in drug development. This focus means that not only very recent papers are discussed.

Most of the examples published deal with the interaction of anesthetics with membranes, but calcium channel openers, β-blockers, antimalarials, anticancer drugs and antibiotics have also been examined, and the NMR signals of atoms and groups of the lipid and of the drug molecules have been followed and described as a function of the interaction. For the interaction of anesthetics and the involved mechanism of action, the reader's attention is directed to a review [45].

11.3.4.1 ^{31}P-NMR to Study Changes in Orientation of Phospholipid Head Groups

One important parameter which can be derived from ^{31}P-NMR spectra of phospholipids is the chemical-shift anisotropy parameter $\Delta\sigma$. This is defined as the width of the resonance signal of the low-frequency "foot" at half height (Figure 11-8). The chemical-shift anisotropy is related to the average orientation of the phospholipid head groups relative to the normal plane of the bilayer and also to the molecular motion of the lipid molecules. As membrane-active compounds often change the average orientation of the lipid head groups as well as the phase transition temperature of the lipids from liquid-crystalline to gel phase, $\Delta\sigma$ is a sensitive parameter to account for such changes. The application of this technique is discussed using the example of the interaction of antimalarial drugs with DPPC [77].

Four cationic antimalarials, chloroquine (**1**), quinacrine (**2**), quinine (**3**) and mefloquine (**4**) have been studied using DPPC bilayer membranes and the ^{31}P- and ^{2}H-NMR technique. In the structures **1-4** the positive charges given by protonated nitrogen atoms are indicated. Figure 11-8 shows the ^{31}P-spectra of DPPC, fully hydrated above T_t, in the absence and presence of three of the drugs. Mefloquine exerts the strongest effect on the chemical shift anisotropy parameter (most intensive broadening of the "foot"), followed by quinine. The corresponding $\Delta\sigma$ values are 57 ppm and 54 ppm. It has previously been shown by computer simulation that changes of $\Delta\sigma$ above T_t are primarily due to changes in the orientation of the phosphate moiety of the phospholipid head groups relative to the bilayer normal. In combination with changes in the phase transition temperature $\Delta\sigma$ is an indicator of membrane-active compounds. The results indicate that mefloquine has a more pronounced effect on the orientation of the head groups than quinine.

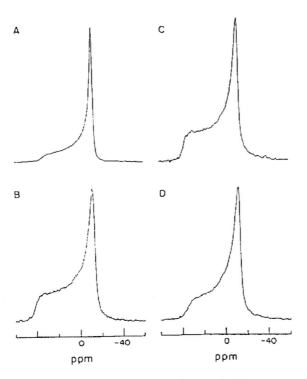

Figure 11-8: ^{31}P-NMR spectra of aqueous dispersions of drug-DPPC mixtures (1:2 mol/mol) at 50°C. **A** DPPC; **B** DPPC+mefloquine; **C** DPPC+Quinine; **D** DPPC+quinacrine (reproduced from [77] with permission from Elsevier Science, Amsterdam, The Netherlands).

Chloroquione (1)

Quinacrine (2)

Quinine (3)

Mefloquine (4)

For chloroquine no interaction was observed. The shapes of the spectra are typical of the bilayer conformation of the lipid molecules, i.e. no change to the hexagonal phase has occurred. The temperature dependence of DPPC in the presence and absence of the antimalarials is shown in Figure 11-9.

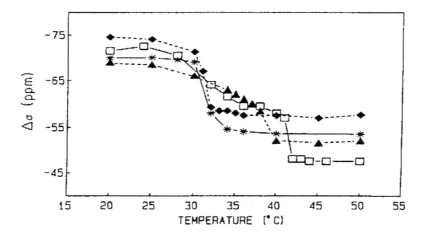

Figure 11-9: Plot of Δσ vs temperature for aqueous dispersions of drug-DPPC mixtures (1:2 mol/mol). □ DPPC; ◆ DPPC + mefloquine; * DPPC + quinine; Δ PC + quinacrine (reproduced from [77] with permission from Elsevier Science, Amsterdam, The Netherlands).

Mefloquine produced an increase in Δσ to 57 ppm and a decrease in T_t of about 10°C. In comparison quinacrine caused a smaller but significant increase in Δσ to 51.5 ppm and the degrease in T_t was 2.5°C. The observed sharp rise in the absolute value of Δσ (see Figure 11-10) upon lowering the temperature corresponds to T_t.

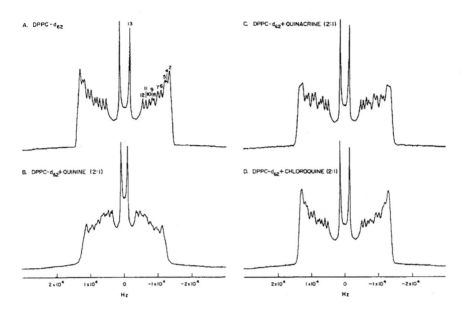

Figure 11-10: ^{2}H-NMR of drug/DPPC mixtures (1:2 mol/mol) at 40°C. **A** control (peaks 1 and 3 are resolved from peak 2 only at higher temperatures); **B** DPPC-d$_{62}$ with quinine; **C** with quinacrine; **D** with chloroquine (reproduced from [77] with permission from Elsevier Science, Amsterdam, The Netherlands).

The drug-induced changes were confirmed by the results of ^{2}H-NMR experiments. For these studies fully hydrated chain perdeuterated phospholipid was used. The spectra in Figure 11-10 arise from the superposition of the axially averaged powder patterns produced by the various deuterons of the C^{2}H$_2$ and the terminal C^{2}H$_3$ group. From each segment the order parameter S_{C2H} can be derived from the peak-to-peak quadrupole splitting, Δν. It corresponds to the perpendicular orientation of the bond relative to the external magnetic field and is described by eq. 1.

$$\Delta \nu^i = (3/4)\,(e^2qQ/h)S^i_{C2H} \qquad (1)$$

e^2qQ/h = 167 kHz is the quadrupole coupling constant of a deuteron in the C^{2}H bond. ^{2}H-NMR spectra of lipids deuterated at specific positions in the side chain show the presence of the order parameter profile along the side chain of phospholipids in a multilayer conformation. A plateau of relatively higher S_{C2H} is observed for C^2H_2 segments near the glycerol backbone [78].

Viewing the spectra in the absence and presence of the drugs confirms that the DPPC-d_{62} remains in the basic bilayer structure. Quadrupole splitting was slightly decreased by quinacrine, strong perturbation of the ^{2}H-NMR line shape was observed, however, in the presence of quinine and mefloquine at low molar ratios. In contrast no interaction with the phospholipid even at high molar ratios was seen for chloroquine. The authors conclude that the results of the ^{2}H- and ^{31}P-NMR show the capability of mefloquine and to a slightly lower degree of quinine to intercalate into the DPPC bilayer which could account for their demonstrated ability to penetrate into the interior of the bilayer. In contrast chloroquine and quinacrine show very little or no or interaction with the head groups. According to the authors, the accumulation of mefloquine in the membrane could also be the reason for the ability of mefloquine to overcome the resistance of chloroquine-resistant strains of *Plasmodia*.

11.3.4.2 ^{31}P-NMR to Study Drug Transmembrane Transport

An interesting application of ^{31}P-NMR is the study of transmembrane transport of drugs. An example is presented in a paper by Tam Huyinh-Dinh and coworkers [79]. These authors studied the transmembrane transport of lipophilic glycosyl phosphotriester derivatives of 3'-azido-3'-deoxythymidine (AZT). The latter is used in the therapy of AIDS. Its application is, however, limited by serious adverse reactions, especially bone marrow suppression. Previous studies of the same authors have shown that lipophilic phosphotriesters of thymidine are transported across unilamellar vesicles and that the length of the alkyl chain is of importance for drug-membrane interaction. The nucleoside and the hexadecyl moiety cannot be changed because they are essential for transport and antiviral effect. Therefore it was assumed that changes at the sugar moiety could alter the physicochemical properties to improve selectivity. The experiments were performed with 6-substituted mannosyl phosphotriesters of AZT, derivative **5**, and a 1-substituted phosphotriester **6**. ^{31}P-NMR spectra were obtained in aqueous solution in the presence of large unilamellar vesicles (LUV) and in the presence of LUV and Mn^{2+}-ions. These ions cannot permeate the LUV membranes and their paramagnetism led to a large broadening of the resonance signals of those drug molecules which remain in the extravesicular volume. In consequence only the resonance signals of molecules were observed which had crossed the membrane and were located within the vesicles. It is obvious from Figure 11-11 that only molecules of derivative **6** were able to cross the membrane of the vesicles.

(5)

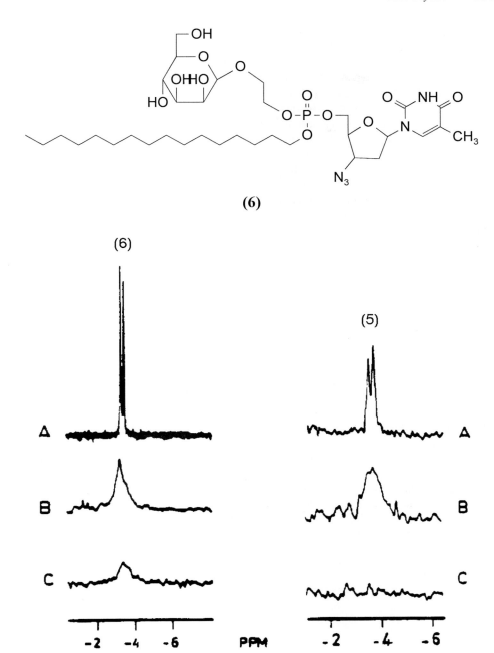

Figure 11-11: ^{31}P-NMR (121 MHz) of phosphotriesters **5** and **6** in the absence **A** and presence **B** of large unilamellar vesicles and after addition of Mn^{2+}ions (reproduced from [79] with permission from American Chemical Society, Washington, DC, USA).

In Figure 11-11A the two phosphorus resonances of **5** and **6** in the absence of LUV correspond to the two diastereoisomers. The chemical shifts of **5** and **6** are similar. The signal line width of **5** is, however, twice that of **6**. The line width of **6** is similar to that observed for glycosyl phosphotriesters of thymidine or 5-fluoro-2'-deoxyuridine (data not shown). It indicates that these molecules form similar small micelles in aqueous solution. In contrast, the 6-mannosyl derivative **5** forms larger aggregates. The signals of both derivatives are broadened upon addition of LUV, that of **5** by a factor of 10 (Figure 11-11B). When Mn^{2+}-ions were added, the resonance signals of **5** were completely broadened so that no signal could be observed. In contrast, the resonance signals of derivative **6** could still be observed (Figure 11-11C). The intensity of the low field resonance was decreased by a factor of 3, for the high field resonance by a factor of 2 and the line width of the signals were not affected in the presence of Mn^{2+}-ions. This shows that the phosphotriester **6**, where the resonance signals are of reduced intensity but not influenced by Mn^{2+} ions, is partly transported into the intravesicular water-membrane interface, whereas derivative **5** interacts only with the outer layer of the membrane.

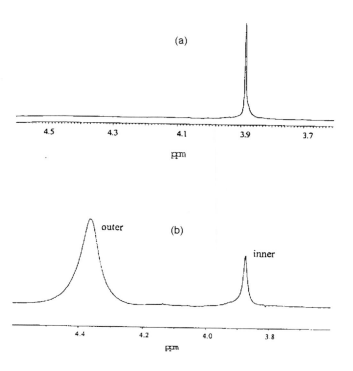

Figure 11-12: Resonance of glycolic acid before **a** and after **b** addition of $PrCl_3$ as shift reagent to the vesicle system. pH 3.94, 30°C, 100 nm vesicle diameter (reproduced from [80] with permission from Elsevier Science, Amsterdam, The Netherlands).

A combination of ^{1}H-NMR and the paramagnetic shift reagent $PrCl_3$ has been used to study the permeation of glycolic acid through LUVs of phosphatidylcholine at various cholesterol concentrations and at different pH values [80]. Figure 11-12a shows the ^{1}H-NMR spectrum of glycolic acid in the presence of DPPC LUVs. A single resonance at about 4 ppm can be observed. The addition of PrCl$_3$ led to a downfield shift of the glycolic acid resonance signal outside of the vesicles. Both peaks are broadened because of the slow exchange regime (Figure 11-12b). This increase in linewidth of the inner resonance as a result of exchange is described by eq. 2.

$$\Delta v = k_{app}/\pi \qquad (2)$$

where Δv is the broadening in Hz arising from the exchange process and k_{app} (in s^{-1}) is the pH-dependent, vesicle concentration-independent rate constant of transport across the membrane. Δv is obtained by subtracting the line width in the absence of exchange, v_0, from the observed line width, v. With decreasing temperature the line width decreased but became constant below 283 K ($v_0 \approx 5$ Hz). The observed pH dependence of the permeation rate is given in Figure 11-13. and is described by eq. 3.

$$k_{app} = k_a \alpha + k_b(1 - \alpha) \qquad (3)$$

k_a is the rate constant of the uncharged acid, k_b, of the negatively charged base and α the fraction of undissociated acid. This example on glycolic acid shows the ability of NMR techniques to evaluate the rate constants for permeation in LUVs under equilibrium conditions and to study the effects of pH, temperature, vesicle size and membrane composition (cholesterol) on the permeation rate of compounds.

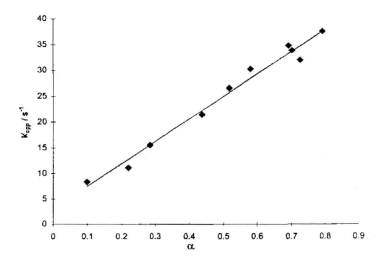

Figure 11-13: Effect of pH (and therefore α, the fraction of undissociated acid) on the apparent rate constant of permeation, k_{app}.
The slope is 43.7 s^{-1} (SD = 2.2 s^{-1}) and the intercept 3.1 s^{-1} (reproduced from [80] with permission from Elsevier Science, Amsterdam, The Netherlands).

11.3.4.3 ^{1}H-NMR in Combination with Pr^{3+} for the Studying of Drug-Location

Often ^{1}H-NMR is used in combination with paramagnetic probes of Praseodymium cations, Pr^{3+}, which allows by dipolar interaction the separation of the resonance signals of the intra- and extravesicular choline head groups of the phospholipids. Extravesicular Pr^{3+} is in rapid exchange between the ^2H_2O and the phosphate sites on the choline head groups in the outer layer of the vesicles. The separation arises from the downfield shift of the extravesicular head group signal (*O*), the second signal originates from the intravesicular choline headgroup resonance (*I*). From the ratio of these signals the size of the vesicles can be calculated [81]. Figure 11-14 shows the ^{1}H-NMR spectrum of DPPC vesicles in the presence of 5 mM Pr^{3+}. It shows the separation of resonance signals of the inner (*I*) and outer (*O*) choline head groups, the proton resonance of the lipid acyl chains (*H*) and of the terminal methyl group (*M*) [82]. Lysed DPPC vesicles were obtained by cycling from 60° to 30° to 60°, i.e. on cycling through T_t. After lysis the interaction of the Pr^{3+} with the inner head groups led to an increase in the intensity of the outside signal and a decrease in intensity of the inside signal so that the *O:I* ratio increases.

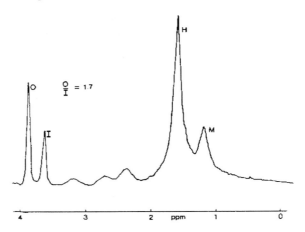

Figure 11-14: ^{1}H NMR spectrum of DPPC vesicles (10mg/ml) in the presence of 5 mM Pr^{3+} showing separate signals from the inner (I) and outer (O) choline head groups, the lipid acyl chains (H) and terminal methyl groups (M). Chemical shifts are with respect to external TMS (reproduced from [82] with permission from Elsevier Science, Amsterdam, The Netherlands).

The ratio can be used to calculate the percentage of lysed vesicles. The authors have studied the effect of normal alcohols on the vesicular permeability induced at the phase transition temperature. The incorporation of alcohols in the vesicles increased the degree of lysis on cycling through T_t. Compared to the control, an increase in the *O:I* ratio is observed after cycling through T_t (Figure 11-15).

In another ^{1}H-NMR study the interaction of β-blockers with sonicated dimyristoyl-phosphatidylcholine (DMPC) liposomes in the presence of Pr^{3+} has been reported [83]. The presence of Pr^{3+} increased the splitting of the choline trimethylammonium group signals which arise from the phospholipid molecules located at the internal and external layer of the bilayer. The downfield shift of the external peak (*E*) is considerably stronger than the upfield shift of the internal peak (*I*). (Figure 11-16).

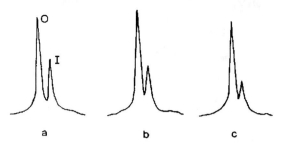

Figure 11-15: ^{1}H NMR spectra showing the increase in the O:I ratio of DPPC vesicles in the presence of 50 mM decan-1-ol. **a** after incubation, O:I ratio = 1.58, **b** after 6 cycles, O:I ratio 1.92 (% lysis = 11.6); **c** after 10 cycles O:I ratio = 2.25 (% lysis = 20.6) (reproduction from [82] with permission from Elsevier Science, Amsterdam, The Netherlands).

The difference in chemical shift of the two signals (Δ Hz) increased linearly with increasing Pr^{3+} concentration up to 10 mM (data not shown). Upon addition of the β-blockers the effect of Pr^{3+} is reversed and propranolol (PPL) has the strongest effect. (β-blockers are presented in Table 11-7 together with their partition coefficients into DMPC liposomes and into octanol).

Table 11-7: Physicochemical properties of β-blockers of the propranolol type (adapted from ref. [83] with permission of Taylor and Francis, UK).

Commpound	pKa	log K'_m[1]	log P_{oct}[2]
Propranolol (PPL)	9.45	2.62	3.56
Alprenolol (APL)	9.70	2.23	3.10
Oxprenolol (OPL)	9.50	1.54	2.18
Toliprolol (TPL)	9.60	1.54	1.93
Metoprolol (MPL)	9.70	1.23	1.88
Atenolol (ATL)	9.55	1.09	0.16

[1] partition coefficient L-α-dimyristoylphospatidylcholine/phosphate buffer pH 7.4
[2] Partition coefficient octanol/buffer added [109].

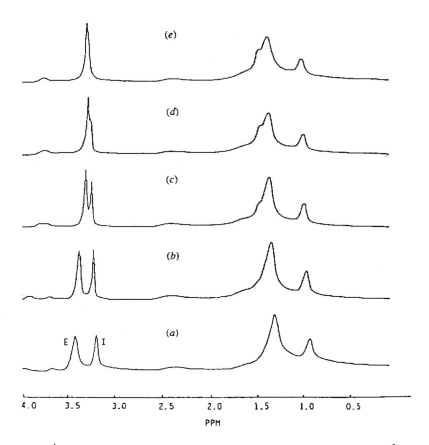

Figure 11-16: ^{1}H-NMR spectra of liposomes of a DMPC (14,4 mM)+ Pr^{3+} (2mM) + propranolol(PPL) at **a** 0 mM, **b** 0.5 mM, **c** 2 mM, **d** 4 mM, **e** 5 mM (reproduced from [83] with permission from Taylor & Francis, UK).

At 5 mM the splitting of the resonance signals is completely reversed by PPL (Figure 11-16) whereas for OPL or TPL the reversal is not completed even at 20 mM. The relative power of the β-blockers to reduce Δ Hz follows a linear relationship with respect to β-blocker concentration and is described by eq. 4.

$$\log \Delta \text{Hz} = D - KC \qquad (4)$$

where the slope K is the displacement constant, C is the concentration (mM) of β-blockers and D is the mean Δ Hz (64.3 ± 2.4) at 2 mM Pr^{3+} in the absence of β-blocker. Propranolol shows the strongest displacement effect. Additionally a slight downfield shift of the methylene group signal of the acyl chain of DMPC is observed, indicating the penetration of PPL into the liposomes. For the *meta*- and *ortho*-substituted β-blockers a highly significant correlation between the displacement constant K and the phospholipid/buffer partition coefficient log K'_m has been found ($r = 0.95$, $n = 4$), compared to the regression coefficient with log P_{oct} ($r = 0.92$). β-blockers with *para*-substitution (MPL, ATL) and low K'_m-values deviate from this correlation. They exerted stronger effects than could be expected on the basis of their low lipophilicity. The authors assume that β-blockers possessing *ortho*- and *meta*-substitution are non-selective whereas those having *p*-substituents on the aromatic ring structure are involved in membrane interactions that lead to cardioselectivity. It is concluded that for the non-selective β-blockers the partitioning into the membrane is decisive; for the more hydrophilic, selective derivatives, however, polar-group interaction at the membrane surface is dependent on the orientation of the β-blocker molecules. It can be concluded that the displacement of Pr^{3+} from membrane surfaces by β-blockers is related to their interaction with the polar head groups of the phospholipid and that *para*-substituted derivatives may adopt a different orientation. The results show that it is risky to rely on partition coefficients only to describe drug-membrane interactions and underlines the great power of NMR techniques to obtain insight into such processes.

11.3.4.4 The Use of ^{2}H-NMR and ^{13}C-NMR to Determine the Degree of Order and the Molecular Dynamics of Membranes

Deuterium NMR is a powerful technique to obtain information on both the degree of order and the molecular dynamics of liquid-crystalline media. It has been extensively used on model as well as on natural membranes. Deuterium nuclear magnetic resonance has been used as a probe to investigate chain packing in lipid bilayers, and the effects of hydrocarbons and alcohols and their location in the membrane have been determined [84]. The deuterium nucleus gives rise to a doublet in the resonance spectrum caused by the interaction between its quadrupole moment ($I = 1$) and the surrounding electrical field gradient. The separation between the two spectral lines can be observed in oriented samples to follow the relationship in eq. 5.

$$v_2 - v_1 = (3/4)(e^2qQ/h)(3\cos^2\theta - 1) \qquad (5)$$

where e^2qQ/h is the ^{2}H quadrupole coupling constant (~ 170 kHz) and θ the angle between the magnetic field and the axis of the molecular ordering. All values of θ are possible in an aqueous dispersion of phospholipids. The spectrum of ^{2}H nuclei is a 'powder pattern', and the two possible peaks correspond to $\theta = 90°$ (perpendicular edges) and the two shoulders to $\theta = 0°$ (parallel edges) [85].

Solid-state ^{2}H experiments with drug/phospholipid preparations can be carried out by observing spectra from either the drug or the phospholipid. Depending on the degree of labeling, orientation and position of the drug in the membrane can be determined. It can also be appropriate to combine ^{2}H-labeling with ^{13}C-labeling, for instance of the drug molecules, and it is known that the effects of anesthetics on glycolipids differ from those seen on phospholipids.

Figure 11-17 shows as an example the ^{2}H-NMR spectra of the glycolipid 1,2-di-*O*-tetradecyl-(β–D-glucopyranosyl)-*sn*-glycerol (β-DTGL) (**7**) labeled at C-1' of the carbohydrate head group (left) and C-3 of glycerol (right) in the absence (top) and presence (bottom) of tetracaine **8** at pH 9.5 and 52°C [85], [86] where the tetracaine is almost totally uncharged. In the absence of tetracaine at 52°C the spectrum describes axially symmetric motion of the lamellar phase. In the presence of uncharged tetracaine the quadrupole splitting of β-DTGL labeled at C-1' of the hydrocarbon head group is reduced by more than a factor of 2 (Figure 11-17). A non-lamellar phase is induced. It can be assumed that the lipid is in the hexagonal phase, as observed for pure β-DTGL at elevated temperatures (58°C), i.e. the lamellar structure of β-DTGL is very sensitive to tetracaine. In contrast, the interaction of tetracaine with phosphatidylcholine with or without cholesterol results in a reduction of the lipid order parameters in the plateau and in the tail regions of the acyl chains, and the effect is larger with the charged form of tetracaine [87]].

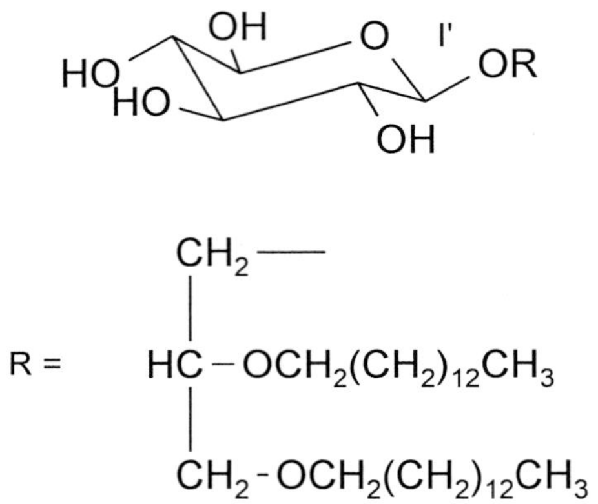

β-DTGL (7)

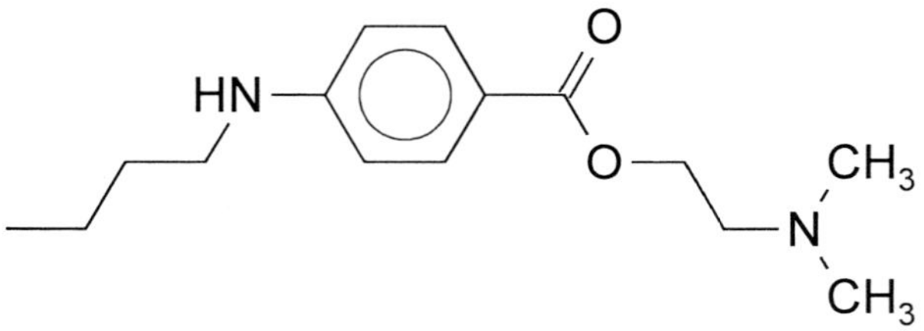

TTC (8)

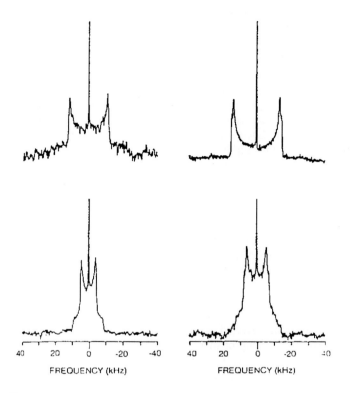

Figure 11-17: ^{2}H-NMR spectra (30.7 MHz) of β-DTGL (**7**) labeled at C-1' of the carbohydrate headgroup (*left*) and at C-3 of glycerol (*right*) in the absence (*top*) and presence of tetracaine (**8**) (*bottom*) at pH 9.5, 52°C (reproduced from [86] with permission from Elsevier Science, Amsterdam, The Netherlands).

11.3.4.5 Change in Relaxation Rate $1/T_2$; a Method of Quantifying Drug-Membrane Interaction

^{1}H- and ^{13}C-NMR spectra of small molecules in the presence of proteins or phospholipids can lead to changes in the spin-lattice relaxation rate $1/T_1$ and the spin-spin relaxation rate $1/T_2$, the latter expressed as the line width of the resonance signal. Changes in $1/T_2$ can be related to a decrease in the rotational freedom of small molecules in the presence of a "receptor" with which they can interact [88], [89]The transverse relaxation time, T_2, depends on both the internal and overall molecular mobility and contributions of several mechanisms as dipole-dipole, spin-rotation, quadrupole interaction and chemical shift anisotropy can be involved [90].The changes in peak line width, v, at half peak heights are proportional to the change in $1/T_2$. Expressed as function of the drug/protein or drug/phospholipid ratio in the vesicles, $1/T_2$ can be used to quantify the degree of interaction, provided care is taken that no other factors have led to signal broadening given in eq. 6.

$$T_2 = 1/(\pi v_{1/2}) \quad (6)$$

In a large range of phospholipid concentrations the broadening depends linearly on lipid concentration. The slope of such plots is proportional to the "affinity" of the drug molecules or of molecular substructures to the phospholipid and can be taken as a measure of the degree of interaction. Such studies have been performed to elucidate the molecular basis for drug-induced phospolipidosis, and it has been found that for drugs of the chlorphentermine type and other catamphiphilic drugs a relation between the degree of interaction and the disturbance of the phospholipid metabolism does exist [91], [97].

The mobility of anesthetic steroids (Table 11-8) in model membranes has been examined by ^{1}H- and ^{2}H-NMR spectroscopy [92]. Line width broadening in the ^{1}H-NMR and ^{2}H-NMR proton signal of the C(O)CH$_3$ substituent of the steroids in the presence of phosphatidylcholine has been determined at various temperatures, and a correlation with their anesthetic effect is assumed (Table 11-9). Resonance signals with relatively narrow line widths were observed at 37°C for the active anesthetics 1–4. Upon raising the temperature the resonance signals narrowed which is consistent with an increase in mobility. For the membrane preparations with the inactive derivatives 5–7, resonances (especially ^{1}H) could not be detected even at the highest temperature. This suggested that the intense line width broadening of these steroids was due to their extreme restriction in mobility in the phospholipid bilayer. Only relatively narrow ^{1}H-NMR resonances of the C(O)CH$_3$ group of the steroids could be observed in the presence of the very intense lecithin signals. To make the observation of the steroid signals of all derivatives possible they were selectively deuterated (>95%) in the acetyl methyl position and the ^{2}H-NMR spectra recorded. Again the resonance signals become smaller with increasing temperature, indicating increasing mobility, but remain relatively broad for the derivatives with low biological activity (Table 11-9). In this example the interaction with the phospholipids leads to a decrease in biological activity.

Another example is depicted in Figure 11-18 showing the broadening of the proton resonance signals of the –N–CH$_3$ group and the aromatic protons of trifluoperazine — a compound which also shows activity in the reversal of multidrug resistance in tumor cell lines — as a function of bovine brain phosphatidylserine (BBPS) concentration. Figure 11-19 shows plots of increasing interaction ($1/T_2$) of several catamphiphiles with increasing BBPS concentration [93].

The interaction of trifluoperazine and other catamphiphiles is so strong that the interaction can only be reversed upon addition of Ca^{2+} in case of lidocaine and verapamil showing the weakest interaction i.e. the lower slope. In case of amiodarone and some other derivatives, however, (see Figure 20) even an increase in interaction upon addition of Ca^{2+} is observed. This effect of increasing ion concentration is due to the strong hydrophobic interaction forces found for amiodarone and supports the results of NMR, X-ray diffraction and modeling studies that amiodarone is deeply buried in the hydrocarbon chains of the bilayer [47], [94].

Table 11-8: Steroid structure for the pregnane analogues 1-7.(Adapted from Table 1 of Makriyannis et al., ref. [92] with permission from American Chemical Society, Washington, DC, USA)

Steroid	R1	R2	R3
1	3α-OH	5α-H	=O
2	3α-OH	5α-H	H2
3	3α-OH	5β-H	=O
4	3α-OH	5β-H	H2
5	3β-OH	5α-H	H2
6	3β-OH	5β-H	H2
7	3α-OH	5α-H	=O

Table 11-9: ^{1}H-(left) and ^{2}H-NMR line width (right) ($\nu_{1/2}$), respectively, of resonances corresponding to the C(O)CH$_3$ protons of steroids incorporated in lecithin vesicles at different temperatures.(adapted from [92] with permission from American Chemical Society, Washington, DC, USA)

Steroid	$\nu_{1/2}$ Hz 37°C	$\nu_{1/2}$ Hz 50°C	$\nu_{1/2}$ Hz 70°C	$\nu_{1/2}$ Hz 37°C	$\nu_{1/2}$ Hz 50°C	$\nu_{1/2}$ Hz 70°C	Anestetic activity (mg/kg)[1]
1	8.0	5.0	2.5	40	17	12	3.1
2	8.0	4.0	2.5	-	-	-	3.1
3	8.0	5.0	2.5	47	35	20	6.3
4	5.5	4.0	2.5	26	19	14	3.1
5	[2]	[2]	[2]	[2]	[2]	30	100
6	[2]	[2]	[2]	60	42	27	25
7	[2]	[2]	[2]	[2]	[2]	[2]	inactive

[1] Lowest dose producing loss of righting reflex; [2] resonance not detected

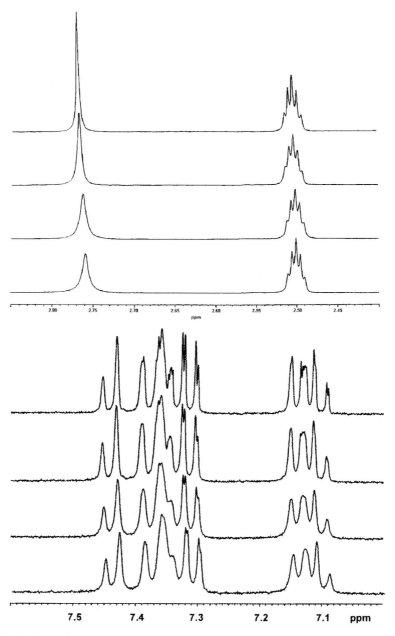

Figure 11-18: Broadening of proton resonance signals of trifluoperazine (*top*, methyl groups ~ 2.9 ppm, standard DMSO 2.65 ppm; *bottom*, aromatic protons) as a function of increasing bovine brain phosphatidylserine concentrations (BBPS) (control, 0.01–0.03, 0.05 mg/500 ml) (reproduction from ref.[93] with permission from Wiley-VCH, D-69451 Weinheim, Germany).

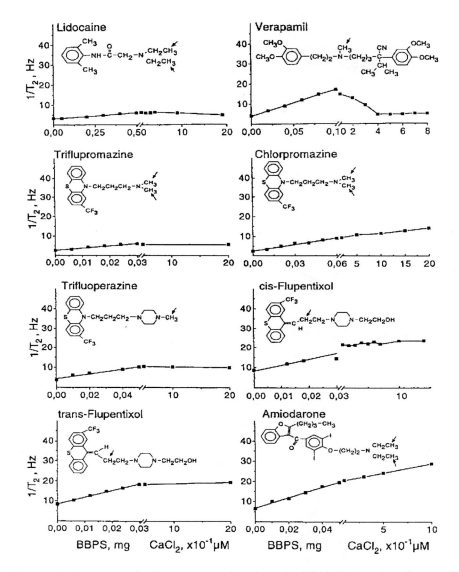

Figure 11-19: Increase in drug proton relaxation rate ($1/T_2$) for a series of catamphiphiles as a function of increasing BBPS concentration alone and after addition of increasing CaCl$_2$ concentration. Note the change in the abscissa. The spin systems measured are indicated with arrows on the corresponding structures (reproduced from [93] with permission from Wiley-VCH, D-69451 Weinheim).

Detailed interaction studies have a been performed with the neuroleptic drugs *cis*- and *trans*-flupentixol. These stereoisomers also act like trifluoperazine as modifiers of multidrug resistance in tumor cell lines where it has been observed that the *trans*-isomer is about 5 times as potent as the *cis*-form. Surprisingly it was found that upon addition of phospholipids the slopes ($1/T_2$ vs BBPS concentration) as a measure of strength of interaction differ for the two stereoisomers.

The steeper slope for *trans*-flupentixol suggests that it interacts about twice as strongly with lecithin as does *cis*-flupentixol (Figure 11-20) [47], [95].This indicates a stereospecific interaction of these catamphiphiles with phospholipids. The strength of interaction of the stereoisomers with the negatively charged BBPS is more extensive than with the neutral DPPC [91], and the chemical shifts of the H1 signals were different for *trans*- and *cis*-flupentixol.

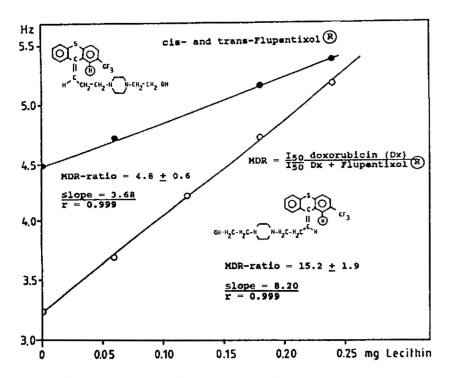

Figure 11-20: Relation between MDR reversing activity (reversal of resistance against doxorubicin) by *cis*- and *trans*-flupentixol and the degree of interaction with phospholipid liposomes (slope of the plot of changes in $1/T_2$ as a function of lecithin concentration) (reproduced from [95] with permission from Elsevier Science, Amsterdam, The Netherlands).

The largest difference was observed for the H1 proton, indicating a reduction in electron density around this nucleus in the case of *trans*-flupentixol. To indicate the main feature in the behavior of the hydrogens of these stereoisomers when interacting with the phospholipid, the slopes of *cis*- were plotted versus the slopes of *trans*-flupentixol in the DPPC and BBPS interaction. The structure of flupentixol and numbering of the protons is given in formula **9**.

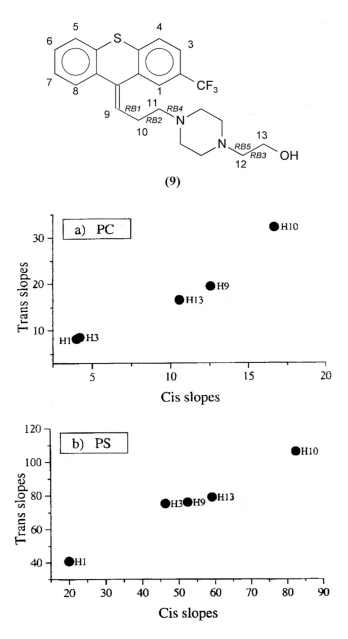

Figure 11-21: Plots of proton $1/T_2$ slopes of trans-flupentixol vs $1/T_2$ slopes of cis-flupentixol in **a** presence of phosphatidylcholine (DPPC) and **b** phosphatidylserine (DPPS) lipid environment (reproduced from [96] with permission from Wiley-VCH, D-69451 Weinheim, Germany).

Figure 11-21 shows that H1 has the lowest slope, H10 the highest and H9 is in the middle range between H1 and H10 on both plots. These hydrogens correspond to that part of the molecule related directly to the stereoisomeric differences between cis- and trans-flupentixol.

The coupling constants of H9 (7.3 Hz) and H10 (7.2 Hz) can be related to the preferred conformations of cis- and trans-flupentixol and were used as constraints to calculate with the PCMODEL program the conformations which best fit the experimental data (Figure 11-22). Experimental and computational results support the assumption that cis- and trans-flupentixol could interact in a stereodependent manner with both phospholipids [96].

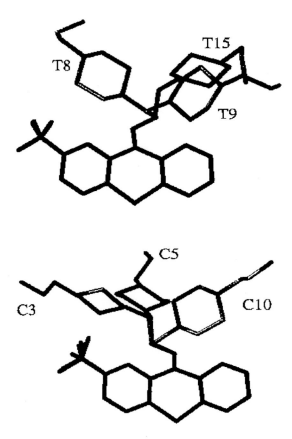

Figure 11-22: Conformation of **a** trans-flupentixol and **b** cis- flupentixol as obtained from the conformational analysis after energy minimization and selected in accordance with the NMR experimental data (hydrogen atoms omitted) (reproduced from [95] with permission from Wiley-VCH, D-69451 Weinheim, Germany).

Similar studies on n-alkyl substituted benzylamines show a sudden increase in line width broadening above a chain length of C_6 (Figure 11-23). The sudden increase in strength of interaction corresponding to a π-value of about 3 is paralleled by a sudden and strong increase in negative inotropic activity [97].

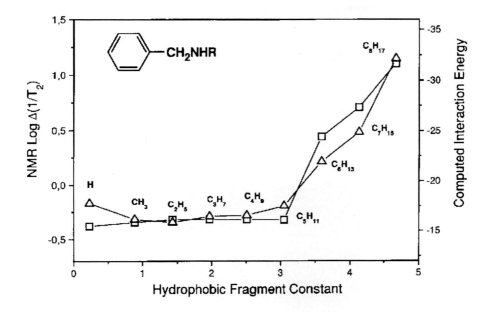

Figure 11-23: Plot of NMR binding ($1/T_2$) (-△-) versus N-alkyl hydrophobic fragment constants compared with an overlay plot of computed interaction energies vs N-alkyl hydrophobic fragment constants (-□-) for benzylamines R = H to R = C_8H_{17} (reproduced from [100] with permission from Wiley-VCH, D-69451 Weinheim, Germany

The observations made in the last two examples lead to another possible effect of drug-membrane interaction, the change in the preferred conformation or orientation of the drug molecule under the environmental conditions of the membrane.

11.3.4.6 NOE-NMR in the Study of Membrane Induced Changes in Drug Conformation

It is well known that the preferred minimal-energy conformations of drug molecules can differ in the crystalline state, in solutions and in the state bound to protein receptors. An example of this is provided by the different conformations of acetylcholine bound to the nicotinic acetylcholine receptor derived from two-dimensional NOE spectra (Figure 11-24) [98]. Major conformational differences between the free and bound acetylcholine were observed. This applies especially to the arrangement of the N-C-C-O backbone which changes from *gauche* in solution to nearly *trans* in the bound state. In the bound state the two electronegative oxygens are positioned on the same side of the acetylcholine molecule and the hydrophobic acetyl methyl and choline methyl form a continuous hydrophobic area over the rest of the molecule. It should therefore not be surprising that a chiral matrix such as a phospholipid membrane can also act as a specific receptor and lead to a change in the conformation or orientation of a ligand molecule. An example for this to take place is provided by the above mentioned simple N-alkyl substituted benzylamines where a bilinear dependence of the change in phase transition temperature, displacement of Ca^{2+} ions and change in $1/T_2$-relaxation rates on the chain length of the N-alkyl substitution was observed. There was no effect or even a negative effect on the interaction with phospholipids up to a chain length of about C_6. Thereafter a strong increase in interaction was observed. We had speculated on a change in conformation from the extended to a folded conformation [98]. TOE-NMR experiments performed subsequently fully support this speculation. With transient Overhauser effect, (TOE), the time-dependent changes of the magnetization of a particular nucleus after the inversion of the magnetization of a neighboring nucleus can be measured.

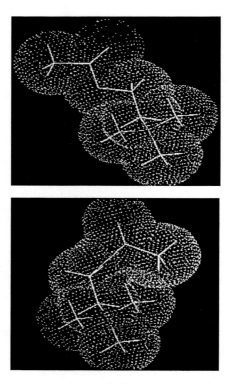

Figure 11-24: Computer representations of the structure of acetylcholine. In its free solution-state conformation (*top*), oxygen atoms appear on opposite sides of the molecule. In its receptor-bound state (*bottom*), oxygen atoms appear on the same side (reproduction from [98] with permission from PNAS, Washington DC, 20418, USA).

The 2D-TOE experiments with the derivatives were performed in the presence and absence of lecithin vesicles, and the results are shown in Figure 11-25. Whereas up to a chain length of *n*-butyl and pentyl the extended conformation exists in the absence and presence of the phospholipid, this form exists for the longer chain length only in the absence of phospholipid (lecithin).

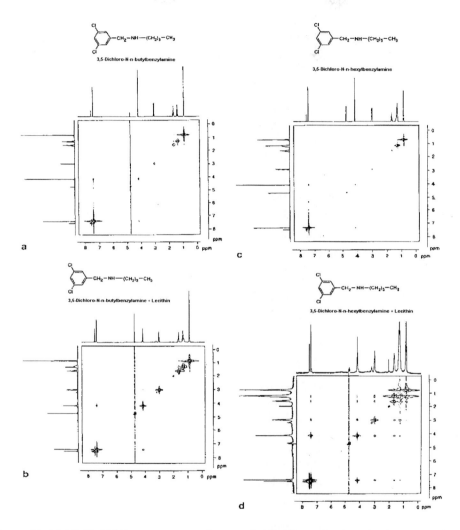

Figure 11-25: NMR-NOE spectra of 3,5-dichlorobenzyl-*N*-butylamine and 3,5-dichlorobenzyl-N-hexylamine in deuterated phosphate buffer in the absence **a**, **c** and presence **b**, **d** of lecithin (reproduced from [99] with permission from Wiley-VCH, D-69451 Weinheim, Germany).

A cross peak occurs only between the *ortho* protons of the dichlorophenyl and the benzylic methylene in the absence of lecithin. In the presence of phospholipid, however, additional cross peaks indicating proximity of the *N*-hexyl protons and the aromatic ring are clearly present for the hexyl derivative whereas for the butyl derivative no NOE effect between the *N*-butyl and the aromatic protons can be detected [99].

Molecular modeling studies correlate very well with the experimental observations, [100]. The plot of computed interaction energies versus lipophilicity proceeds parallel to the plot of $1/T_2$ versus lipophilicity (Figure 11-23). The result confirms the proposed conformational change with lengthening of the N-alkyl groups of the benzylamines. It is an example of the general model proposed by Herbette et al. [54] for possible consequences of drug-membrane interactions on conformation and/or orientation (see Section 10.3.2.2.3.). The examples discussed provide support for the assumption that the active conformation can be induced by the lipid environment, but also that inactivity may be related to such changes in other examples. Changes in conformation have also been shown by NOE-NMR for verapamil in the presence of lipid mimetic solvents such as acetonitrile [101] and on binding of verapamil and verapamil analogs to amphiphilic surfaces (phospholipid-coated polystyrene-divinylbenzene beads) [102].

11.3.4.7 Combination of Methods and Techniques for the Studying of Drug-Membrane Interactions

The different methods discussed differ in their ability to describe the various aspects of membrane-drug interaction. HPLC techniques can be used to quantify the overall affinity of drugs to membranes and to determine log P_{oct} or log k'_{oct} proportional log k'_{membr} values. Studies on monolayer and Ca displacement are also techniques for studying the relative affinity of drugs to phospholipid head groups and effects on membrane surface tension. Calorimetry especially can supply data on changes in membrane fluidity, phase transition, phase separation and domain formation. The spectroscopic techniques supplement information on drug and phospholipid conformation, order and dynamics as well as on drug orientation, location and transfer rates. To get detailed insight into drug membrane interaction a combination of these techniques is advocated and several applications have been published and are discussed in a few examples.

Besides freeze-fracture electron microscopy a combination of DSC, NMR and monolayer techniques has been applied to study the various aspects of the interaction of the class IV Ca^{2+} antagonist flunarizine [103] with various phospholipids. DSC shows only limited interaction of flunarizine with DPPC; it destabilizes the L_β-phase without stabilizing the L_α-phase. In contrast not only the onset of phase transition is influenced by flunarizine but also the phase transition temperature and the completion of the transition in the case of DPPS, indicating a significant interaction. With DPPE, flunarizine causes no effect on the L_β- to L_α-phase change, but a strong concentration dependent induction of the H_{II} phase is observed, which points to an expansion of the acyl chain region of the DPPE bilayer by the lipophilic molecules. The limited effect on the L_β to L_α transition of DPPC, the complete lack of this in case of DPPE, and induction of the H_{II} phase in PPE indicate the location of flunarizine in the hydrophobic core of these phospholipids. The strong effect on phase transition of DPPS, suggests a more specific interaction of flunarizine with DPPS probably with the negatively charged head groups (Figure 11-26). This specific interaction follows also from studies on monolayers, where the cut-off surface pressure for the interaction of flunarizine with the phospholipids has been evaluated.

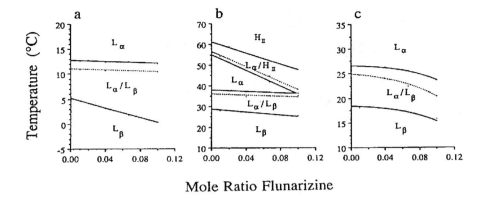

Figure 11-26: Phase diagrams of flunarizine-dielaidoylphospholipid mixtures; **a** phosphatidylcholine DPPC, **b** phosphatidylethanolamine DPPE and **c** phosphatidylserine DPPS (reproduced from ref. [103] with permission from Academic Press Ltd., San Diego, USA).

The cut-off point is the surface pressure at which flunarizine in the aqueous phase (0.5 mM) is no longer penetrating into the monolayers of the three phospholipids. It shows that the strongest interaction occurs with DPPS (44.0 mN.m^{-1}) and the weakest with DPPE (33.4 mN.m^{-1}), which is in good agreement with the DSC results. ^{31}P-NMR was then used to follow the displacement of flunarizine from the head groups by Ca^{2+}-ions. ^{31}P-NMR allows us to determine the mobility of the phosphorus atoms in the head groups. The interaction of Ca^{2+} with the head groups of DPPS leads to an immobilization of the head groups and in consequence to a broadening of the resonance signal so that the signal almost disappears. Figure 11-27 demonstrates this event in the absence and presence of flunarizine. It shows the ability of equimolar flunarizine concentrations to almost completely prevent the binding of Ca^{2+} ions. An interesting study has been published in which the four complementary techniques, DSC, FT-IR, ^{1}H-NMR and deuterium NMR of side-chain perdeuterated DMPC (DMPC-d$_{54}$) have been used to evaluate the interaction of three reversible H$^+$/K$^+$-ATPase inhibitors [104]. The study has been undertaken to characterize the interaction between DMPC model biological membranes and two non-covalent inhibitors of gastric H$^+$, K$^+$-ATPase. Compound **10**, (SK&F 96079) partitions readily into the bilayer of DMPC, causes a slight disordering of the lipid hydrocarbon side-chain motion, a reduction in cooperativity and a decrease in phase transition temperature. FT-IR and the deuterium NMR studies showed, however, that the bilayer structure remains intact up to high molar compound/lipid ratios. ^{1}H-NMR NOE-measurements performed for **10** and analogue **11** led to an insight into the orientation of the two compounds within the bilayer. Both drugs are largely cationic under physiological conditions (pK_a 9.54 for **10** and 8.63 for **11**). The nature of the spin probe data and the intermolecular NOE were unspecific for compound **10** but specific for compound **11**, (SK&F96464). For compound **10** the NOEs in the spectral region especially between 4 and 4.8 ppm, could not be reliably assigned to effects of compound-lipid interactions. Only the resonances of the β-carbon of the glycerol moiety of DMPC at about 5.2 ppm were observed to be involved in NOEs with **10**. NOEs from the glycerol-β protons to several signals of **10** were observed, indicating considerable mobility of the compound in relation to the head group region.

SK&F 96079 (10)   **SK&F 96464 (11)**   **SK&F 95018 (12)**

In contrast, derivative **11** showed increased solubility in water and rather sharp resonance signals also in the presence of DMPC, due to rapid exchange between free and membranous environment. The conformation of **11** was varied interactively using an Evans-Sutherland PS-300 to take into account the observed NOEs and the considerable shielding of the H2 and H3 protons and deshielding of H5 and H6 of the quinoline ring.
Irradiation of H2, H3 and H5 produced slight NOEs to the envelope of signals at about 7.3 ppm which correspond to H7 and the protons attached to the *ortho*-toluidino group. Upon irradiation H3 resonance produced also an NOE to the 2'-CH$_3$ signal (see **11**).
By a conformation with a torsion angle of about 80° for the angle C3-C4-N-Cl' and about – 105° for C4-N-Cl'-C2', the protons H2 and H3 are placed inside the ring current shielding cone centered in the *ortho*-toluidine ring and the protons H5 and H6 outside in accordance with the NMR results. Rotation of the *ortho*-toluidine ring by 180° about the Cl'-N bond yields a conformation which is also significantly populated. For the two conformations assumed the authors concluded on the basis of the only slight NOEs observed between compound and membrane signals that a significant fraction of compound **11** is in the aqueous phase. Typical NOE difference spectra resulting from the irradiation of protons of compound **11** are shown in Figure 11-28.
Under the conditions of the experiment **11** can form two favorable electrostatic interactions with DMPC head groups. Hydrogen bonds can occur between the amino protons of **11** and the *sn*-1 side chain carbonyl of a DMPC molecule and between the protonated quinoline ring of the molecule and a primary phosphodiester oxygen of the same DMPC molecule. Figure 11-29 shows the stereo plot of the two molecules in the bound conformation consistent with the intramolecular NOEs. These conformations are also consistent with the NOEs between the compound and the glycerol β-proton of DMPC. The two geometrically favorable electrostatic interactions possible for compound **11** seem to be the precondition for its specific orientation in the membrane. The authors concluded further that compounds of type **10** which show only a slight effect on cooperativity of the gel to crystalline phase transition should not lead to deleterious toxic properties, especially

hemolytic, which have been observed for the amphiphilic derivative **12**, (SK&F 95018) for which disruption of model bilayers to non-ordered systems has been observed [105].
Another interesting example to be mentioned is the interaction of positively charged antitumor drugs with cardiolipin-containing DPPC vesicles [106]. The example allows us to compare ^{31}P-NMR with DSC measurements. The authors concluded from the results of ^{31}P-NMR studies on dioleoylphosphatidylethanolamine/cardiolipin (2:1) liposomes that adriamycin and 4'-*epi*-adriamycin "are cabable of inducing phase separation under liquid-crystalline conditions". In contrast, celiptium, 2-*N*-methylelipticinium and ethidiumbromide could not produce phase separation. This finding would be of special interest because of the considerable evidence that cardiolipin-adriamycin interaction plays an important role in the inhibition of mitochondrial function, both *in vitro* and *in vivo*, and this could be related to the anthracycline induced cardiotoxicity.

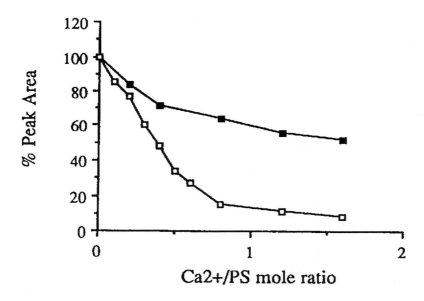

Figure 11-27: ^{31}P-NMR study of the effect of flunarizine on the Ca^{2+}-induced reduction in the head group signal of dielaidylphosphatidyl-serine (DPPS); DPPS □ and DPPS + flunarizine (1:1) ■ (reproduced from [103] with permission from Academic Press Ltd., San Diego, USA).

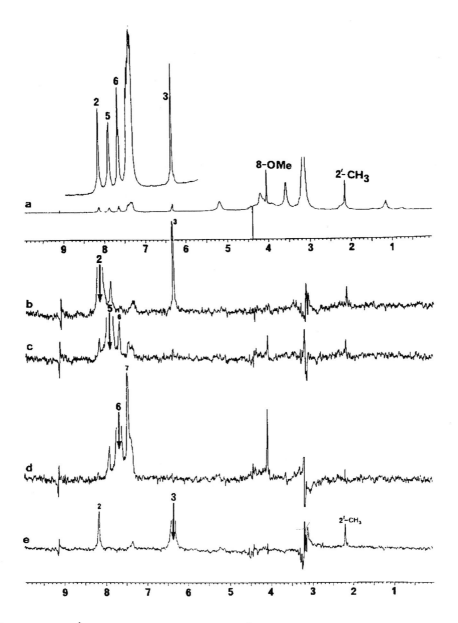

Figure 11-28: **a** ^{1}H-NMR spectra (360 MHz, 25°C) of a sonicated aqueous dispersion of 5 mg DMPC-d$_{54}$ and SK&F 96464 **11** in a 4:1 molar ratio. **b** to **e** correspond to NOE difference spectra resulting from 250 ms irradiation at the **b** H2, **c** H5, **d** H6 and **e** H3 signals of the compound (reproduced from [104] with permission from Elsevier Science, Amsterdam, The Netherlands).

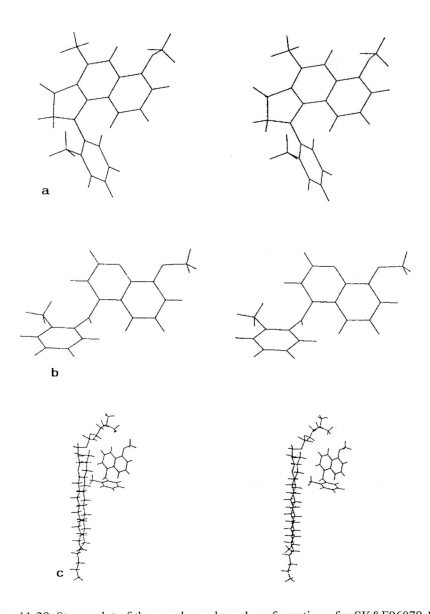

Figure 11-29: Stereo plot of the membrane bound conformation of **a** SK&F96079 **10** and **b** SK&F 96464 **11**, consistent with the intermolecular NOEs and ring current perturbation of chemical shifts, **c** a hypothetical complex between a DMPC molecule and SK&F 96464 **11**, consistent with fatty acid spin probe data. The structure is according to the authors only intended to be representative and not unique (reproduced from [104] with permission from Elsevier Science, Amsterdam, The Netherlands).

A later publication from this group [107] dealing with ^{31}P-, ^{2}H-NMR DSC studies on zwitterionic and anionic phospholipids in the absence and presence of doxorubicin (adriamycin) did not agree with the former results [107]. Doxorubicin had a stronger disordering effect on the membrane of lipid mixtures enriched with anionic lipids. However, "extensive segregation of DOPE and DOPA or DPS was not observed even under conditions of H_{II} phase formation". The reason for this discrepancy, according to the authors, was that in the earlier paper phase separation was obtained with membranes "subject to gel-liquid to crystalline phase transition". However, in the first paper this was denied.

The example shows the great importance of the experimental conditions (lipid composition, temperature, ratio of lipid/drug concentration etc.) which have a decisive influence on the quality, type and strength of interaction. Any further interpretation with respect to biological activity *in vitro* and *in vivo* can become risky.

11.4 Concluding remarks

The several methods available for the studying of drug-membrane interactions are used preferably on artificial membranes. One has, however, to be careful in assuming a direct pharmacological or biological relevance of the measured effects of drugs on artificial membranes.

The changes in phase transition of artificial bilayers for example upon addition of drugs is only a parameter which allows an interpretation of strength and type of interaction with the studied phospholipid and is useful for the derivation of quantitative structure-interaction relationships.

Most of the biomembranes will be in the liquid-crystalline phase under physiological conditions and the complex composition will decrease the chance of measuring phase transitions. It should be mentioned that the effects of drug molecules on fluidity and degree of order of a membrane can even be oppositey below and above T_t [108]. NMR-measurements of drug-membrane interactions are normally performed above T_t.

Other effects caused by the intrusion of drug molecules into membranes such as changes in surface charge, transmembrane potential, allosteric conformational changes of membrane integrated proteins etc. could very well be based on biological membrane effects.

It can also be assumed that cataphiles, because of the microheterogeneity of biomembranes, induce phase transitions of biological importance in special restricted membrane regions (domains, lipid anuli).

Strongly interacting drugs such as flupentixol or amiodarone lead to a complete separation of the control transition signal and the drug-induced phase transition signal. In between the two signals the thermogram returns to the base line. This indicates that not all transition states between maximal drug-induced and unaffected phospholipid molecules do exist.

The reduction in T_t is independent of the phospholipid/drug ratio. This means that the phospholipid/drug domains possess a constant composition. The domain becomes enlarged upon addition of drug (increase in area under the signal of the thermogram), but no change in structure or "quality" occurs.

Especially the various NMR techniques alone or in combination with other analytical techniques are most valuable for following the interaction of drug molecules with the hydrocarbon chains and the polar head groups of the bilayer. They allow us to obtain information on preferred conformation and orientation of ligands and membrane constituents. The examples discussed show that amphiphiles can interact with polar head groups and that the preferred conformation of drugs and head groups may change. This could be decisive for drug activity or inactivity. The various NMR methods are also useful to determine transport kinetics and partitioning of drug molecules into bilayers. Type and strength of interaction depends on the structure and physicochemical properties of both the drug molecule and the phospholipids and have a direct influence on drug accumulation within a certain membrane and therefore on drug activity and selectivity.

It is hoped that this chapter will help to stimulate and intensify research into the effects of the composition of membranes and structure of drugs on their interaction processes [109].

References

1. H Lüllmann, K Mohr, A Ziegler, Taschenatlas der Pharmakologie, Thieme Verlag, Stuttgart (1990)
2. P J Quinn,The Molecular Biology of Cell Membranes, The MacMillan Press Ltd, London (1976)
3. R M Epand, *Biochemistry* **24**, 7092-7095 (1985)
4. A A Spector, M A Yorek, *J Lipid Res* **26**, 1015-1035 (1985)
5. P Yeagle, The Structure of Biological Membranes, CRC Press Boca Raton, Ann Arbor, London (1992)
6. Phospholipids Handbook, edited by G Cevc, Marcel Dekker, INC New York, (1993)
7. M G Lee and M Jacobson, *Curr Top Membr* **40**, 111-142 (1994)
8. T E Redelmeier, M J Hope, P R Cullis, *Biochemistry* **29**, 3046-3053 (1990)
9. J E Rothman, D K Tsai, E A Dawidowicz, J Lenard, *Biochemistry* **15**, 2361(1976)
10. A J Verkleij, R F A Zwaal, B Roelofsen, P Comfurius, D Kastelijn, L L M van Deenen, *Biochim Biophys Acta* **323**, 728 (1973)
11. F P Devaux, A Zachowski, *Chem Phys Lipids* **73,** 107-120 (1994
12. P Williamson, R A Shlegel, *Mol Membr Biol* **11**, 199-216 (1994)
13. A van Hellvoort, A J Smith, H Sprong, I Fritzsche, A H Schinkel, P Borst, G van Meer, *Cell* **87**, 507-517 (1998)
14. D Kamp, C W M Haest, *Biochim Biophys Acta* **1372**, 91-101(1996)
15. C Diaz, A J Schroit, *J Membr Biol* **151**(1), 1-9 (1996)
16. L G Herbette, *Dev Cardiovasc Med* **68**, 353-365 (1987)
17. S J Singer, G L Nicholson, *Science* **175**, 720-731 (1972)
18. G Bobel, *Proc Natl Acad Sci U S A* **77**, 1496-1500, (1992)
19. O Y Polovinikova, T N Simonova, Y S Tarahkovsky, T N Nikonova, *Biol Membr* **12**, 260-278 (1995)
20. R Schwyzer, *Biochemistry* **25**, 4281-4286 (1986)
21. M Chow, Ch J Der, J E Buss, *Curr Op Cell Biol* **4**, 629-636 (1992)
22. R M Epand, S Yechiel, J P Segrest, G M Anantharamaiah, *Biopolymers* **37**, 319-338 (1995)
23. M Murata, S Kagiwada, S Takahashi, S Ohnishi, *J Biol Chem* **266**, 14353-14358 (1991)
24. K Chiba and T Mohri, *Biochemistry* **28,** 2995-2999 (1989)
25. R Winter, *Chem In unserer Zeit* **24**, 71-81 (1990)
26. M Edidin, *Annu Rev Biophys Bioeng* **3**, 179-201 (1974)

27. R D Klausner, D E Wolf, *Biochemistry* **19**, 6199-6203 (1980)
28. M K Jain, H B White, *Adv Lipid Res* **15**, 1-60 (1974)
29. P Siekevitz, *Annu Rev Physiol* **34,** 117-140 (1972)
30. P K J Kinnunen, *Chem Phys Lipids* **57**, 375-399 (1991)
31. O G Mouritzen and K Jorgensen, *BioEssays* **14**(2), 129-136 (1992)
32. H Ching-hsien, in: Phospholipid Binding Antibodies, 4-27, edited by E N Harris, T Exner, G R V Hughes, R A Asherson, CRC Press Boca Raton, Ann Arbor, Boston (1995)
33. D D Lasic, *J Liposome Res* **5,** 431-441 (1995)
34. F Szoka and D Papahadjopoulos, *Annu Rev Biophys Bioeng* **9**, 467-508 (1980)
35. A D Bangham, M W Hill and N G A Miller, *Methods Membr Biol* **1**, 1-68 (1974)
36. F Szoka and D Papahadjopoulos, *Proc Natl Acad Sci U S A* **75**, 4194-4198 (1978)
37. Szoka, F Olso, T Heath, W Vail, E Mayhew and D Papahadjopoulos, *Biochim Biophys Acta* **601**, 559-571 (1980)
38. R C MacDonald, R I MacDonald, B Ph M Menco, K Takeshita, N K Subbarao, H Lan-rong, *Biochim Biophys Acta* **1061**, 292-303 (1991)
39. M Riaz, N D Weiner, J J Schacht, *J Pharm Sci* **78,** 172-175 (1989)
40. J T Buboltz, G W Feigenson, *Biochim Biophys Acta* **1417**, 232-245 (1999)
41. D Chapman, Physical Chemistry of Phospholipids, 117-142, in: Form and Function of Phospholipids, edited by G B Ansell, J N Hawthorne, R M C Dawson, BBA Library Vol 3, Elsevier, Amsterdam (1973)
42. E Sackmann, *Ber Bunsenges Phys Chem* **82**, 891-909 (1978)
43. H K Kimmelberg, The Influence of Membrane Fluidity on the Activity of Membrane-Bound Enzymes, 205-293in: Dynamic Aspects of Cell Surface Organization, edited by G Poste, G L Nicolson, North-Holland Publishing Comp Amsterdam, New York, Oxford (1977)
44. A Caruthers, D L Melchior, *Trends Biochem Sci* **11**, 23-34 (1986
45. Drug and Anesthetic Effects on Membrane Structure and Function, edited by R C Aloia, C C Curtain, L M Gordon Wiley-Liss, New York (1991)
46. Methods for Studying Membrane Fluidity, edited by R C Aloia, C C Curtain, L M Gordon, Wiley-Liss, New York (1989)
47. J K Seydel, E A Coats, H -P Cordes, M Wiese, *Arch Pharm (Weiheim)* **327**, 601-610 (1994)
48. L G Herbette, D W Chester, D G Rhodes, *Biophys J* **49**,91-94 (1986)
49. D W Chester, L G Herbette, R P Mason, A F Joslyn, D J Triggle, D E Koppel, *Biophys J* **52**, 1021-1030 (1987)
50. R P Mason, D W Chester, *Biophys J* **56**, 1193-1201 (1989)
51. S-J Bae, S Kitamaru, L G Herbette, J M Sturtevant, *Chem Phys Lipids* **51**, 1-7 (1989)
52. R P Mason, J Morling, and L G Herbette, *Nucl Med Biol* **17**, 13-33 (1990)
53. D G Rhodes, R Newton, and L G Herbette, *Mol Pharmacol* **42**, 596-602 (1992)
54. R P Mason, D G Rhodes and L G Herbette, *J Med Chem* **34**, 869-877 (1991)
55. D Chapman and S A Penkett, *Nature* **211**, 1304-1305 (1966)
56. P R Cullis and B de Kruijff, *Biochim Biophys Acta* **559**, 399-420 (1979)
57. B de Kruijff, P R Cullis, A J Verkleij, M J Hope, C J A van Echteld, T F Taarschi, P van Hoogevest, J A Killian, A Rietveld and J J H H M van der Steen, 89-142

in:Progress in Protein-Lipid Interaction, Vol 1 edited by A Watts and J H M de Pont, Elsevier, Amsterdam (1985)
58. E Oldfield, Structural and Dynamic Properties of Lipids and Membranes, *Ser Portland Press Res Monogr* **3**, 119-135 (1992)
59. H Fribolin, Ein- und zweidimensionale NMR-Spektroskopie, VCH-Verlagsgesellschaft Weinheim (1988)
60. G E Martin and A S Zektzer, Two-Dimensional NMR Methods for Establishing Molecular Connectivity, VCH Publishers, Weinheim (1988)
61. A Watts, *Biochim Biophys Acta* **1376**, 297-318 (1998)
62. R J Wittebort, A Blume, T -H Huang, S K Das Gupta and R G Griffin, *Biochemistry* **21**, 3487-3502 (1982)
63. A Watts and P J R Spooner, *Chem Phys Lipids*, **57**, 195-211 (1991)
64. J Seelig, Magnetic Resonance of Membranes and Micelles, 261-273 in: Modern Trends of Colloid Science in Chemistry edited by H -F Eicke, Birkhäuser Verlag, Basel (1985)
65. E J Dufourc, Chr Mayer, J Storer, G Althoff, G Kothe, *Biophys J* **61**, 42-57 (1992)
66. M P Milburn and K R Jeffrey, *Biophys J* **58**, 187-194 (1990)
67. A Watts, *Stud Biophys* **127**, 29-36 (1988)
68. P M MacDonald, J Leise and F Marassi, *Biochemistry* **30**, 3558-3566 (1991)
69. T P W MacMullen, R N A H Levis, R N McElhaney, *Biochim Biophys Acta* **1416**, 119-134 (1999)
70. M -A Krajewski-Bertrand, A Milon Y Nakatani and G Ourisson, *Biochim Biophys Acta* **1105**, 213-220 (1992)
71. H de Boeck and R Zidovetzki, *Biochemistry* **28**, 7439-7446 (1989)
72. H de Boeck and R Zidovetzki, *Biochemistry* **31**, 623-630 (1992)
73. J A Hamilton, J M Vural, Y A Carpentier, R J Deckelbaum, *J Lipid Res* **37**(4) 773-782 (1996)
74. D P Siegel, J Banschbach, P L Yeagle, *Biochemistry* **28**, 5010-5019 (1989)
75. R M Epand, R F Epand and C R Lancaster, *Biochim Biophys Acta* **945**, 161-166 (1988)
76. R M Epand, T J Lobl, H E Renis, *Biosci Rep* **7**, 745-749 (1987)
77. R Zidovetzki, I W Sherman, A Atiya, H de Boeck, *Mol Biochem Parasitol* **35**, 199-208 (1989)
78. J Seelig, *Q Rev Biophys* **10**, 353-418 (1977)
79. Y Henin, C Gouytte, O Schwartz, J -C Debouzy, J -M Neumann, T Huynh-Dinh, *J Med Chem* **34**, 1830-1837(1991)
80. R G Males, F G Herring, *Biochim Biophys Acta* **1416**, 333-338 (1999)
81. W C Hutton, P L Yeagle, R B Martin, *Chem Phys Lipids* **19**, 255-265 (1977)
82. G R A Hunt, M Kaszuba, *Chem Phys Lipids* **51**, 55-65 (1989)
83. G V Betagery, Y, Theriault, J A Rogers, Membrane *Biochemistry* **8**, 197-206 (1989)
84. N Boden, S A Jones, F Sixl, *Biochemistry* **30**, 2146-2155 (1991)
85. M D Rice, R J Wittebort, R G Griffin, E Meirovich, E R Stimson, Y C Meinwald, H J Freed, H A Scheraga, *J Am Chem Soc* **103**, 7707-7710 (1981)
86. M Auger, I C P Smith, H C Jarrell, *Biochim Biophys Acta* **981**, 351-357 (1989)
87. M Auger, H C Jarrell, I C P Smith, *Biochemistry* **27**, 4660-4667 (1988)

88. O Jardetzki, G C K Roberts, NMR in Molecular Biology, Academic Press, New York (1991)
89. J K Seydel, *Trends in Pharmacol Sci* **12**, 368-371 (1991)
90. D J Craig and K A Higgins, 61-138 in: Annual Reports on NMR-Spectroskopy, edited by A G Webb, Academic Press, London (1990)
91. J K Seydel, O Wassermann, *Naunyn-Schmiedeberg's Arch Pharmacol* **279**, 207-210 (1973)
92. A Makriyannis, Chr M DiMeglio, W Fesik, *J Med Chem* **34,** 1700-1703 (1991)
93. I K Pajeva, M Wiese, H -P Cordes, J K Seydel, *J Cancer Res Clin Oncol* **122,** 27-40 (1996)
94. L G Herbette, M Trumbore, D W Chester, A M Katz, *J Mol Cell Cardiol* **20**, 373-378 (1988)
95. J K Seydel, H -P Cordes, M Wiese H Chi, K -J Schaper, E A Coats, B Kunz, J Engel, B Kutscher, H Emig, 367-376 (1991), in: QSAR: Rational Approaches to the Design of Bioactive Compounds, edited by C Silipo and A Vittoria, Elsevier, Amsterdam
96. I K Pajeva, M Wiese, *Quant Struct -Act Relat* **16**, 1-10 (1997)
97. J K Seydel, H -P Cordes, M Wiese, H Chi, N Croes, R Hanpft, H Lüllmann, K Mohr, M Patten, Y Padberg, R Lüllmann-Rauch, S Vellguth, W R Meindl and H Schönberger, *Quant Struct -Act Relat* **8,** 266-278 (1989)
98. R W Behling, T Yamane, G Navon and L W Jelinski, *Proc Natl Acad Sci U S A* **85**, 6721-6725 (1988)
99. J K Seydel, M Albores Velasco, E A Coats, H -P Cordes B Kunz, M Wiese, *Quant Struct -Act Relat* **11**, 205-210 (1992)
100. E A Coats, M Wiese, H Chi, H -P Cordes, J K Seydel, *Quant Struct -Act Relat* **11**, 364-369 (1992)
101. St Tetreault, V S Ananthanarayanan, *J Med Chem* **36,** 1017-1023 (1993)
102. G S Retzinger, L Cohen, S H Lau, F J Kezdy, *J Pharm Sci* **75**, 976-982 (1986)
103. P G Thomas, *Cell Biol Int Rep* **14** (4), 389-397 (1990)
104. D G Reid, L K MacLachlan, R C Mitchell, M J Graham, M J Raw, P A Smith, *Biochim Biophys Acta* **1029**, 24-32 (1990)
105 H B Jones, D G Reid, J S Luke, *Toxicol In Vitro*, **3**, 299-309 (1989)
106 K Nicolay, A -M Saureau, J -F Tocanne, R Brasseur, P Huart, J -M Ruysschaert, B de Kruijf, *Biochim Biophys Acta* **940**, 197-208 (1988)
107. F A de Wolf, K Nicolay, B de Kruijff, *Biochemistry* **31**, 9252-9262 (1992)
108. W K Surewicz, W Leiko, *Biochim Biophys Acta* **643**, 387-397 (1981)
109. Comprehensive Medicinal Chemistry, Vol 6, edited by C J Drayton, Pergamon Press, Oxford, New York (1990)

Chapter 12

I. Wawer

12 Solid-State NMR in Drug Analysis

12.1 Fundamentals and Techniques of Solid State NMR

NMR spectra cannot be measured in solids in the same way in which they are routinely obtained in solutions because NMR lines from solids are too broad. In solution all interactions apart from chemical shift and indirect coupling are averaged to zero by thermal motions of molecules. Magnetic interactions in the solid state are described by a Hamiltonian H [1], which is a sum of several contributions: Zeeman interaction (the same as in solution), direct dipole-dipole interaction, magnetic shielding (giving chemical shifts), scalar spin-spin coupling to other nucleus, and for nuclei with $I > 1/2$ also quadrupolar interactions.

$$H = H_{Zeeman} + H_{dipolar} + H_{chemical\ shift} + H_{spin\ coupling} + H_{quadrupolar}$$

The most important is the dipole-dipole interaction with neighboring nuclei. Magnetic fields of nuclei i (^1H) in the neighborhood of the nucleus j (for example ^{13}C, ^{15}N) under observation generate local fields:

$B_{loc} \sim \mu / r^3_{ij} (3\cos^2 \theta_{ij} - 1)$, which depend on the magnetic moment μ of the nuclei and the mutual orientation of the nuclei ($3\cos^2 \theta_{ij} - 1$), where θ_{ij} is an angle between internuclear vector and the direction of static magnetic field.

The interaction is significant for ^1H and ^{19}F with large magnetic moments, but decreases rapidly with the distance r^3_{ij}. Such dipolar nuclear interactions are orders of magnitude larger than chemical shifts, and, since many mutual interactions are present simultaneously, the result is a wide resonance line. To obtain a high-resolution spectrum the line-narrowing procedures, such as magic angle spinning (MAS) or special pulse sequences, have to be applied. The dipolar interactions vanish for $\cos^2 \theta = 1/3$ (because $3 \times 1/3 - 1 = 0$); thus, if the sample is rotated around an axis inclined at an angle $\theta = 54° 44''$ (the magic angle) to the magnetic field, the line broadening is significantly suppressed. How fast should the sample rotate?

To illustrate the effect of rotation speed two examples were selected with only one type of nuclei under observation, and thus a single resonance line was expected in every case. The ^{77}Se MAS NMR spectrum of the selenium analog of amino acid D,L-methionine is shown in Figure 12-1.

232   Solid-State NMR in Drug Analysis

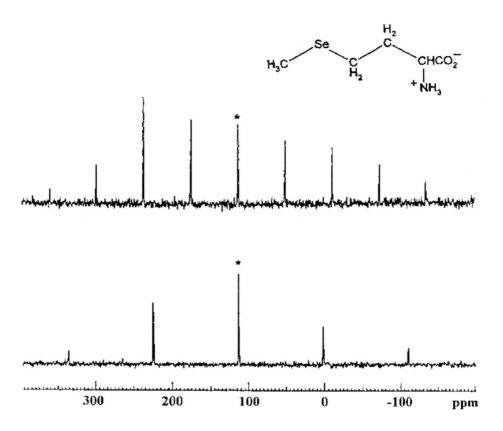

Figure 12-1: ^{77}Se MAS spectra of selenium analog of methionine; sideband patterns as a function of spinning speed (Courtesy of NMR Laboratory, Center of Molecular and Macromolecular Studies, Lódz, Poland).

The ^{77}Se is a ½ spin nucleus of 7.7% abundance and the recording time for the experiment is as for ^{13}C NMR. For slow spinning (3530 Hz), one observes a set of lines spaced at the rotation frequency at the left and right side (upper trace): the lines decrease in intensity and move away from the center resonance as the spinning rate increases to 6367 Hz (lower trace).

^{31}P is a nucleus of 100% abundance with spin $I=1/2$; a ^{31}P NMR spectrum of solid racemic cyclophosphamide, carcinostatic agent, is illustrated in Figure 12-2 According to the X-ray data, the molecules of cyclophosphamide are linked by NH...O=P bonds into chains and there is one molecule in the crystal unit. The spectra were recorded with two slightly different rotation frequencies in order to find the resonance (at 30.7 ppm) between sidebands. Chemical shift was referenced to K_2HPO_4 (1.6 ppm with respect to 80% H_3PO_4). In the spectra from Figure 12-1 and Figure 12-2 the central line is marked with an asterisk. This line does not move upon increasing rotational speed; its position is given by $\sigma_{iso}=1/3(\sigma_{11}+\sigma_{22}+\sigma_{33})$. The three components σ_{ii} (i=1,2,3) of the shielding tensor (and three angles which specify their orientation with respect to the direction of magnetic field) describe the anisotropy of the shielding. According to the convention, $\sigma_{11} < \sigma_{22} < \sigma_{33}$, the anisotropy is defined as: $\Delta\sigma = \sigma_{33} - \frac{1}{2}(\sigma_{11}+\sigma_{22})$, and the asymmetry parameter $\eta = (\sigma_{11}-\sigma_{22})/(\sigma_{33}-\sigma_{iso})$ [1]. The anisotropy parameter reflects the distortion of the geometry from ideal (tetrahedral, octahedral), and the asymmetry parameter characterizes the environment of the nucleus under observation.

The rotational speed should be greater than the anisotropy $\Delta\sigma$, given in Hz; if this is not the case the signal is flanked by rotational side bands on both sides. The effect is particularly often observed for ^{31}P and ^{77}Se (Figure 12-1 and Figure 12-2) which exhibit large chemical-shift anisotropy.

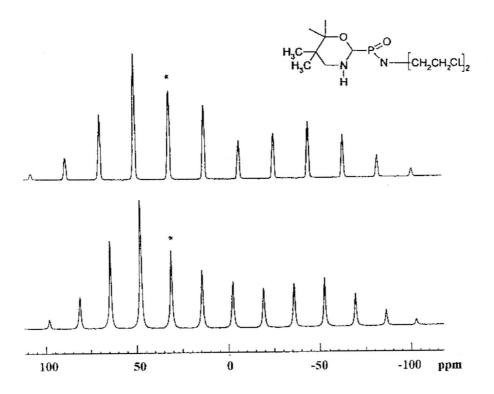

Figure 12-2: ^{31}P MAS spectra of cyclophosphamide; rotation speed 2300 Hz (upper trace) and 2030 Hz (lower trace); (Courtesy of NMR Laboratory, Centre of Molecular and Macromolecular Studies, Lodz, Poland).

The intensities of side-band lines reflect the shape of the static (without rotation) spectrum of solids. If a large single crystal could be obtained and installed (stacked) in the probehead of an NMR spectrometer, a narrow signal would be observed for a particular nucleus; the chemical shift will depend on the orientation of the crystal axes with respect to magnetic field. Thus, an assignment of three tensor components with respect to the molecular geometry may be made. In the case of a polycrystalline sample (placed in a small vessel constructed of zirconium oxide), there is a random distribution of all possible orientations, and a broad line results from a superposition of multiple resonances. The principal elements of the shielding tensor may be obtained directly from a spectrum, but their orientation with respect to the molecular coordinate system is not obvious. Nevertheless, chemical-shift anisotropy is an important source of information on molecular structure and bonding.

Chemical shift anisotropies for aliphatic carbons are 15–50 ppm, but as much as 120–200 ppm for aromatic and carbonyl sp^2-type carbons (since directional variation in electron density is larger within multiple bonds).

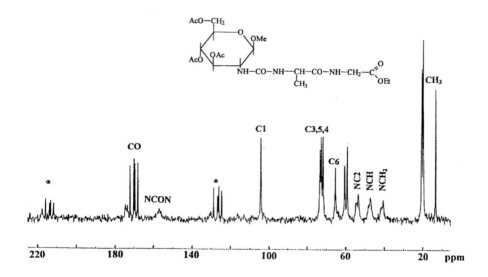

Figure 12-3: ^{13}C CP MAS spectrum of ureido sugar, rotation speed 3 kHz.

The ^{13}C MAS NMR spectrum of ureido sugar is shown in Figure 12-3. The rotation speed of ca. 3 kHz was too small to average the anisotropy of C=O; significant side bands are present on both sides of the carbonyl resonances. The spectrum of trolox, a water-soluble analog of vitamin E, recorded at rotation speed of 10 kHz, is sideband free (Figure 12-4).

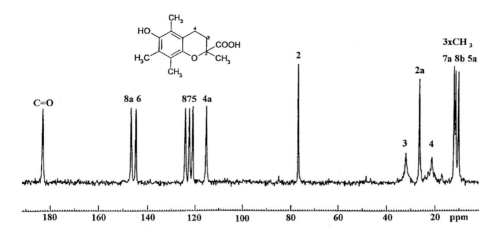

Figure 12-4: ^{13}C CP MAS spectrum of trolox, rotation speed 10 kHz.

Although we will not discuss the solid-state NMR of quadrupolar nuclei, some interesting features of their presence should be mentioned. If the rare spin ½ nuclei are coupled to a quadrupole moment-bearing nucleus, characteristic splittings are observed as a consequence of the fact that the ^{13}C-^{14}N or ^{13}C-^{2}H coupling is not completely averaged out. Theoretical aspects involved in this phenomenon and the equation correlating the value of the splitting with a number of variables (quadrupole parameters, ^{13}C-X distances, and angles which describe the orientation of ^{13}C-X with respect to the electric field gradient) are given in [2]. Valuable structural information regarding the quadrupole tensor at the ^{14}N nucleus can be obtained from the spectra of nitrogen containing compounds recorded under CP MAS conditions. Asymmetric doublets (or broadening) such as those in the spectrum of ureido sugar allowed easy assignment of carbons directly bonded to nitrogen (Figure 12-3).

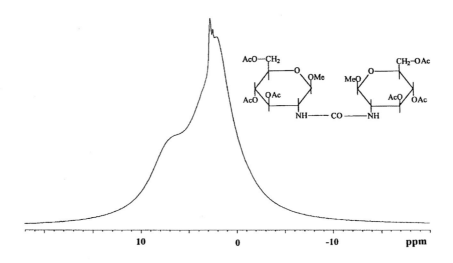

Figure 12-5: ^{1}H MAS spectrum of ureido sugar, rotation speed 16 kHz.

In solution the most popular method is ^{1}H NMR, however this is not the case for solid state. Unfortunately, the dipolar ^{1}H - ^{1}H interactions are of the order 10–50 kHz, and it is not possible to rotate the sample fast enough to remove them. The solid state ^{1}H NMR spectrum, illustrated in Figure 12-5, was measured at rotational speed of 16 kHz and showed a broad line covering the whole region of 10 ppm. High-resolution ^{1}H spectra cannot be obtained with the magic-angle-spinning (MAS) technique alone (^{1}H-^{1}H dipolar couplings can be removed by multiple-pulse techniques such as WAHUHA, MREV-8), and therefore are of little practical use for the characterization of pharmaceutical solids.

Solid state ^{13}C spectra are obtained by a combination of MAS and high-power decoupling, to remove dipolar interaction with ^{1}H. However, ^{13}C are rare atoms (abundance 1.2%), their sensitivity is low and relaxation times T_1 are long. A cross-polarization method is thus applied — a double-resonance technique which improves the observation of the less sensitive nuclei. The advantages of this technique can be seen in the case of solid adamantane (Figure 12-6). A broad contour with high noises (Figure 12-6a) is separated into two resonances when MAS is included (Figure 12-6b), however good signal-to-noise ratio can be achieved with cross-polarization sequence (Figure 12-6c).

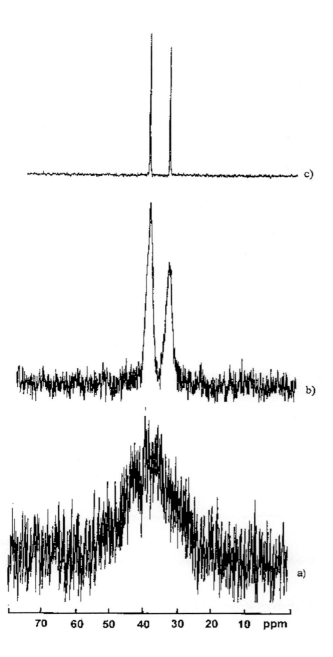

Figure 12-6: ^{13}C spectra of solid adamantane: [a] without MAS and CP, [b] with MAS, [c] with MAS and CP.

The Hartmann-Hahn condition: $\gamma_H \cdot B_{1H} = \gamma_C \cdot B_{1C}$ (γ_H and γ_C are gyromagnetic ratios of both nuclei) should be fulfilled. A 90° pulse of power B_{1H} (irradiated at ^{1}H frequency) aligns the spins (spin locking); simultaneously a pulse of power B_{1C} (^{13}C frequency) is given and the transfer of proton polarization occurs during time t_{cp} (contact time, called also cross-polarization time), increasing the polarization of ^{13}C nuclei. The polarization of carbons can increase by a factor of 4 (because $\gamma_H / \gamma_C \approx 4$), giving significant improvement in sensitivity. Then, the B_{1C} field is turned off and free induction decay (FID) is recorded; the B_{1H} field is still „on" because it serves as high-power decoupling. The sequence can be repeated with short recycle time; it is governed by proton relaxation times, not by the long T_1 of ^{13}C. The rate of cross-polarization of particular carbons is, in first approximation, proportional to the number of directly bonded protons, and the following relative rates are usually observed: CH$_3$ (static) >CH$_2$>CH>CH$_3$ (rotating)>C (nonprotonated) [3].

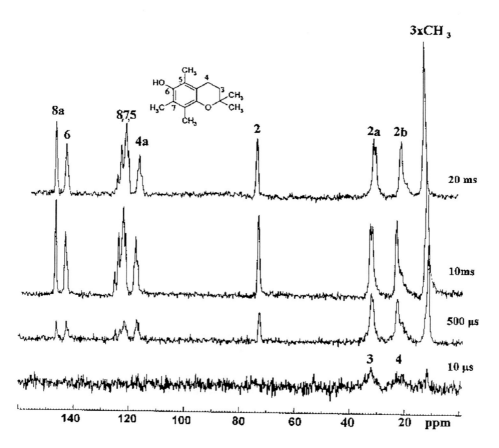

Figure 12-7: ^{13}C CP MAS spectra of chromanol recorded with various contact times.

The series of spectra of chromanol were recorded with various contact times (Figure 12-7). In the spectrum with very short contact time t_{CP} =10 is only the intensity in the region of methylene carbon resonances can be seen. Rotating methyls, quaternary carbons but also aromatic CH carbons need longer t_{CP}. The increase of intensity in the range of shorter contact times is due to heteronuclear cross-polarization rates (described by rate constant T_{CP}).

In general, the intensity I is a function of the contact time t_{CP} where A is the intensity amplitude:

$$I(t_{CP}) = A (1-T_{CP}/T_{1\rho}^H)^{-1} [\exp(-t_{CP}/T_{1\rho}^H) - \exp(-t_{CP}/T_{1\rho}^H)]$$

Long contact times are useless; the intensity of resonances decreases due to the proton spin relaxation (in the spin-locking conditions, i.e. in rotating frame, it is the proton spin relaxation time $T_{1\rho}^H$). Usually, a contact time in the range 2–6 ms was the most frequently applied in cross-polarization experiments. However, according to the plot of intensity versus t_{CP} illustrated in Figure 12-8, the optimal value for chromanol is ca. 10 ms.

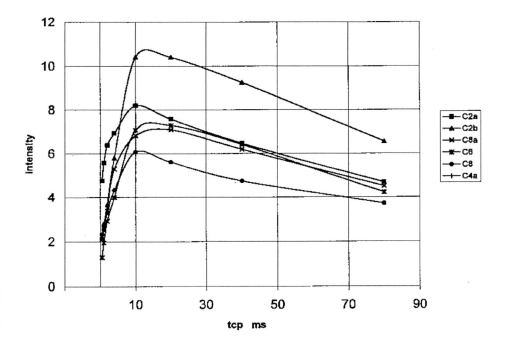

Figure 12-8: Selected results of the variable-contact cross-polarization experiments. The plot of signal intensity versus contact time for chromanol carbons.

It is worth stressing that cross-polarization kinetics and relaxation processes (mainly proton relaxation) govern ^{13}C resonance intensities of CP MAS NMR spectra, and therefore the intensities measured in routine experiments cannot be used for quantitative analysis. In order to obtain quantitatively reliable results a series of spectra should be recorded as a function of varied cross-polarization time and repetition time (to determine proton spin-lattice relaxation times in laboratory frame T_1^H and in rotating frame $T_{1\rho}^H$). Further details of quantitative ^{13}C CP MAS NMR have been reported in [4],[5],[6].

As mentioned above, in the spectra recorded with short cross polarization time ("short contact"), only signals of carbons bearing hydrogen appear, because quaternary carbons take longer to cross-polarize; this allows for selective observation of protonated carbons. Introducing a short (50-µs) delay just before acquisition results in loss of intensity (dephasing) of signals of carbons linked to hydrogens because of very effective proton relaxation whereas the resonances of nonprotonated carbons remain intense. The technique of "delayed decoupling" (known also as „dipolar dephased") enables observation of nonprotonated carbons. Both methods are helpful for assignment of signals in solid-state ^{13}C NMR spectra and are illustrated in Figure 12-9 for a flavonoid galangin, as an example. Quaternary carbon C10 resonance at 103.0 ppm could be distinguished from C6-H (99.5 ppm) and C6-H (94.8 ppm).

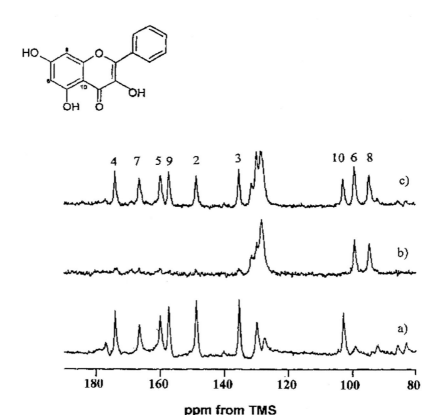

Figure 12-9: ^{13}C CP MAS spectra of the flavonoid galangin recorded with (a) dipolar dephased, (b) short contact and (c) standard CP sequence.

12.2 Solid State Conformations of Drugs and Biologically Active Molecules

Obviously, the spectra recorded for solid state are compared with those obtained for solution. A motionally averaged value of shielding anisotropy σ_{av} is observed in solution and σ_{iso} in the solid state; comparison of the σ_{iso} from the solid and σ_{av} from solution brings important information related to the phase and conformational effect. For solution spectra, information on conformation is lost due to averaging of chemical shifts through fast intramolecular motions. In solids the conformations are locked, revealing specific chemical environments with different chemical shifts. Therefore, similar chemical shifts are usually observed for such molecular fragments, which are rigid in both phases. Significant differences occur for fragments flexible in solution and rigid in the solid phase. These differences are indicative of conformational equilibration. For example, in methoxybenzenes there is fast rotation around the C1-O bond in solution, whereas in the solid state the methoxy group is locked in one particular position with respect to the aromatic ring. In the ^{13}C spectra of solids the differences of chemical shifts for carbons *ortho* to the OCH$_3$ reached 9.2 ppm. It enabled the assignment of resonances of synthetic estrogen, *meso*-hexestrol and its ether derivative [7].

^{13}C NMR signals of the *ortho* carbons C2, C6 of phenol derivatives are split into a doublet, the separation of which is 1.6–2.7 ppm. Higher electron density is induced at C2 when the C1-O-H hydrogen is close to C2-H; in contrast, a lower electron density is induced at C6 and results in larger chemical shift [3]. Such simple means of peak assignment was applied in the analysis of ^{13}C spectra of solid flavonoid, quercetin [8].

Solids cannot be considered as static, with molecules in frozen conformations, since many dynamic processes with rates within the NMR time scale are also observed in the solid phase.

Fast molecular motions in the solid state have been investigated through NMR measurements of the second moment (mean square width of the line) and relaxation. As stated above, dipolar interaction averages to zero for rapid, isotropic motion, such as in solution. In the case of limited motion there is only partial averaging of the interaction, and, as a result, only some narrowing of the resonance line takes place. Some molecules that are globular in shape reorient fast in the solid state; extremely fast motions and thus a significant reduction of the second moment were observed for adamantane and D-camphor. Many systems of interest for pharmacy such as polymers, resins, and surface-immobilized drugs lack crystalline order, and the ^{1}H spectral measurements of broad resonance line allow for the detection of molecular motion in solid samples. The measurement of relaxation times also provides a suitable technique for the detection of molecular motions. Assuming the dipolar mechanism of relaxation, the spin-lattice relaxation time T_1 can be related to the correlation time for the motion τ_c. The dependence of log T_1 on $1/T$ (T is temperature) exhibits a minimum at $\omega_0 \cdot \tau_c = 0.62$ (ω_0 is the spectrometer frequency); the slope of the curve gives activation energy. Such measurements were widely exploited in studying molecular motions in a variety of substances, including polyaminoacids and other biological materials [1]. For example, a reduction in the second moment and minima in the spin-lattice relaxation time were attributed to the reorientation of methyl groups on carbons C18 and 19 in solid steroids: cortisone, cholesterol, progesterone and testosterone [9].

Dynamics in the solid state include processes of an intramolecular proton transfer; the tautomerism of porphyrin [10] or the intramolecular proton transfer in cyclic *NN*-bisaryl formamidine dimers [11] was studied by means of ^{15}N CP MAS NMR.

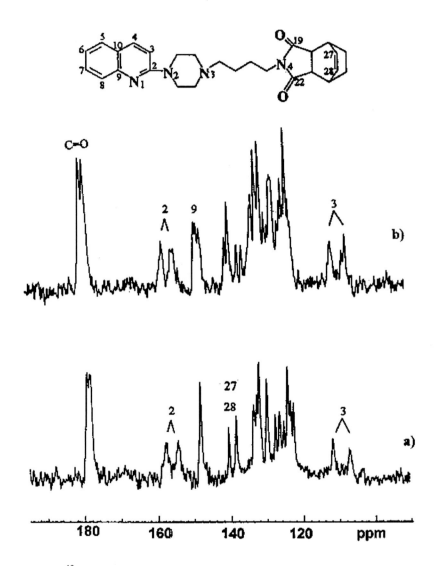

Figure 12-10: ^{13}C CP MAS spectra in the range of aromatic carbons for buspirone analog, (a) solid hydrochloride and (b) solid hydrobromide.

Reorientation of molecular fragments, such as the aromatic ring (as in penicillins, see below) can be investigated in the solid state; the processes are followed by variable-temperature studies. In the solid-state spectra the changes of line shapes (from separate

resonances through the broad contour near coalescence to a single narrow line) are observed, as in solution spectra; the activation parameters can be obtained by line shape analysis.
Interesting conformational effects were observed for solid arylpiperazine derivatives. The ^{13}C CP MAS spectra of the buspirone analog 3a,4,7,7a-tetrahydro-2-{4-[4-(2-quinolinyl)-1-piperazynyl]butyl}-4,7-ethane-1H-isoindole-1,3(2H)-dione hydrochloride and hydrobromide were recorded and analyzed [12]. Two sets of signals appeared, in agreement with single-crystal X-ray diffraction data, indicating that in each of the salts two independent cations are present in the crystal unit. The largest shielding differences of 3.2–4.6 ppm between two sets were found for quinoline aromatic carbons C3 and C2. The splittings look similar (Figure 12-10); however the reasons for separate resonances are quite different.

The X-ray data showed that in the hydrochloride there are molecules with an axial and equatorial quinoline unit; in hydrobromide the quinoline fragment is differently twisted around the C2-N2-piperazine bond, and, additionally, one molecule forms hydrogen bonds with crystalline water. The GIAO-CHF calculations were performed for structural fragments of this buspirone analog. Linear correlations between theoretical and solid state results were obtained, thus enabling a reasonable assignment of carbon resonances of the conformations present in the solid state. It is worth stressing that successful assignments were possible because the geometry for theoretical calculations was directly adapted from X-ray data.
Solid-state NMR studies were performed on various biomolecules: nucleosides [13] D,L-methionine [14] and selenomethionine [15], aminoacids and peptides [16], [17], [18], [19] natural and synthetic melanins [20], flavonoids [8], β-carotene [21], α- and β-D-glucose [22], lactose [23], lactulose [24], sugar derivatives of tocopherol [25] and numerous other sugar derivatives [26].
^{13}C NMR spectra of solid nucleosides: uridine, cytosine, adenosine, guanosine were measured in 1982 by Chang and co-workers [13]. The most distinct spectral difference between the spectrum in solution and that in the solid state appeared for uridine, this was the splitting of C2 (67.6/69.3 ppm) and C5 (58.9/62.4 ppm) resonances of sugar carbons. According to the X-ray data there are two molecules in an asymmetric unit, which have similar gross conformations and anti-conformation about the glycosyl bond; however, a number of bond angles and the torsional angles of the ribose ring are significantly different. These differences in angles as well as the differences in intramolecular hydrogen bonding accounted for the splittings.
The ^{13}C spectra of crystalline α-D-glucose·H$_2$O, α-D-glucose and β-D-glucose were recorded with cross-polarization and a spinning rate of 3 kHz [22]. During the accumulation of the spectrum of α-D-glucose·H$_2$O the hydrate was partially transformed to the anhydrous form. Neither ^{1}H irradiation nor rapid spinning alone resulted in this structural modification; however a combination of centrifugation and the heating produced by the decoupling coil gave an effect of liberation of water from crystals of α-D-glucose ·H$_2$O. The use of ^{13}C labeling at C1, C2, C3 and C6 was helpful in the assignment of resonances. The label yielded increase in intensity of the labeled position but also line broadening due to homonuclear coupling to adjacent carbons in the molecule. Significant changes in chemical shifts were observed upon going from solution to solid. Configuration, hydrogen bonding and crystal field effects determined the exact positions

of ^{13}C resonances of D-glucose. In the ^{13}C CP MAS spectra of lactose [23] the splitting of the C1 and C6 resonances into two peaks was observed (ratio of 3:1), implying that there are at least two nonequivalent glucose molecules per unit cell. The spectra of the anhydrous form were more complicated than those of the solid monohydrate. The ^{13}C spectra of solid lactulose (4-O-β-D-galactopyranosyl-D-fructose) indicated the presence of five isomers [24]. The interpretation of solid-state NMR data for sugars is not obvious even for simple monosaccharides because of conformational and configurational effects, intra- and intermolecular hydrogen bonds and interactions in the crystal lattice (molecular packing effects) but also polymorphism and phase transitions. These problems were recently reviewed by Potrzebowski [24].

Commercial D,L-methionine contains two crystalline forms, which could be separated by crystallization from water or water-ethanol. The ^{13}C spectrum of solid D,L-methionine showed two chemical shifts for the methyl carbons (18.7 and 15.8 ppm); the conformational difference is also reflected in the carbon resonances of C-β and C-γ [14].

The selenoproteins nowadays attracte much attention, and selenomethionine is the predominant form of selenium in food. ^{77}Se and ^{13}C CP MAS experiments were employed to study the structure and dynamics of D,L-and L-selenomethionine [15]. The ^{77}C CP MAS spectrum of solid D,L-selenomethionine showed one signal at 111.5 ppm (see Figure 12-1); in the case of L-selenomethionine two resonances at 122.9 and 84.5 ppm indicated the presence of two molecules in the unit cell. L-selenomethionine undergoes a phase transition at 305 K and isomerization when the sample is rotated rapidly (at a spinning rate of 14.5 kHz).

The important provitamin A, *trans*-β-carotene, is widely studied as a chemopreventive agent against cancer. It is supplemented in vitamin complexes, recommended in cases when the daily diet contains insufficient amounts of fruits and vegetables. The ^{13}C CP MAS spectra of solid *trans*-β-carotene were recorded and assigned on the basis of variable contact time and dipolar dephasing experiments [21]. The data were discussed in terms of molecular motion. The XRD data showed particularly high isotropic temperature factors for C2, C3 and C4, indicating a large amount of thermal motion. The cross-polarization results are in agreement with such conclusion. There is a discrepancy between the NMR spectra in the alkene region and the XRD structure, since the alkene spectral region contains more resonances than expected from the crystal structure. This could originate from some local disorder in the β-carotene chain, which was overlooked in the old (1964) XRD study.

Tocopherols are important components of biological membranes and act as antioxidants. The structure of vitamin E derivatives has been widely studied, although neither by X-ray diffraction nor by solid state NMR, because tocopherols are liquids at room temperature. Fortunately, 6-O-glucopyranosyl and 6-O-mannopyranosyl derivatives are solid, and ^{13}C CP MAS spectra were recorded [25]. The signals in the spectra of solids were assigned by comparison with those taken for solutions. The signals of sugar carbons bearing hydroxyls were broader than those of the respective acetyl derivative, which suggests some differentiation of hydrogen bonds. Remarkable changes in the chemical shifts were observed for C2 and C3 of the chroman ring, which is flexible in solution. Frozen conformation of the long alkyl chain results in deshielding of C1' (of ca. 4 ppm) to C11'. The changes of shielding can be explained by inversion at the C2 carbon of the heterocyclic chroman ring; two conformers, with axially and equatorially oriented alkyl chain, probably exist in solution. The MO calculations and the analysis of shielding indicated that the conformer with axial methyl group and equatorial phytyl chain is preferred in the solid state.

12.3 Polymorphism/Pseudopolymorphism of Drugs

The Food and Drug Administration in the USA requires that appropriate analytical procedures be used in drug guidelines in order to detect polymorphic, hydrated or amorphous forms of the drug substance. Polymorphic change affects the pharmaceutically important physical properties, such as solubility, dissolution rate, powder flow and tableting behavior. Different polymorphs usually show different drug release *in vivo*, and therefore polymorphism of drugs and excipients has been extensively studied [useful Internet resources of references can be found on the Pharmaceutical Polymorphs web site http://www.cbc.umn.edu/~jhan/refl.html].

The greatest advantage of the NMR method lies in the assignment of a minor polymorphic form in the dominant environment of the other form, and, additionally, the analysis is carried out in a non-destructive manner. The majority of pharmaceutical applications of solid state NMR reviewed in 1991 by Brittain et al. [27] and later by Bugay [28] were the investigations of polymorphs.

However, there have been few studies focused on the quantitation of relative amounts of polymorphic forms. Solid-state ^{13}C NMR was used to determine the amounts of carbamazepine anhydrate and dihydride [29]; more recently the methodology of quantitative analysis was applied to delavirdine mesylate mixtures [30]. These works demonstrated that unambiguous identification of the polymorphic forms was possible with empirical detection limits about 2–3 %.

Delavirdine mesylate, a potent non-nucleoside reverse transcriptase inhibitor (developed for the treatment of AIDS), crystallizes in a variety of modifications. The solid state ^{13}C NMR spectra of three selected solid forms (VIII-polymorphic, XII-pseudopolymorphic containing non-stoichiometric amounts of water and XI—the preferred solid form for the formulation because of its superior stability and acceptable bioavailability) revealed distinct differences in chemical shifts and peak splitting [30]. Resonances of isopropyl methyl carbons at 17.3 ppm (VIII), 19.1, 21.2 and 22.2 ppm (XII) and 20.2 ppm (XI) were diagnostic for each form; their resonance intensities were utilized to determine the composition of solid mixtures. Several parameters of the system were determined in order to obtain quantitative results: proton spin lattice relaxation times T_{1H} – to adjust the repetition time of the CP MAS NMR experiments (1.34 s for XI, 1.87 s for VIII, 1.03 s for XII), cross polarization rate constant T_{CH} and proton spin-relaxation time $T_{1\rho}$ which are important for setting the optimal contact time. Statistical results of the least-squares analyses were discussed; the estimated standard deviation, ESD, which provided an estimate of the precision, was larger when the peak height ratio was used for quantitation and smaller when using the peak area. However, for overlapping resonances, measurements of peak height appear to provide better precision due to less subjectivity.

There are many examples where structural diversity was detected by crystallography (X-ray diffraction, XRD), and solid-state NMR evidenced the effects of crystal packing, mainly various hydrogen bonding patterns. Solid-state ^{13}C NMR is a complementary technique to crystallography for obtaining structural information about solids, since the number of resonances in the spectrum is equal to the number of non-equivalent carbons in an asymmetric structural unit. Since NMR uses powdered compounds, one can avoid the difficulties of growing good quality single crystals suitable for XRD analysis. The method is particularly useful in cases where crystallographic examination is not possible because the sample is not crystalline or forms complex polymorph. NMR provides an immediate indication of the presence of any configurational or conformational multiplicity that may exist in the solid state. The ^{13}C spectrum of the well-known drug exhibiting polymorphism [31], phenobarbital, may serve as an illustration (Figure 12-11).

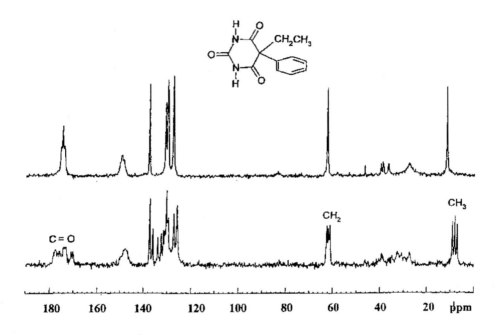

Figure 12-11 ^{13}C CP MAS spectrum of phenobarbital; (*bottom*) polymorphic, without preparation, (*top*) recrystallized from ethyl acetate.

Three forms are present in the commercial solid, evidenced by the three components of methylene and methyl carbon resonances. Recrystallization from ethyl acetate results in only one form. Polymorphism was also recognized in the case of caffeine [3232]. The ^{13}C CPMAS spectrum of solid caffeine (Figure 12-12) shows five carbonyl resonances whereas there are three C=O groups. The assignment of signals in the spectra of phenobarbital or caffeine is not straightforward, it requires X-ray diffraction structural data and theoretical calculations of shielding constants for the forms present in the crystals.

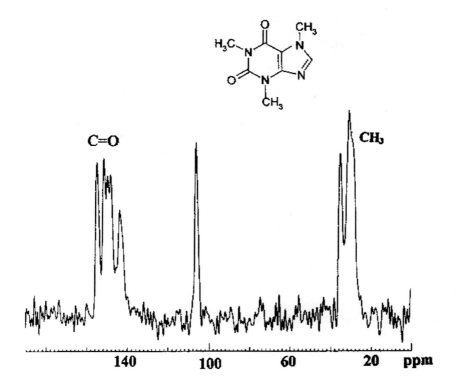

Figure 12-12 ^{13}C CP MAS spectrum of caffeine; five resonances of carbonyl carbons indicate polymorphism.

Numerous drugs exist as hydrates with various amounts of water molecules incorporated in the crystal lattice. Significant differences in the solid state NMR spectra result from the interaction of the drug molecule with crystalline water.

Different polymorphic forms of solid benoxaprofen, nabilione and pseudopolymorphic (i.e. including solvent) crystal forms of cefazolin were analyzed in 1985 by Byrn et al. [33]. Solid state ^{13}C NMR spectra of crystal forms I and II of benoxaprofen were recorded; a spectrum of the pharmaceutical granulation exhibited the signals of excipients but was identical to that of form II. The spectra of nabilone I and II allowed for a determination of which polymorph is present in a binary mixture. Hydrated crystal forms of cefazolin, the sesquihydrate, pentahydrate and monohydrate, gave different signal positions; the amorphous form can be distinguished from crystalline ones because it showed broadened resonances.

Various polymorphs of steroids were found: three pseudopolymorphic forms of testosterone [34], anhydrous and monohydrate forms of androstanolone [35], six crystal forms of cortisone acetate [36] and five forms of prednisolone *tert*-butylacetate [37].

Prednisolone *tert*-butyl acetate, which is a clinically useful glucocorticoid, exists in crystalline forms with different crystal packing; the disordered side chain and water of crystallization were found by XRD. Since conformation of steroids in the solid state is related to their biological activity it seemed appropriate to investigate these forms by solid-state NMR. The spectra indicated [37] that conformational differences in the side chain do not cause major changes. The largest changes in chemical shifts occur with the carbonyl carbons and the neighboring unsaturated carbon atoms, and correlate with hydrogen bonding; as the O···O distance in the hydrogen bonding system decreases the difference between the solid state and solution, chemical shifts increase.

Cortisone in the form of the acetate ester, administrated as an anti-inflammatory agent, exists in polymorphic and pseudopolymorphic forms (solvate crystals). Six crystalline forms were distinctly found by the analysis of solid-state ^{13}C NMR spectra of nine various samples [36]. Crystal structures were known for three forms and, therefore some correlation between NMR and XRD results could be made. Larger chemical shifts of C3=O are related with the presence of hydrogen bonding; other chemical shift differences between solid forms appear in the side chain and the conformations of ring A.

In numerous studies, solid-state NMR was used as an additional source of structural information together with crystallography, IR spectroscopy and differential scanning calorimetry (DSC). The characterization of solid-state structures of losartan [38], furusemide [39], cyclopenthiazide [40], chiral leukotriene antagonists [40], [41], fosinopril sodium [42], [43], [44] and the antibiotics cefaclor (dihydrate) [45] and penicillins [46], [47] was possible. Losartan, an antihypertensive agent (in 1993 in clinical development) was found to exist in two enantiotropic polymorphic forms, I a low temperature stable and II a high temperature stable form. In the absence of single-crystal XRD data, useful information was derived from solid-state ^{13}C spectra [38]. Splitting of the resonances in the spectrum of form I indicated the presence of more than one orientation for the *n*-butyl side chain and the imidazole ring. No peak splitting was observed for form II. It is obvious, therefore, that heating changes the packing in the crystal in such a way that only one molecule is present in the unit cell.

Solid-state ^{13}C NMR analysis of two frusemide crystal forms revealed differences in chemical shifts [39]. Marked differences between solution and solid state appear for the carboxyl group and for pyrane ring carbons; however, most likely the conformation in solution is not the same as those present in the solid-state crystal forms.

Fosinopril sodium, an angiotensin-converting enzyme inhibitor, has been found to exist in two anhydrous polymorphic forms. No single crystal of sufficiently high crystallographic quality was obtained for the polymorphic phases and polymorphism was characterized by solid state ^{13}C and ^{31}P NMR. The analysis of chemical shifts in ^{13}C spectra suggested that polymorphism originates from conformational differences in the acetal side chains and/or *cis-trans* isomerization about the peptide C-N bond. The difference in chemical shift between ^{31}P resonances in the two spectra is 2.2 ppm and demonstrates the existence of various environments of phosphorus nuclei in each polymorph [43].

Aspartame (L-aspartyl-L-phenylalanine methyl ester), a dipeptide sweetener, is increasingly used in foods and pharmaceuticals. Aspartame hemihydrate undergoes polymorphic transitions, hydration and dehydration. The crystal structure of form I of the hemihydrate was known earlier but the commercial form II is a different polymorph. The differences in the ^{13}C spectra of the two solid hemihydrates allow for identification of these polymorphs [48]. The spectrum of I exhibits one peak per carbon, in agreement with the one crystallographic site in the unit cell. In the spectrum of II several resonances are split, indicating that there are at least three crystallographically inequivalent sites. Chemical shifts of backbone carbons remain almost constant, but those of side-chain carbons are not. The apparent differences in conformation are probably limited to the carboxylic acid and phenyl side chains. The comparison of ^{13}C NMR, FTIR and crystallographic results suggests that the crystal structure of II is less symmetric, although the packing arrangement of the two forms may be similar.

A novel crystal form II of the benzodiazepine chlordiazepoxide was reported [49]; the standard form I and the new one II are similar with respect to the cell volume or crystal symmetry. Both structures consist of a pair of hydrogen-bonded dimers in the asymmetric unit. The major difference is the crystal packing of the dimers; in the form II the dimers are displaced with respect to each other. Both forms were characterized by solid-state ^{13}C NMR and the obtained spectra were different. The splitting of the methyl group signal was observed in II. Notable differences were also found for aromatic and olefinic carbons. The differing chemical environments in the two crystal structures can explain these effects.

Bisphosphonates (used to treat various bone, teeth and calcium metabolism diseases), contain the P1-C-P2 group, and ^{31}P CP MAS NMR was used to characterize the solid-state properties and stability of disodium clodronate tetrahydrate [50]. The MAS spectra were recorded with cross-polarization sequence and proton decoupling. The P···P intermolecular interaction produces dipolar broadening at 640 Hz (when the distance r_{PP} is 0.3125 nm), which is fully reduced at a rotation speed of 4500 Hz. The ^{31}P spectrum exhibits two isotropic lines A and B at 14.3 ppm and 9.0 ppm, respectively, surrounded with spinning side bands. The intensity of these side bands was used to estimate the anisotropy of shielding. The geometrical properties of the clodronate anion were obtained from XRD data. Assuming that the ^{31}P chemical shift correlates with the P-O bond length, phosphorus P1 was assigned to the high-field signal at 9.0 ppm because the mean value of the P1-O bond length is slightly shorter. Comparison of the tetrahedron regularity with the smaller shielding anisotropy supports the assignments. A fast rise in temperature leads to loss of crystal water and an average structure with single ^{31}P resonance appeared; slow dehydration converts the crystalline form to the anhydrous structure with two non-equivalent ^{31}P atoms. Dehydration is fully reversible; when the dehydrated sample was allowed to equilibrate with moisture the spectrum was again that of the original sample of disodium clodronate tetrahydrate.

Meropenem, a promising agent for combating resistant bacterial infections, is marketed as a powder blend of the crystalline drug and sodium carbonate. NMR spectra recorded for solution and solid state showed [51] that the carbonate in the formulation was the cause of a chemical modification of meropenem in solution. The ^{13}C CPMAS spectrum of the reconstituted formulation showed the signal of bicarbonate at 159 ppm and a broader one, centered at 164 ppm of meropenem-CO_2 adduct. It is interesting that such a carbon dioxide adduct could be observed in the solid state, obtained after lyophilization.

The potassium salt of phenomethylpenicillin was studied by the combination of ^{13}C CP MAS NMR and XRD [46]. The X-ray diffraction study indicated that there are four molecules in the crystallographic asymmetric unit, which differ in the orientation of the phenoxy side chain with respect to the penam unit. The variable temperature spectra indicate that phenyl rings of all molecules perform 180° flips around their C_2 axes. The aromatic region of the spectrum showed a number of resonances of varying intensity and line width. On increasing the temperature from 180 to 240 K the majority of the aromatic carbon peaks broadened and at higher temperatures progressively narrowed. The equivalence of the *ortho* (C2′, C6′) pair and *meta* (C3′, C5′) pair of resonances whereas *ipso* (C1′) and *para* (C4′) are unaffected was interpreted as indicative of a 180° aromatic ring reorientation. Similar spectral behavior has been observed in solution and called „chemical exchange"; frequently met dynamic processes in the solid state are proton transfer; dynamic change involving larger structural fragments is less common because of the insufficient space left in the crystal lattice.

The study of potassium penicillin was extended [47] to the crystalline Li, Na, Rb, CS and free acid-penicillin. The X-ray diffraction data on these crystals yielded information on their structure, bonding, intermolecular contacts, librational oscillations or positional disorder. There are one, two or four molecules in the asymmetric unit. The crystals have layer structures composed of hydrophilic fragments (cation coordinated by the penV oxygen atoms) and hydrophobic (phenyl groups of the side chains). The principal change in the series of crystals is the variation in the radius of the cation; with the increase in size of the cation the number of ligators at the cation increases from 2 to 7.

The ^{13}C CP MAS spectra of penicillins at lower temperatures showed discrete resonances for each of the six non-equivalent aromatic carbon atoms (Figure 12-13). As the temperature is increased the resonances of *ortho* and *meta* carbons display typical two-site exchange lineshape changes. Rates of the ring flip exchange process and activation energies were determined. The flip occurs at very different rates and activation energies. One can expect that expansion of the unit cell would lead to the decrease in the energy barrier to the spatially demanding motion of the phenyl ring. There is no specific correlation between librational amplitudes and the rates of the ring flip process although a qualitative correlation between these rates and the structure of the crystal in which the rings are located does exist. In Li-penV, which has the highest activation barriers, the phenyl rings lie between pairs of amide groups of penV anions from the other molecule type. H-penV also has a high activation barrier to ring flips, and in this structure it is the C2-methyl groups of molecules from a neighboring layer that provide the most rigid obstacle to flip. In the Na-penV the extended conformation of the side chain leads to the closest contacts to phenyl rings of molecules in the adjacent layer. It is quite understandable that a single parameter from a structure determination is inadequate to summarize the quality of the local environmental with respect to relative ease of the ring flips. More probably it is a complex interplay of different structural factors. The analysis of such motions in molecular crystals of drugs is possible using a combination of X-ray diffraction and solid-state NMR techniques.

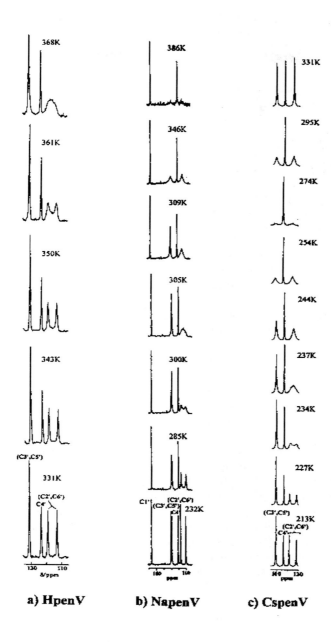

Figure 12-13 Parts of aromatic regions of the ^{13}C CP MAS spectra of phenoxymethyl penicillin (a) free acid HpenV and salts (b) NapenV, (c) CspenV at selected temperatures (reproduced with permission from [47]).

12.4 Studies of Tablets and Excipients

It is of interest to the pharmaceutical industry to be able to characterize drugs directly in their final dosage forms. Two of the first tablets studied were aspirin [52], [53] and acetaminophen [54].

Several commercial aspirin samples were used [52]: Bayer aspirin, extra-strength Bufferin, Empirin, and maximum-strength Anacin. There was no significant difference in the spectral pattern, except for the different content of the carbohydrate used in the tablet formation. The results showed that there were no interactions between the drug and the components of the buffer in the solids. The tablets were then suspended in aqueous solution and lyophilized. The spectrum of lyophilized Bufferin changed significantly indicating the interactions with the buffer components such as magnesium carbonate. The solid-state spectrum of soluble aspirin [53] showed the signals of citric acid and tartaric acid in polymorphic forms; the buffer system used in the formulation of the drug also contained sodium bicarbonate.

The spectrum of a paracetamol tablet is illustrated in Figure 12-14. Broad signals of excipients are in the range 60–100 ppm (carbohydrates) and 30–40 ppm (alkyl carbons of polyethylene glycol). Aromatic C2 and C6 as well as C3 and C5 carbons of the paracetamol molecule show separate resonances, due to the frozen conformation of -NH(CO)CH$_3$ substituent at C1 and the -OH group at C4.

^{13}C CP MAS NMR spectra of tablets or capsules of prednisolone, enalapril maleate, lovastatin, simvastatin, and, the non-steroidal anti-inflammatory agents ibuprofen, flurbiprofen, mefenamic acid, indomethacin, diflunisal, sulindac and piroxicam were recorded [55]. All samples contained small amounts of drug in the presence of large amounts of excipients. The results showed that solid-state NMR can detect the drug in low dose tablets as well as differentiate between tablets containing drug or placebo. The technique of delayed decoupling appeared to be very useful. Comparison of the spectra of drug tablet and placebo confirmed that the large peaks in the range 50–120 ppm are due to excipients. During the 50-µs delay before acquisition the magnetization of carbons directly attached to protons rapidly dephases; the resulting spectrum showed the signals of quaternary and rotating methyl groups, but the signals of other carbons are absent. This is therefore a convenient method to suppress the signals of excipients, which, as carbohydrates, have a number of methine and methylene carbons.

Polymers or amorphous solids used as pharmaceutical excipients have been also studied by means of solid-state NMR. The investigations included celluloses [56], [57], amyloses and starches [58], [59], hydrous and anhydrous lactose [27] and cyclodextrins [58], [59], [60] — easily available natural host compounds, increasingly popular for encapsulation of drugs [61].

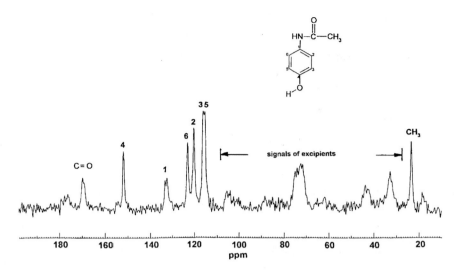

Figure 12-14 ^{13}C CP MAS spectrum of the paracetamol tablet.

^{13}C CPMAS NMR spectra of V-amyloses showed resolved resonances assigned to C1, C4 and C6 sites by comparison with solution chemical shifts for (1-4)-D-glucans [58]. In aqueous solutions of α- and β-cyclodextrin single resonances were obtained for C1 and C4 sites due to conformational averaging. Solid cyclodextrin hydrates are obtained by aqueous recrystallization. In solid-state spectra for C1 and C4 sites there are six and seven discrete resonances for α-cyclodextrin hexahydrate and β-cyclodextrin undecahydrate [58]. The signals are better resolved for the crystalline sample [54] than amorphous, dehydrated substances. The chemical shifts of the C1 and C4 resonances can be correlated with the conformation about the (1-4) linkage; chemical shifts of C6 are sensitive to hydrogen-bonding conformation [60]. ^{13}C CP MAS NMR spectra of the solid α-,β- and cyclodextrin are illustrated in Figure 12-15. In principle, a separate peak for each carbon atom in the molecule might be expected, and the number of resonances is larger for γ- than for α-cyclodextrin. It is worth noting that the resonances of cyclodextrins appear from 55 to 105 ppm, leaving free the outside spectral region where aromatic or aliphatic carbon resonances occur. This enables cyclodextrin/drug interactions to be studied, since the resonances of the included drug molecules can be observed.

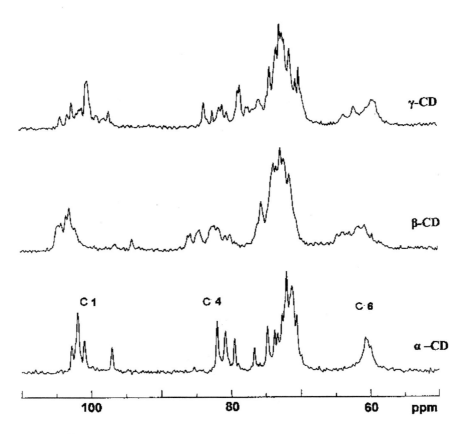

Figure 12-15 ^{13}C CP MAS spectra of α-, β- and γ-cyclodextrin.

References

1. C A Fyfe, Solid State NMR for Chemists, 1983, C F C Press, Guelph, Ontario, Canada
2. A C Olivieri, L Frydman, L E Diaz, *J Magn Reson* **75,** 50-62 (1987)
3. L B Alemany, D M Grant, D M Pugmire, T D Alger and K W Zilm, *J Am Chem Soc* **105,** 2133-2142 (1983)
4. L B Alemany, D M Grant, D M Pugmire, T D Alger and K W Zilm, *J Am Chem Soc* **105,** 2133-2142 (1983)
5. R K Harris, *Analyst* **110,** 649-655 (1985)
6. D G Retwisch, M A Jacintha and C R Dybowski, *Anal Chim Acta* **283**, 1033-2194 (1995)
7. H Saito, M Yokoi, M Aida, M Kodama, T Oda and Y Sato, *Magn Reson Chem* 155-161 (1988)
8. I Wawer and A Zielińska, *Solid State NMR* **10,** 33-38 (1997)

9. E R Andrew and M Kempka, *Solid State NMR* **2**, 261-264 (1993)
10. J Braun, M Schlabach, B Wehrle, M Köcher, E Vogler, H-H Limbach, *J Am Chem Soc* **116**, 6593-6604 (1994)
11. R Anulewicz, I Wawer, M K Krygowski, F Maennle, H H Limbach, *J Am Chem Soc* **119**, 12223-12230 (1997)
12. A Szelejewska-WoŸniakowska, Z Chilmonczyk, A Lecę, I Wawer, *Solid State NMR* **13**, 63-70 (1998)
13. C J Chang, L E Diaz, W R Woolfenden and D M Grant, *J Org Chem* **47**, 5318-5321 (1982)
14. L E Diaz, F Morin, Ch L Mayne, D M Grant and Ch Chang, *Magn Reson Chem* **24**, 167-170 (1986)
15. M J Potrzebowski, R Katarzyñski, W Ciesielski, *Magn Reson Chem* **37**, (1999)
16. P S Belton, A M Gil, S F Tanner, *Solid State NMR* **1**, 67-72 (1992)
17. W F Schmidt, A D Mitchell, M J Line and J B Reeves, *Solid state NMR* **2**, 11-20 (1993)
18. Ando S, I Ando, A Shoji and T Ozaki, *J Am Chem Soc* **110**, 3380-3386 (1988)
19. A Shoji, S Ando, S Kuroki, I Ando and G A Webb, in Annual Reports on NMR Spectroscopy, ed G A Webb, Academic Press, vol 28, pp 55-98 (1993)
20. C J Chang, L E Diaz, W R Woolfenden and D M Grant, *J Org Chem* **47**, 5318-5321 (1982)
21. W Kolodziejski T Kasprzycka-Gutman, *Solid State NMR* **11**, 177-180 (1998)
22. P Pfeffer and K Hicks, *J Carbohydr Chem* **3**, 197-217 (1984)
23. W L Earl and A F W Parish, *Carbohydr Res* **115**, 23-32 (1983)
24. G A Jeffrey, R A Wood, P E Pfeffer and K B Hicks, *J Am Chem Soc* **105**, 2128-2133 (1983)
25. S Witkowski, P Walejko, I Wawer, *Solid State NMR* **10**, 123-128 (1998)
26. M J Potrzebowski in Recent Advances in Analytical Chemistry, ed Atta-Ur Rahman, Gordon and Breach Publishers, pp 359-404 (1998)
27. H G Brittain, S J Bogdanowich, D E Bugay, J DeVincentis, G Lewen and A W Newman, *Pharm Res* **8**, 963-973 (1991)
28. D E Bugay, *Pharm Res* **10**, 317-327 (1993)
29. R Suryanarayanan and T S Wiedmann, *Pharm Res* **7**, 184-187 (1990)
30. P Gao, *Pharm Res* **13**, 1095-1104 (1996)
31. Y Kato, F Watanabe, *J Pharm Soc Jap* **98**, 639-648 (1978)
32. F Sabon, S Alberola, A Terol and B Jeanjean, *Trav Soc Pharm Montpellier* **39**, 19-24 (1979)
33. S R Byrn, G Gray, R R Pfeiffer and J Frye, *J Pharm Sci* **74**, 565-568 (1985)
34. R A Fletton, R K Harris, A M Kenwright, R W Lancaster, K J Packer and N Sheppard, *Spectrochim Acta* **43A**, 1111-1120 (1987)
35. R K Harris, B J Say, R R Yeung, R A Fletton and Lancaster R W, *Spectrochim Acta* **45A**, 465-469 (1989)
36. R K Harris, A M Kenwright, B J Say, R R Yeung, R A Fletton, R W Lancaster and G L Hardgrove, *Spectrochim Acta* **46A**, 927-935 (1990)
37. S R Byrn, P A Sutton, B Tobias, J Frye and P Main, *J Am Chem Soc* **110**, 1609-1614 (1988)

38. K Raghavan, A Dwivedi, G C Campbell, E Johnston, D Levorse, J McCauley and M Hussain, *Pharm Res* **10**, 900-904 (1993)
39. C Doherty and P York, *Int J Pharm* **47**, 141-155 (1988)
40. J J Gerber, J G van der Watt and A P Lötter, *Int J Pharm* **73**, 137-145 (1991)
41. Ball R G and Baum M W , *J Org Chem* **57**, 801-803 (1992)
42. E B Vadas, P Toma and G Zografi, *Pharm Res* **8**, 148-155 (1991)
43. H G Brittain, K R Morris, D E Bugay, A B Thakur and A T M Serajuddin, *J Pharm Biomed Anal* **11**, 1063-1069 (1993)
44. A B Thakur, K Morris, J A Grosso, K Himes, J K Thottathil, R L Jerzewski, D A Wadke and J T Carstensen, *Pharm Res* **10**, 800-809 (1993)
45. H Martinez, S R Byrn and R R Pfeiffer, *Pharm Res* **7**, 147-153 (1990)
46. J Fattah, J M Twyman, S J Heyes, D J Watkin, A J Edwards, K Proust and Ch M Dobson, *J Am Chem Soc* **115**, 5636-5650 (1993)
47. M Wendeler, J Fattah, J M Twyman, A J Edwards, Ch M Dodson, Heyes S J , K Prout, *J Am Chem Soc* **119**, 9793-9803 (1997)
48. S S Leung, B E Padden, E J Munson, D J W Grant, *J Pharm Sci* **87**, 501-506 (1998)
49. D Singh, P V Marshall, L Shields, P York, *J Pharm Sci* **87**, 655-661 (1998)
50. J T Timonen, E Pohjala, H Nikander, T T Pakkanen, *Pharm Res* **15**, 110-115 (1998)
51. Ö Almarsson, M J Kaufman, J D Stong, Y Wu, S M Mayr, M A Petrich and J M Williams, *J Pharm Res* **87**, 663-666 (1998)
52. C Chang, L E Diaz, F Morin and D M Grant, Magn *Res Chem* **24**, 768-771 (1986)
53. L E Diaz, L Frydman, A C Olivieri and B Frydman, *Anal Lett* **20**, 1657-1666 (1987)
54. N R Jagannathan, *Current Science* **56**, 827-830 (1987)
55. P J Saindon, N S Cauchon, P A Sutton, C Chang, G E Peck and S R Byrn, *Pharm Res* **10**, 197-203 (1993)
56. G E Maciel, Kolodziejski W , M S Bertran and B E Dale, *Macromolecules* **15**, 686-687 (1982)
57. P T Larsson, Wickholm K and T Iversen, *Carbohydr Res* **302**, 19-25 (1997)
58. M J Gidley and S M Bociek, *J Am Chem Soc* **110**, 3820-3829 (1988)
59. K R Morgan, R H Furneaux and R A Stanley, *Carbohydr Res* **235**, 15-22 (1992)
60. S J Heyes, N J Clayden and Ch M Dobson, *Carbohydr Res* **233**, 1-14 (1992)
61. V J Stella, R A Rajewski, *Pharm Res* **14**, 556-567 (1997)

Chapter 13

I. Wawer

13 MR Imaging and MR Spectroscopy

13.1 *In vitro* and *in vivo* Measurement Methods

MR imaging is the realization of an old dream of physicians — how to look inside the body of a patient without cutting it. MR spectroscopy goes one step further — it attempts to explain the disease on a biochemical, molecular level. Both are of high interest to pharmacy since the response to treatment can be followed.
One can even suppose that the need to ask the patient " how do you feel?" will soon be reduced to a mere matter of courtesy; the qualitative and quantitative answer could be supplied by MRI and MRS diagnostic methods (including feeling, *vide* functional MRI).

Before we describe magnetic resonance imaging, a summary of the standard NMR experiment is needed. If a static magnetic field B_0 is applied to a sample containing water, magnetic moments of protons (rotating with frequency ω_0) line up either parallel or antiparallel to the applied field. A small excess magnetization M_0 is created, parallel with the field direction. For reorientation of magnetic moments a radiation B_1 with radiofrequency $\omega = \omega_0$ is necessary.
The relationship: $\omega = \gamma B_1$ (γ = gyromagnetic ratio of nuclei) holds. The RF radiation is applied in pulses; if the pulse is applied for t sec., the excess magnetization M_0 is tipped through an angle: $\varphi(rad)=\gamma B_1 t$ and gives rise to a transverse component, which is detected by the spectrometer. The magnetization returns rapidly to equilibrium position; its decay can be characterized by the spin-spin (T_2) and the spin-lattice (T_1) relaxation times. Free induction decay (FID), converted to NMR signal, is observed between RF pulses. The signal is collected from the whole volume and the sample should be homogeneous.
Neither solution nor solid-state NMR techniques give spatial information on a sample.
Basic principles and theory of spatially resolved NMR, called MR imaging or tomography, can be found in [1], [2], [3], [4], [5], we will give here only brief comments on measurement methods.
MRI uses a spin-echo sequence, i.e. the sequence of two pulses, usually 90° followed by 180°, which produces spin-echo in an inhomogeneous magnetic field (the transverse magnetization can be measured at time 2TE, where TE is the echo-time). Inhomogenety of magnetic field can result from the presence of deliberately applied field gradients. Consider magnetic field B_0 applied along a z direction together with an additional field, producing a linear gradient. The resonance frequency is proportional to the magnetic field and the condition: $\omega_0=\gamma B_0$ has to be replaced by: $\omega_0(z) = \gamma(B_0+G_z z)$, where G_z is the field gradient at position z. Resonance frequencies of nuclei placed along the z axis have to be changed accordingly; now the FID containing information on position of nuclei, i.e. frequency-encoded FID, will be obtained. Adjusting the duration of the pulse, and thus the spectral width, it is possible to excite the nuclei at the selected position in the sample.

Such „slice selection" allows the imaging, although it is only a one-dimensional experiment.
Commercial medical scanners use imaging sequences based on the „Two-Dimensional Fourier Transform" (2DFT) method or spin-warp method [4], [5]]. One transverse gradient is applied for phase-encoding then orthogonal transverse gradient follows and the FID is sampled. The process is repeated for n phase-encoding; increments of gradients to obtain $n \cdot n$ data set, which is Fourier transformed to give the $n \cdot n$ image. The imaging pulse sequences are usually more complicated, but are composed of common steps: excitation (slice selection)→evolution (phase encoding)→detection (frequency encoding) and image reconstruction. Three parameters: the amplitude, the frequency and the phase can be used for encoding the information. In a two-dimensional experiment two of these parameters (i.e. phase and frequency) serve for localization, the third (intensity) should give the image contrast. In the three-dimensional experiment, phase encoding must be two-fold in order to create the third encoding dimension.

Many preprogrammed sequences are now available and are routinely applied. The spin-echo technique is frequently used as the RARE sequence (Rapid Acquisition with Relaxation Enhancement, for fast T_2 contrast images), and the gradient-echo technique in the version of GEFI (fast scanning using variable RF excitation angles or spoiler gradients to acquire differing contrast) or SNAP (ultrafast extension). Short pulse sequences are of particular interest for fast imaging, in this case the flip angle of the excitation pulse is much smaller than 90°, and much shorter repetition times can be used. A fast version, known as FLASH (Fast Low Angle Shot) technique was developed by Haase; it allows good quality images to be obtained within seconds [6] or even a fraction of a second (in the order of 100 ms) rather then minutes or hours necessary for conventional spin-echo methods. Even faster data acquisition can be achieved by the multiple signal read-out technique called echo-planar imaging. The coherences excited by a 90° RF pulse are refocused many times with increasing phase-encoding increments. The total data acquisition time which can be obtained is in the order of 10 ms; this is really a very good time resolution, valuable in the cases when the images of moving objects have to be recorded.

The images look realistic (see Figure 13-1), like those in a black-and-white photograph (or even like the color one, if colors are assigned to intensities by a computer program). Nevertheless, one should keep in mind that such picture could never be seen because is comes from the inside of the object. If the experiment is performed in such a way that one parameter dominates the contrast we describe it as, for example, relaxation- or diffusion-weighed representation. However, the gray shades in the image are not reflecting only this single parameter but are combination of many others as well. Systematic variation of pulse intervals permits obtaining T_1 or T_2 maps to be obtained. Relaxation contrasts can also be produced artificially by treating the object with relaxation contrast agents based on paramagnetic species (iron micro-particles, gadolinium complexes). Blood Oxygenation Level Dependent (BOLD) contrast is based on differences in magnetic susceptibility of hemoglobin in oxygenated and deoxygenated states and has been extensively used in functional MRI studies of human brain activation. Since BOLD images are sensitive to the oxy- /deoxy-hemoglobin ratio, this technique can also be used to monitor changes in tissue oxygenation and oxygen consumption.

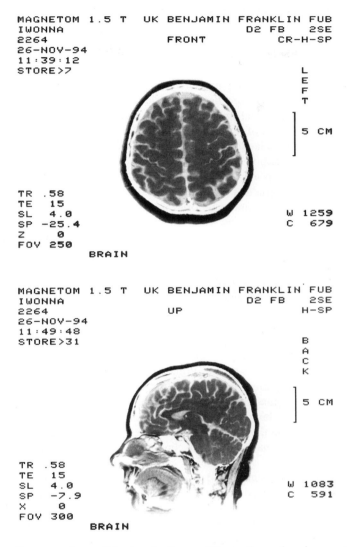

Figure 13-1: Cross-sections of the head of one of the authors showing anatomical details of the brain, obtained by ^{1}H MRI at 1.5 T.

Localized MR spectroscopy was developed to obtain chemical as well as spatial information. Frequency offsets (caused by the inhomogeneous susceptibility or chemical shift distribution) can be recorded by introducing a further measuring domain, called the spectral-encoding domain. Thus, the dimension of the experiment is increased by one. The spatial distribution of compounds with distinct chemical shifts can be obtained as chemical-shift-selective maps. The spatial distribution of frequency offsets can be acquired by imaging pulse sequence, such as echo-time encoding. The method is suited for the ing of maps of frequency offsets due to chemical shift, magnetic susceptibility differences or magnet inhomogeneities.

The two methods most widely used for acquiring spatially localized spectra *in vivo* are Point-REsolved Spectroscopic pulse Sequence (PRESS) and Stimulated Echo Acquisition Mode (STEAM), they are frequently implemented on commercially available 1.5 T whole-body systems. Pulse sequences are based on a localization sequence using three slice-selective RF pulses with nominal tip angles α, β, γ and relative pulse phases that depend on the location of the volume of interest. The PRESS sequence is characterized by setting a=90o, b=g=180o, careful selection of the relative sizes of gradients produces the sensitive short echo-time localization sequence needed to record MRS spectra. In the STEAM localization sequence the tip angles of all three slice-selective pulses are equal to 90°; a pair of de and re-focusing gradient pulses is inserted in the TE/2 period and also a strong crusher gradient in the middle interval. Both methods have advantages and suffer from several drawbacks. PRESS achieves worse suppression of contaminating signals and requires more RF power for slice-selective 180° pulses. Shortening of the echo time TE leads to the increased contamination by signals (water) from outside of the volume of interest. Hybrid sequences, combining slice selective chemical shift imaging (CSI) and STEAM or PRESS to reduce lipid contamination and spatial aliasing, are also used. For better characterization of the metabolite content *in vivo*, the combined 3D spatial localization and 2D NMR pulse sequences (J-resolved, COSY, SECSY) have been developed [7], [8], [9].

NMR systems for investigations *in vivo* use horizontal-axis superconducting magnets (1.5-9.4 T) with bore diameters of 15-90 cm. Probes for sample diameters of 7-20 cm are designed for small animals whereas larger bores and magnetic field strength lower than 4 T are used for humans. The imaging technique is capable of giving images of cross sections through the human/animal body with good resolution in all three dimensions. As mentioned above, depending on the dimensionality of the region from which the signal is observed, one can obtain information by point-, line-, plane- or volume-selective techniques.

MRI yields anatomical information and is mainly used for clinical diagnostic. As an example, sections through the head of one of the authors showing the brain, obtained by ^{1}H MRI at 1.5 T, are illustrated in Figure 13-1. Early studies concentrated on head and extremities (hand, leg) because MRI of fast moving organs such as the heart, or inner organs moving with breathing, is more demanding experimentally. A modification of the spin-echo sequence for correcting motion artifacts was necessary. Navigator echo, NE, can correct phase errors from small motions during the imaging sequence because for each image echo a correction echo without phase encoding is recorded [10], [11]. Most MR imaging studies on humans are performed at field strengths of 1.5 T, however measurement at lower fields has also some advantages. The sensitivity of the MR spectroscopic signal, which depends to the field strength, is approximately three times higher at 1.5 T than at 0.5 T; better homogeneity of the magnet results in a line width approximately one third (at 0.5 T) of that at 1.5 T. The type of MR-visible nuclei, the concentration and the relaxation times determine resonance intensity (signal-to-noise ratio) at any field strength.

The signal-to-noise ratio, the spatial resolution and the total measuring time are closely related. The total measuring time is $N \cdot T_R$; where N is the number of times the measuring sequence with duration T_R is carried out. In a typical 2D experiment with slice selection a total measuring time is $256 \cdot T_R$ when the sequence is performed once. After data processing one can obtain a 16·16 matrix of pixels, each with a spectrum, yielding a mapping of the desired metabolite distribution in the plane (slice). Small voxel dimensions require extensive time averaging; for one 2D MRS imaging experiment, dimensions $k \cdot k$ pixel it is $k^2 \cdot T_R$ [12]. An acceptable measuring time for a patient is ca. ½ h, and signal-to-noise ratios determine in practice the spatial resolution, which is about 1 cm on a whole-body system and can be 1 mm on an animal system, when time of experiment is less restricted.

MRS is far more demanding than MRI (mapping water) because the metabolites detectable by MRS are present *in vivo* in millimolar concentration (concentration of water in the body fluids is above 40 M). MRS investigation of metabolites may last for one to several hours. The results are most frequently presented as ratios between metabolite signals. Absolute quantitation of the metabolite content *in vivo* is difficult, as it is also in the case of ^{1}H MRS. Problems with localization, pulse sequences, relaxation rate, spectral resolution, signal intensity calibration, etc. have to be overcome. ^{1}H MRS is usually limited to the spectral range 2–4.3 ppm, exhibiting signals of the most important low molecular metabolites. Overlapping of signals and the splitting of resonances (^{3}J coupling constants) complicated the assignments and the quantitation procedures. The high-frequency spectral region containing mainly signals of aromatic and amide protons (5–9 ppm) is heavily disturbed by the water resonance (variation of bulk susceptibility and water suppression sequences produce problems with base line correction). In the low-frequency region 0–2 ppm broad signals of macromolecules occur, which are not easy to interpret.

^{1}H, ^{31}P and ^{19}F MRS have been applied in a number of studies with the aim of exploring the potential of the technique for differential diagnosis, and therapy, and to study biochemical aspects associated with drug therapy. The endogenous and exogenous compounds as well as the reaction of an organism to the drug can be monitored, either by imaging or by localized spectroscopy. However, because of low sensitivity only the effect of drugs on the endogenous metabolism may be followed *in vivo* and not the metabolism of the drug itself. Research applications also involve ^{13}C and ^{15}N with isotopically enriched substances. This provides new possibilities in many areas of medicine and pharmacology.

A number selected applications of *in vitro* and *in vivo* MRI/MRS will be discussed in order to illustrate the power of the technique. Our aim is to give the readers an insight into present and future applications in various areas of pharmacy.

13.2 ^{1}H MR Imaging of Tablets

NMR signals for solids not subjected to magic angle spinning exhibit broad signals of low intensity, and the solid substances cannot be visualized with spin-echo or gradient-echo methods usually applied in Magnetic Resonance Imaging for taking images. However, if a liquid such as water or oil, producing an intense signal, is introduced into the cavities of a solid the inner structure can be displayed. Resolution at microscopic level is achieved (better than 10 µm), and therefore the method is also called NMR microscopy. The size of the smallest object that can be distinguished in the image depends on: (i) the magnetic properties of nuclei used for imaging, (ii) the strength of the imaging gradient used, (iii) the spin-spin relaxation time T_2 and (iv) the time that the signal can be sampled for. High resolution requires large field gradients and such large gradients are more easily produced when the sample is small. Studies on tablets are carried out on small-bore, high-field magnets. Strong magnetic fields and some heating effect are of no consequence for a tablet, which is not a living creature; the advantage of using strong magnetic fields and small RF coils is that good signal-to-noise ratio can be achieved in a reasonable time. Sometimes it is necessary to acquire the image in a very short time. The imaging can be speeded up by reducing resolution or dimensionality. Frequently it is possible to select the slice and follow the process in time in a one–dimensional model.

For the studies of fluid displacement, the fast imaging technique FLASH [6] can be used to follow the water ingress into tablets. The images shown in Fig. 10.2 a-b were obtained in 0.4 s each using the home-built NMR spectrometer (^{1}H frequency 271.3 MHz, 6.4 T magnet) in the Institute of Nuclear Physics, Cracow, Poland. The time of the experiment was short enough to follow the dissolution of the tablet in water (with 0.1 M HCl added to imitate the environment of gastric juice). The tablet, at the bottom of the glass NMR tube, was placed inside the probehead and the first image was obtained after 3 min (Figure 13-2a). The place of cross-section is marked with a white line and at the top of the picture is the plot of the water signal intensity. The tablet dissolves rapidly and after 20 min a part of coating was separated, seen as black shape on the right hand side (Figure 13-2b).

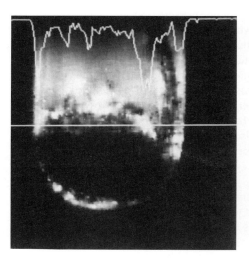 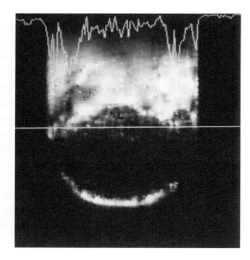

Figure 13-2: Images of tablets immersed in 0.1 M HCl solution, each obtained in 0.4 s using home-built NMR spectrometer, ^{1}H frequency 271.3 MHz, 6.4-T magnet.

at the same time. Several reports provide indications that biological phenomena such as drug distribution, drug-receptor interaction and gastrointestinal drug absorption (which is frequently heterogeneous) can be also interpreted on the basis of their fractal nature and follow the fractal kinetics [13]. One may suppose that such descriptions will be even more important for understanding processes that drive drug release *in vivo* (see below).

The change in intensity of water signals with time (such as plotted in Figure 13-2) can be considered as a measure of drug release rate and may lead to different dosage form characteristics. MRI creates a potential for studies and optimization of controlled release dosage forms. The processes such as hydration and diffusion can be monitored, which are important for the mechanism of drug release.

The MRI method was utilized in 1991 [14] to follow the progress of hydration in poly-(L-, DL)lactic acid tablets. The samples were prepared from low-molecular mass poly-lactic acid polymers and various amounts of theophylline and compressed into tablets of 9 mm diameter. The tablets were hydrated in the buffer solution and then placed inside the probehead of the Bruker MSL 400 spectrometer for imaging. The 2DFT imaging with spin-echo pulse sequence were applied to obtain slice-selective images. The process of diffusion, described by water penetration into the tablet, was not influenced by buffer pH (in the range 1.2– 4.4) in the case of L-poly-lactic acid tablets whereas for DL-poly-lactic acid tablets both the influence of buffer pH and drug content were found. Images of DL-poly-lactic acid tablets showed qualitative and also quantitative alterations in matrix structure caused by diffusion of the buffer. Porosity of a tablet is an important parameter in the production of solid dosage forms because it is connected with mechanical stability, disintegration, wetting properties and the release of drug. Pore size distribution and porosity have been estimated usually by mercury porosimetry, but the number of obtained data points is low and the object is destroyed.

In 1995 ^{1}H-NMR microscopy was used for the first time to study the porosity of intact tablets filled with silicone oil [15]. The images of the tablets filled with silicone oil (doped with gadolinium) were obtained with a Bruker AM300 NMR spectrometer, wide bore magnet and probehead specially designed for micro-imaging, allowing sample diameters from 1 to 25 mm. The cross-sectional signals through the tablet were transformed by computer into contour plots and color images. Spatial resolution was about 95 µm for an edge length of the cube. Two-dimensional images of tablets can be obtained in about 30 min; for three-dimensions 4 h or more are necessary, depending on the required resolution and signal-to-noise ratio. Porosity distributions within tablets, cracks and cavities were seen in physically intact tablets. The different states of densification during compaction of powders and different compaction mechanisms with plastic-type or brittle-type tablets could be observed. Thus, the method could be very useful for studying and controlling compression of tablets in the pharmaceutical industry.

In the formulation of hydrophilic gel-forming matrices, water-soluble and swellable cellulose derivatives have been widely applied. Hydroxypropylmethylcellulose (HPMC), hydroxypropyl- cellulose (HPC) and carboxymethylcellulose (CMC) form gels and swell on contact with water. There are several reports dealing with water mobility in such a polymer matrix using NMR Imaging. The MRI technique was applied [16]for studying the formation of gel layer in hydrating HPMC tablets. The tablets were mounted and hydrated for up to 3 h. The in-plane resolution was 100 µm and each image was acquired in ca. 2 min. Diffusion weight images showed different degrees of polymer hydration at different depths within the gel, and revealed additionally that HPMC matrices expand on hydration much more in the axial direction than in the radial direction. This work demonstrated that NMR imaging could be used to measure non-invasively diffusion of water, thus obtaining valuable information on dynamics within various dosage forms.

Micronized low-substituted hydroxypropylcellulose (LH41), HPC and HPMC tablets during the hydration process were investigated recently [17]. The tablet was hydrated and then placed inside the core of the MRI, and the measurements were carried out on a Bruker spectrometer operating at 200-MHz with 4.7-T magnet. Transverse images of the LH41 tablet showed that swelling, deformation and cracking occurred at the edge of the tablet. The amount of absorbed water and overall swelling of the tablets decreased in the order: HPMC>HPC>>LH41. The three kinds of tablets expanded in the transverse section much more than in the coronal direction. The apparent self-diffusion coefficient and the spin-spin relaxation time of water in the LH41 tablet were smaller than the respective values of free water indicating that water is restricted and interacts strongly with the polymer. Thus, gel layer properties may be different among the three polymers. According to the analysis of spin-spin relaxation time, one type of restricted water exists in the LH41 tablets and two types of water with different mobility are in the HPC and HPMC tablets.

The results discussed above confirm that the imaging methods open new fields of investigation of drugs dissolution and release from various types of tablets, now tested *in vitro* but in the near future to be tested also *in vivo*. The bioavailability of drugs from solid oral dosage forms is influenced by various physiological factors such as peristalsis, stomach residence time or intestinal transit. Therefore it is of great interest to study *in vivo* behavior of pharmaceutical dosage forms. To our knowledge only a very few MRI studies on *in vivo* localization of dosage forms have been carried out. Monitoring of the gastrointestinal transit is possible with γ-scintigraphic methods but the radiolabeled drugs or dosage forms are necessary. MRI is a non-invasive and non-radioactive technique and allows the visualization of the digestive tract as well. Simultaneous visualization of the tablet and the gastrointestinal tract of the rat was performed recently [18]. The anesthetized animal was positioned in the birdcage-type resonator (60 mm diameter, 90 mm length). The resonator was tuned to 85.54 MHz ^{1}H and 80.47 MHz for ^{19}F, and the maximum gradient strength was 200 mT/m. The T_1-weighed spin-echo images of the rat abdomen were recorded in eight contiguous coronal ^{1}H MRI slices of 2 mm thickness and one ^{19}F MRI slice of 10 mm thickness. Normally, fluorine is not present in the gastrointestinal tract but 100% abundance and high sensitivity make it very good label. For visualization purposes perfluorinated compounds have to be used. Polytetrafluoroethylene (PTFE powder) microtablets as a model for solid dosage form and minicapsules (made of PTFE and polyethylene) filled with hexafluorobenzene as surrogates for pharmaceutical capsules were prepared; for the investigations in rat the toxic hexafluorobenzene was sealed in capsules. Oral contrast agents (Gd-DTPA and ferric ammonium citrate, mixed with lipids) were applied to depict the lumen of the digestive organs. In the ^{1}H MRI images the microtablets appeared as „dark spots" on the brightly enhanced digestive lumen. Visualization in the rat intestines was more difficult due to fast disintegration of the microtablet after stomach passage (thickening of the coating was necessary to reduce swelling) and the irregular distribution of the contrast media. Solid dosage forms can also be labeled by incorporating trace amounts (less than 0.1% w/w) of non-toxic iron oxide particles. The super-paramagnetic particles may be incorporated in various types of pharmaceutical dosage forms and enable the *in vivo* behavior to be investigated.

An alternative approach can be the labeling of tablets with spin probe, 3-carboxy-2,2,5,5-tetramethyl-pyrrolidine-1-oxyl (PCA) radical [19]. PCA loaded tablets were implanted into the neck of an anesthetized Wistar rat. EPR and NMR measurements were performed on the same animals. MR images were obtained on a 7.0-T horizontal bore magnet; rats were placed supine over RF surface coil. The spin-echo T_1 and T_2 weighted sequences were used to obtain 7 slices of 2 mm thickness with 4 acquisitions. The purpose of the study was to compare drug release and polymer erosion from biodegradable poly(-fatty acid dimer-sebacic acid), P(FAD-SA) polyanhydrides. Biodegradable polymers are used clinically, but knowledge about their degradation and drug delivery is limited. The comparison of polymer degradation processes *in vitro* and *in vivo* is of special interest. The P(FAD-SA) 50:50 and 20:80 tablets were studied; for both polymers the thickness of the tablets decreased with time. Polymer 50:50 was almost entirely absorbed by day 44 and a residue of polymer 20:80 remained after 65 days. The processes such as edema, deformation of the implant, encapsulation and bioresorption were observable by MRI *in vivo*. The EPR studies gave confirmation that water penetrates inside the tablet; in P(FAD-SA) 20:80 implants EPR signals were detectable after 65 days (in agreement with MRI results) whereas *in vitro* the nitroxide radical was released within 16 days.

MRI provides the pharmaceutical scientist with a powerfull noninvasive method for picturing events inside controlled-release dosage forms [20]; with MRI it is possible to observe, follow and measure the processes of hydration and diffusion. Thus, the process of drug release can be monitored *in vitro* and *in vivo*. The potential of MR imaging is only beginning to be exploited.

13.3 ^{1}H Imaging of Plants

In biological tissue the most abundant source of protons is water, and in most cases MR imaging means obtaining maps of water density. Differences between T_1 and T_2 relaxation times provide a contrast mechanism. Both relaxation times are sensitive probes of molecular motions. As NMR imaging is able to acquire maps of water distribution and to monitor water flow, it is clear that this method has potential as a valuable tool in the study of plants. Several images of fruits and nuts have been presented earlier; that of an intact lemon exhibiting stones is widely known (see the book by Günther H. [21]).
Sensitivity and resolution in NMR imaging was discussed in 1986 [22], and the MRI method was applied to image water distribution in plant steam. The image of the flower *Taraxacum officinale* is shown in Figure 13-3 and was obtained using NMR spectrometer (^{1}H frequency 271.3 MHz, 6.4 T magnet) at the Institute of Nuclear Physics, Cracow. The flower was placed inside the 7 mm diameter glass tube. Long preparation and imaging time caused loss of water — the plant withers away.

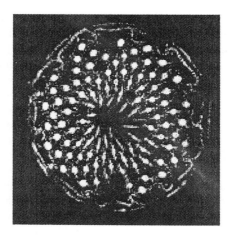

Figure 13-3: ^{1}H MR image of the flower *Taraxacum officinale*; placed in NMR tube

With the decrease in the amount of water, the plant changes shape, and the leaves roll up. It creates a problem with the selection of slice, especially when the most suitable place for quantitative measurements is to be found. Nevertheless, one can observe which parts of the plant retain the largest amount of water, as determined after 1 or 2 h of drying. In the in completely dried parts of plant, mold or bacteria develop, and microbiological contamination has to be removed, for example, by radiation sterilization. It is therefore of interest for the technology of plant sources to follow the process of drying in detail.

According to our experiments with *Urtica dioica, Achilea millefolium,* or *Equisetum arvense* after a 1 h stay inside the probehead the amount of water was not sufficient to provide an imagine of major tissues of the plant with a reasonable signal-to-noise ratio. In order to gain enough time for measurements, the plant should be left in soil or placed in water. The image of *Equisetum* where vascular bundles and parenchyma can be distinguished is shown in Figure 13-4 a, b.

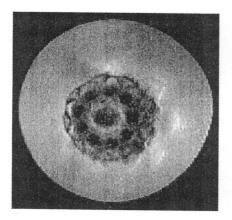 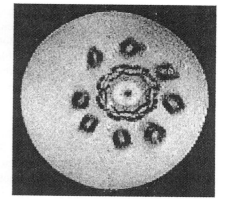

Figure 13-4: a, b ^{1}H MR images of *Equisetum arvense*; NMR tube was filled with water.

The spin-echo technique has been used for many years to measure molecular self-diffusion, but the use of pulsed field gradients in a spin-echo sequence allowed for measurements of displacement of water molecules as a function of position, i.e. for measurements of localized self-diffusion. Wheat grain (*Triticum aestivum*, at the mid grain-filling stage) was inserted in a 5-mm NMR tube with moist cotton wool to maintain humidity during several hours of imaging [23]. Water self-diffusion rates to the transverse 1.3-mm section of grain were measured, the resolution being 150 µm. Different water intensities in the endosperm tissue and vascular bundle area can be seen; by the use of selective excitation it is possible to measure the localized transport in any direction.

NMR imaging offers a unique means of accessing structure, growth and hydrodynamics of roots *in situ* because soils are permeable to static and radio-frequency magnetic fields. An investigation of the roots of the bean *Vicia faba* was described in 1986 [24]. Potted samples were positioned vertically within the receiver coil (14 cm diameter), and ^{1}H NMR images were obtained at 63 MHz on a 1.5 T system; 256 · 256 point spin-warp spin-echo sequence was applied with 0.56 mm · 0.56 mm resolution in the imaging plane. The image intensity was proportional to the mobile water density; signals from bound water or cellulose did not appear because of short T_1 relaxation times. Cotyledons and roots were clearly seen in the images, and the variation of the image quality with soil type was noticed (the effect of paramagnetic particles). Water flow was investigated using water doped with Cu^{+2} as contrast agent. The processes of plant wilt and recovery and light-stressed foliage were monitored.

There is no other non-invasive means of assessing the germination process *in situ*. The information on germination is very important for optimizing growth conditions in seedbeds. Seeds that had germinated were also imaged by Bottomley et. al. [24]. Differences of water content in the various stem tissues of *Pelargonium hortum* were observed using ^{1}H MRI [25]. The images were obtained at 64 MHz on a medical system (1.5-T magnet, gradient coils of 28 cm diameter) with a resolution of 100 µm in the transverse plane. The plant was slowly and actively transpiring. There is a correlation between increased proton density and longer spin-lattice relaxation time. In the slowly transpiring plant, T_1 ranged from an average of 659–865 ms with a proton density variation of 72–100%. In the actively transpiring plant, T_1 ranged from 511 to 736 ms and the proton density was reduced to 62–88% of the value found in the slowly transpiring plant. The results are in agreement with the two-state water model: water bound to macromolecules is characterized by short T_1, while free water has much a longer T_1. As the percentage of free water increases (increase in relative proton density), the weighted T_1 also increases. The values of T_1 reflect an average of free and bound water and also a spatial average across cell walls or intracellular spaces.

A series of NMR micro-imaging experiments carried out in order to obtain an insight into *in vivo* transport in plants were described in 1995 [26]. The castor bean seedlings were grown hydroponically on the top of glass tubes inserted directly into a standard micro-imaging probe head; the roots were continuously aerated. The experiments were performed with Bruker AMX 300 and 500 MHz spectrometers. A bean seedling consists of the root, the hypocotyl and the endosperm with two cotyledons. A T_1-weighted image with a planar resolution of 25 µm · 25 µm in a 1-mm transverse slice through the showed cortex parenchyma, pith parenchyma and eight vascular bundles. The bundles consist of lignified vessels, xylem (conducting the flow of water from the roots to the leaves) and a complex of cells and sieve tubes, phloem (containing phloem sap, rich in sucrose). A 3D chemical-shift imaging (CSI) experiment was performed using spin-echo techniques for determining the sucrose distribution within the seedling. The image of sucrose was obtained by displaying the H6a,b resonance at 3.72 ppm. Eight spots clearly indicated that areas enriched in sucrose in the hypocotyl may be identified as the phloem. Therefore, for the first time a non-invasive determination of the sucrose concentration in the intact plant was possible. Other resonance lines observed in the ^{1}H spectra of the CSI experiment could not be identified due to the signal overlap. A correlation peak-imaging technique was developed in order to obtain a COSY spectrum for each of the 16 · 16 image matrix elements. The spatial distribution of glucose, glutamine, and amino acids could be visualized. Moreover, a difference technique was developed to detect water flow and to suppress the signal of the stationary water present in the tissues. A method, based on q-space imaging, allowed for measurement of flow velocities within a voxel (voxel resolution was 47 μm · 47 μm · 2 mm). The flow image indicated the bi-directional flow of water in vascular bundles; for xylem a volume flow rate of 35 μl/h was calculated and phloem 15 μl/h. for. It is worth noting that approximately 20 μl/h is consumed by growth and lost by evaporation. The study demonstrated that detailed investigation of water transport is possible in an intact plant. The experiment can be repeated with the same plant, and therefore all changes can be monitored in relation to the changes in the environment (nutrient deficiency, drought). Physiological processes can be studied in different parts of the plant, for example photo-synthesis in leaves, uptake of nutrients by roots and transport of metabolites.

Identification and quantitation of metabolites in *in vivo* ^{1}H imaging is not straightforward because of the complexity of the spectra. Line broadening caused by susceptibility effects and the small chemical shift range prevent resolution of metabolite peaks and correct assignments. The 2D experiments on ethanolic extracts from the hypocotyl of castor bean (*Ricinus communis*) seedlings allowed for identification of 14 resonances in the range 0.96-5.5 ppm (from: sucrose, α- and β-glucose, glutamine/glutamate, lysine, arginine, leucine, isoleucine and valine) [27]. In order to improve the spectral resolution, homonuclear 2D correlation spectroscopy has been combined with phase encoding of the spatial dimensions. The method of correlation-peak imaging (CPI) with two spatial and two spectral dimensions was applied to image the metabolites from the hypocotyl of 6-day-old castor bean seedlings. In the *in situ* 2D COSY spectrum of a 4-mm slice of the hypocotyl, the cross peaks arising from amino acids and carbohydrates can be observed, despite the line broadening *in vivo*. It was possible to compare the CPI metabolite images with plant anatomy. High intensity of the sucrose cross peak was found in the vascular bundles, as expected. The pith parenchyma showed a gradient of the sucrose cross-peak intensity toward its center. The glutamine/glutamate cross-peaks appeared as a ring covering vascular bundle and cortex parenchyma regions; arginine and lysine are observed in the vascular bundles.

The knowledge about *in situ* distribution and concentration of metabolites in plant physiology has hitherto been limited, mainly because of the lack of an appropriate, non-invasive experimental technique. The CPI experiments could be used to monitor *in vivo* the concentration and localization of the compounds important for pharmacy which are still obtained from plant sources.

13.4 Studies on Biopsy Samples and Tissues

Once the biopsy sample or tissue fragment is removed from the living organism it can be investigated by the techniques of micro-imaging with high spatial resolution. The inner structure can be displayed without destroying the sample, in contrast to the destructive methods such as histological ones. However, biopsates are preferably analyzed as metabolite extracts and not as intact tissue samples. The reason is the post-mortem decay process, which starts immediately after resection, and also the scientific interest, the main goal being to gain information on the biochemical pathways occurring in cells; the spatial structure is less critical. *In vitro* MR spectroscopy is useful since the spectra can be measured at high fields, which offer better signal resolution than that found with *in vivo* measurements. More precise identification of the different metabolites could be made, enhancing our understanding of the spectral results obtained *in vivo*.

NMR spectroscopy of cell/tissue extracts uses standard high-resolution NMR equipment like that used for biofluids, so that the identification of many metabolites and determination of their concentrations is possible. Usually, the ^{1}H and ^{31}P spectra are recorded, and the low-abundance ^{13}C and ^{15}N nuclei have to be enriched. The enrichment is achieved by supplying labeled substrates, such as glucose or amino acids, into the cell culture medium. The MRS method is noninvasive and nondestructive but rather insensitive, because studies *in vivo* require concentrations in the order of 1 mmol. NMR spectroscopy of cell or tissue extracts can identify compounds at a concentration of a few μmols. The materials for *in vitro* studies are obtained either from stereotactic biopsies or from conventional microsurgical operations. For immediate *ex vivo* studies of intact samples the specimens are sliced to produce fragments which fit into a 5 or 10 mm O.D. NMR tube and are immersed in D_2O. Usually, after dissection, the tissue specimen is rapidly immersed in liquid nitrogen and may be stored. It is weighed in the frozen state, homogenized and extracted with perchloric acid or 70% ethanol. After low-speed centrifugation and lyophilization, the dried powder is dissolved in D_2O and the NMR spectra are recorded.

The viable cells should be immobilized for the time of NMR studies. They are grown on a gel matrix of agarose, carrier beds or kept in suspension within semipermeable hollow fiber. The time resolution is in the range of one to a few minutes for approximately 10^8 cells within the measured volume. The studies performed on intact biopsies with subsequent investigation by standard histological techniques are very interesting; the correlation between metabolite levels and histopathological features is of high significance. Many forms of cancer still presents ambiguous diagnoses, and there is therefore a strong need for a non-subjective method of analysis. We believe that such a method will be NMR provided by spectroscopy. In some cases the spectra from normal and cancerous samples can be recognized visually, but in most cases this is not possible. Therefore a set of procedures should be established to normalize the spectra, identify most sensitive regions and, further, to apply multi-variant analysis. A large data base of spectra *in vivo* is therefore necessary. Some chosen examples of the studies on tissue samples and cultured cells are discussed below.

Biopsies obtained from brain tumors (placed immediately in buffered saline in D_2O and frozen in liquid nitrogen) were later, after thawing, investigated by ^1H MRS [28]. Spectra obtained from samples from different biopsies of the same tumor, and even from samples obtained from the same biopsy differ significantly from each other. Reduced amounts of NAA, Cho and Cr were observed and various amounts of mobile lipids (high intensity of the resonance at 1.3 ppm from $-(CH_2)_n$ and at 5.3 ppm from -CH=CH-). Analysis of histological characteristics suggested that the samples with high amounts of mobile lipids contain more cellular necrosis that samples with low amounts of these lipids. Since necrosis is a prognostic factor in malignant astrocytomas, these lipids are likely to have prognostic significance. Mobile lipids have been observed in tumors but not in normal tissue of the brain.

High-resolution ^1H spectra of brain tissue extracts at 600 MHz (14.1 T) allowed the separate detection of several signals contributing to the *in vivo* resonances; for example the *in vivo* Cho peak between 3.14 and 3.35 include signals from $N^+(CH_3)_3$ of glycerophosphocholine GPC, phosphocholine PCho, choline, H5 of inositol and NCH_2 of taurine [29]. The analysis of signals enabled discrimination between well-differentiated astrocytomas and glioblastoma multiform samples to be performed on the basis of the ratio between the integrated choline-containing resonance and the creatine peaks (creatine + phosphocreatine). Water-soluble metabolites extracted from 60 surgically excised samples of various brain tumors and of 4 normal brains were measured quantitatively using *in vitro* high-resolution ^1H NMR [30]. The concentration of each metabolite in normal brain in μmol/100 g was as follows: NAA 555.6, total creatine 1076.3, choline-containing compounds 64.3, glycine 82.2, taurine 152.3, alanine 91.9, inositol 174.3 and phosphorylethanolamine PEA 115.3. The concentrations of metabolites were determined in nine different tumors. The NAA level was very low in some areas of glioblastoma that showed loss of neurons and axons. The concentration of total creatine in astrocytoma was about 15-40% of that in normal brain and decreased according to the degree of malignancy. A choline peak is associated with the species involved in membrane turnover, and the concentration of choline-containing compounds increased in proportion to the grade of malignancy but decreased according to the extent of necrosis in glioblastomas. Glycine increased remarkably in glioblastoma and might be useful for the differential diagnosis of glial tumors from metastatic brain tumors. Large amount of taurine was detected in medulloblastoma, taurine levels might be the expression of increased cellular proliferation and tumoral aggressiveness. High alanine content was detected in meningioma and glioblastoma. Inositol was increased especially in neurinoma (the inositol peak consists of four compounds: phosphatidylinositol, inositol polyphosphatides, inositol monophosphate and myo-inositol); myo-inositol is probably important for cell growth and is found in high concentrations in the nervous system. The reason for the high inositol content in neurinoma is unknown; nevertheless, the prominent peak of inositol is useful for diagnosis of neurinoma. It can be concluded that ^1H gives valuable information on tumors for their differential diagnosis.

In future, *in vivo* MRS might provide data on tumor metabolism and the response to therapy [31]. First such studies are reported below.

Synthetic ether lipids are a new class of antitumor agents, alkyllysophospholipid (edelfosine) and hexadecylphosphocholine (miltefosine) show pronounced antineoplastic activity *in vitro* and *in vivo*. They are cytotoxic against tumor cells and also are able to induce programmed cell death (apoptosis) in certain tumor cells. Ether lipid derivatives act via cell membranes. It was interesting to observe the membrane composition and metabolism of treated cells. Tumor cells may react to therapy by necrosis, apoptosis, termination of proliferation or do not react at all. The use of ^{1}H, ^{13}C, ^{31}P MRS of cell extracts may give insight into metabolic pathways of tumor cells affected by drug treatment. Two different tumor cell lines were used to investigate the effect of chemotherapy with miltefosine [32]; the human epithelial tumor cell line KB was a model for a responder to treatment and the C6 glioma cell line of rat, in which no cytotoxic action was observed. The uptake of miltefosine was followed by the ^{31}P NMR of lipid extracts. The most pronounced difference was the high content of β-acyl-γ-alkyl-phosphatidylcholine (PCa) in KB cells; membranes of C6 cells contained higher amounts of ethanolamine-phospholipids. After drug treatment in the membranes of KB a marked decrease of PCa was observed. ^{13}C NMR was used to observe phospholipid metabolism (the cells were incubated with 1-^{13}C-glucose for 2 h before extraction with perchloric acid) and signals of several lipid species (phospholipids, neutral lipids, fatty acids, cholesterol) present in the membrane could be identified. Treatment with miltefosine markedly altered the spectra. The signals of neutral lipids triacylglycerol TAG and diglycerol DAG and also those of fatty acids increase, whereas the signals of the labeled carbons of the glycerol backbone of phospholipids decreased. Thus, during the first hours of treatment, a stop in glycogen synthesis along with marked activation of glucose metabolism and pronounced fatty acid synthesis was observed. The TAG and DAG increase seemed to be an early marker of miltefosine-induced apoptosis. The results show that multinuclear NMR is an excellent method for investigation of tumor cells and helps to evaluate the mode of action of an anticancer drug at molecular level.

The effect of gossypol on cultured TM3 Leydig and TM4 Sertoli cells was studied by means of ^{31}P and ^{23}Na NMR [33]. Gossypol, a natural bis-sesquiterpene produced by the cotton plant, was tested earlier as an antispermatogenic agent. However, due to its effects on cellular energy metabolism it is a potential anticancer drug. Gossypol is optically active; the *in vivo* antispermatogenic activity is largely associated with (-)-gossypol. Relative ATP levels and intra- and intercellular sodium levels were determined by ^{31}P and ^{23}Na NMR. In order to observe the toxic effects of gossypol, perfusion experiments with a medium containing 0–100 μM of the compound were carried out. However, relatively high concentrations of gossypol were required to cause short term effects; acute toxicity was observed with a concentration corresponding to 40·IC50, TM3 cells being more sensitive than the TM4 cells. Incubation with a low gossypol concentration markedly affected the energetic status of the cells (increase in the level of intracellular sodium, decrease of ATP level), especially of the TM3 cells. Since the sodium pump is the major energy consumer in the cell, the observed increase in the intracellular sodium level may be a direct consequence of the decreased level of ATP, detected by ^{31}P NMR. The authors [33] conclude that inhibition of energy production in somatic testicular tissue is rather unlikely to be a major cause of the antispermatogenic effect of gossypol.

Colorectal cancer is an important public health problem because of its high incidence and high rates of recurrence after resection. These facts prompted a study [34] with human biopsies in order to find malignancy markers useful in diagnosis and monitoring of the *in vivo* response to therapy. Biopsies were examined *ex vivo* and as perchloric acid extracts by ^{1}H NMR spectroscopy at 9.4 T. Analysis of extracts demonstrated significant changes in the concentration of the endogenous compounds. Among these metabolites, the high myo-inositol and taurine levels are interesting as possible malignancy markers. The observation of taurine in one-dimensional spectra of intact biopsies was easy; however, for the detection and quantification of myo-inositol the use of two-dimensional COSY spectra was necessary. In the two-dimensional spectra the use of a ratio between the cross-peak volumes of both metabolites permits an excellent differentiation between tumors and normal mucosa.

^{31}P NMR spectra can reflect the energetic status of cells (quantitation of ATP, PCr, Pi); the chemical shift of Pi depends on the intracellular pH, which can be measured in this way [35]. By use of ^{13}C-labelled substrates, the kinetics of phosphatidylethanolamine synthesis has been studied by ^{13}C NMR spectroscopy of extracts of lymphomatous mouse liver following the administration of ^{13}C2–ethanoloamine [36].
The effects of various concentrations of 2-fluoro-2-deoxy-D-glucose (FDG) on the aerobic metabolism of glucose and the reciprocal effect of glucose on the metabolism of FDG in glucose-grown repressed *Saccharomyces cerevisiae* cells were studied by ^{1}H, ^{19}F and ^{31}P NMR [37]. The medium contained $5 \cdot 10^7$ cells/ml. The glucose consumption rate was reduced by 57% and 71% in the presence of 5 mM and 10 mM FDG, respectively; ethanol production rate also decreased. When FDG is the unique carbon source, the α- and β-anomers of 2-fluoro-2-deoxy- D-glucose-6-phosphate and a small amount of 2-fluoro-2-deoxy- D-gluconic acid were observed. Apart from trehalose no other disaccharide was observed in the spectra.

The myometrial, decidual, placental and fetal membrane phospholipids were studied by ^{31}P NMR [38]. The tissue samples were pulverized and extracted. The signals of phosphatidylcholine PCho, phosphatidylethanolamine, sphingomyelin SM and phosphatidylinositol were found. The ratio PCho/SM decreased during pregnancy in the decidua, placenta and fetal membranes but not in the myometrium. The fetal tissues, fetal membranes and placenta contained about twice as much phospholipid as the maternal tissues.

Perfusion of skin with transdermal-penetration-enhancing agents such as ethanol, DMSO-d_6 and propylene glycol was studied by *in vitro* ^{31}P NMR [39]. Epidermal strips from abdominal pig skin were placed in a 10 mm O.D. NMR tube modified for continuous perfusion with buffered salt solution and a serial spectra were recorded on a Bruker AMX500 spectrometer. Signal intensities for phosphomono- and di-esters PME and PDE, phosphocreatine PCr, inorganic phosphate Pi and nucleotide triphosphorate, β-NTP were followed in time. Additional spectra were recorded when the perfusion medium contained dexamethasone. The dexamethasone perfusion resulted in a dose-dependent decrease in PCr and NTP levels and had an effect on PME metabolism.
The quality of the ^{31}P spectrum of epidermis encourages further studies. Investigations of drugs applied on skin as well as skin care cosmetics, which are of importance for the pharmaceutical industry, may be carried out.

13.5 Multinuclear Preclinical Measurements in Animals

MRI and MRS are increasingly applied in preclinical pharmaceutical research. The MRI technique is noninvasive and its advantages compared with invasive pharmacology are obvious. As mentioned above, MRS enables important endogenous metabolites to be determined and gives an insight into tissue metabolism. Therefore the morphology and physiology of the healthy and diseased body can be studied as well as the action of drugs. Horizontal magnets with larger bore are needed for the studies on animals. Lower magnetic field strength and larger bore result in a decrease in sensitivity and spectral resolution in comparison to the tissue studies *in vitro*. Magnetic field gradients in combination with spin-echo sequences are used for localization of the volume of interest. The most frequently used method is ^{1}H MR, because it uses the best nuclei for detection; however, in MRS studies one has to suppress the dominating water signal, which is significantly stronger than the signals of metabolites.

Several human diseases have been studied on animal models. MRI and MRS have been applied to study the status of isolated perfused pig hearts and brains *in vivo* in order to develop better methods of protecting these organs during surgery on the heart and large blood vessels. Similar techniques have been also used on tissue isolated from human hearts to estimate optimum conditions for the preservation of hearts destined for transplantation. Using ^{31}P MRS, the effects of preservation conditions on metabolite (ATP, PCr and Pi) levels and on intracellular pH have been monitored [40]. Very interesting was the experiment in which the consequences of surgery on the heart were followed by imaging the brain (an anesthesized pig was in the magnet). MRI with contrast agent has been used to assess the efficacy of tissue perfusion provided to the heart and brain through the antegrade (normal) and retrograde routes of warm blood supply.

The usefulness of MRI and MRS for the development of drugs for cerebral ischemia was reported in 1992 [41]. MRI and MRS experiments were carried out on a Biospec Bruker spectrometer with a 15-cm horizontal-bore magnet; localization was achieved using a surface coil. The techniques have been routinely used since 1985 in the Preclinical Research laboratory of Sandoz Pharma Ltd; with a daily throughput of 35–40 rats. The cross-sections through the head of a rat 24 h after occlusion of the left middle cerebral artery showed the infarcted area. Determination of the infarcted volume gave information concerning the drug effects. It has been found that a dihydropyridine calcium antagonist reduces the size of infarction when given up to 6 h after onset of stroke. The maximal effect was obtained with an immediate subcutaneous injection of 2.5 mg/kg isradipine. A putative mechanism of action of calcium antagonist leading to cytoprotection may be the improvement of collateral blood supply to the ischemic areas. The effects of those calcium antagonists which easily cross the blood-brain barrier were also observed on cerebral energy metabolism. In isradipine-treated rats the brain ATP levels decay after global ischemia at a slower rate than in controls; the effect is due to the anaerobic glycolysis. The level of ATP after ischemia showed correlation with pH and the lactate level. Isradipine and related dihydropyridine calcium antagonist seem to reduce the ATP consumption in brain, resulting in higher ATP levels during ischemia.

The cell depolarization connected with a transient shift of water from the extracellular to the intracellular compartment occurs during the evolution of infarcts; it is observable as a decrease of the diffusion coefficient in a diffusion-weighted image (DWI). The technique of DWI was applied to the observation of therapeutic intervention during the acute ischemic phase [42], [43]. The effect of the sodium-calcium ion channel modulator RS-87476 was studied [43] in a cat model; the infarct size was reduced by up to 88% when the drug was applied 15 min post-occlusion. Levelopamil, calcium and serotonin antagonist, applied either pre-occlusion or within 2 h after, led to smaller lesion size during the acute phase and reduced the area of edema by ca. 20% [44]. The effect of mannitol on focal cerebral ischemia was studied in cats [45]. Mannitol is known to reduce intracranial hypertension, the images recorded 2 h and 24 h after vascular occlusion showed reduction of an infarct area in the treated group in comparison to controls, and the beneficial effect was attributed to improvement of cerebral blood flow.

Whereas considerable effort is being devoted to developing treatments for ischemic stroke, much less is directed to hemorrhagic stroke. The development of therapies relies on the use of an animal model. A reproducible method of causing bleeding into the brain tissue was developed in which intrastriatal injection of bacterial collagenase disrupts the basal lamina of cerebral capillaries. MRI was performed [46] on rats anesthetized with pentobarbital; rats received collagenase or a collagenase/heparin combination. The injection into rat striatum gives rise to a rapidly forming hepatoma of uniform shape and reproducible size; much of the edema is confined to white matter. The effects in T_2-weighted MR images were attributed to hemoglobin degeneration. As the erythrocytes (containing paramagnetic deoxyhemoglobin and methemoglobin) lyse, the release of methemoglobin causes increased T_2, producing isointensity or hyperintensity on the images. The macrophages ingest extracellular iron and form hemosiderin, and the T_2 in the surrounding tissue decreases. The MR images clearly showed that widespread white matter edema accompanies the hemorrhage and resolves over about 1 week. Brain injury may be due not only to the space occupying effect of hepatoma, but also to the intense inflammatory response and to ischemic injury at the periphery (due to tissue compression and generalized edema). MR imaging can be used to investigate development and resolution of injury following treatment with anti-inflammatory agents, antioxidants, inhibitors of free radicals or neuroprotective drugs [47].
Excitotoxicity is considered as important factor in the pathogenesis of several cerebral disorders. Persistent activation of excitatory amino acid receptors, which are linked to ion channels, initially results in a shift of ions and water causing cellular swelling; long excitation of receptors will lead to neuronal damage and death. The status of the neonatal rat brain was studied after injection of the excitotoxin *N*-methyl-D-aspartate (NMDA) [48]. Rats were anesthetized with ether and immobilized in a 4.7-T horizontal magnet. Anesthesia was accomplished by continuous ventilation with a mixture of N_2O, O_2 and halothane through a face mask. A small hole was drilled in the skull and NMDA was injected in the striatum, and the needle was then replaced by a catheter for remote administration of the non-competitive NMDA-antagonist MK-801 (dizocilpine) or the competitive NMDA-antagonist -CDPene (D-(E)-4-(-phosphono-2-propenyl)-2-piperazine-

carboxylic acid). The injection resulted in a decrease in the apparent diffusion coefficient of brain tissue water, as measured by with diffusion-weighted MRI. The early diffusion changes were accompanied by only mild changes in the overall metabolic status as measured by ^{1}H MRS (single-voxel-localized, with PRESS sequence), ^{31}P MRS and subsequent metabolic imaging (rats were rapidly immersed in liquid N_2, and the brains were cut into small sections). Minimal decreases in the high-energy phosphate levels and small acidosis with modest lactate accumulation were observed in the first 6 h after injection. An overall decrease in ^{1}H MRS-detected metabolites was found after 24 h. Treatment with the non-competitive antagonist within 90 min after NMDA injection rapidly reversed the NMDA-induced changes in the entire ipsilateral hemisphere. The effect of the competitive antagonist was restricted to the cortical areas and was accomplished in a slowed time scale. The results indicate that early excitotoxicity is accompanied by only mild metabolic changes and can be entirely normalized. Thus, excessive activation of EAA receptors alone does not automatically lead to acute and severe irreversible neuronal damage. It is worth noting that such interesting results could be obtained by the combined application of various NMR techniques, notably diffusion-weighted MRI and both ^{1}H and ^{31}P MR spectroscopy.

The multi-slice technique of ^{1}H MRI was used to study regional effects of amphetamine on rat brain [49], a two-coil system enabling three-dimensional perfusion imaging. Amphetamine caused a significant increase in perfusion in many areas of the brain, including the cortex, cingulate and caudate putamen.

Interesting investigations of drug distribution *in vivo* were possible using ^{19}F NMR imaging. With ^{19}F MRI/MRS, the uptake, distribution and elimination of sevoflurane (fluorinated anesthetic) of rat brain was observed [50]. Based on T_1 relaxation time measurements, the sevoflurane signals were separated into two components, attributable to sevoflurane in a mobile and immobile microenvironment. There was a delayed accumulation of the anesthetic in the mobile microenvironment (adipose) and a rapid elimination from the immobile one (brain tissue). At anesthetizing concentrations sevoflurane distributes heterogeneously in the brain.

The supply of oxygen is crucial to the health of tissues but is impaired in numerous diseases. NMR provides an alternative, noninvasive approach for monitoring pO_2. ^{31}P NMR of endogenous high-energy metabolites can be used to infer hypoxia, and ^{1}H MRI, based on BOLD method, can follow tissue contrast changes. However, a linear relationship was found between the ^{19}F spin-lattice relaxation rate of fluorocarbons and pO_2, and hexafluorobenzene was considerably more sensitive to the changes in pO_2 then other fluorochemicals. The method was tested on rats under anesthesia, placed in the horizontal bore of 4.7-T magnet [51]; hexafluorobenzene (20 µl) was injected into the tumor for oximetry study. ^{1}H and ^{19}F images were obtained and the precision of the study should facilitate future investigations on hypoxia.

A well-known radiopharmaceutical 2-fluoro-2-deoxy-D-glucose (FDG) is widely used for positron emission tomography (PET) diagnosis of brain activity. It was also evaluated as an NMR pharmaceutical for cancer detection [52]. ^{19}F chemical shift images of FDG and its metabolites were obtained at 376 MHz, the mice with MH134 hepatoma having been placed under pentobarbital anesthesia in the bird-cage type RF coil. Prolonged retention of DFG and its metabolites over 2 days was observed in the tumor cells and in the heart. The 6-phosphate of DFG was converted reversibly to its epimer the 2-deoxy-2-fluoro-D-mannose in these tissues. The authors conclude that DFG could be used as an ^{19}F NMR pharmaceutical for tumor diagnosis for both *in vivo* MRS and imaging since the signal of 2-deoxy-2-fluoro-D-mannose can be used as a marker of tumor.

Fluorinated representative of antifolates (C2-desamino-C2-methyl-N10-propargyl-2′-trifluoromethyl-5,8-dideazafolic acid), which act as inhibitors of thymidylate synthase and are being evaluated for the treatment of human cancer) was followed *in vivo* in mice. ^{19}F images were measured from the abdomen of mice after intravenous injection (500 mg/kg). A time resolution of 4 min was obtained for two-dimensional and 20 min for three-dimensional imaging. The drug was found in gall and urinary bladders [53]. The results confirm the potential value of ^{19}F MRI in pharmacokinetic studies.

Fast ^{19}F MRI with FLASH technique was applied *in vivo* and the imaging time amounted to only 470 ms [54]; this sequence enabled a pharmacological study to be performed on the clearance of perfluorooctylbromide in the internal organs of rats [55].

MRI of moving organs was possible using a modified navigator-echo sequence. The heart rate is more than 300 beats per minute. It was possible, however, to synchronize the MR data acquisition with the electrocardiogram and obtain high resolution images of the rat heart with pixel dimensions of 0.2·0.2 mm^3 [56]. Using this technique the beneficial effect of the angiotensin-converting enzyme inhibitor spirapril on the established left ventricular hypertrophy in rats [57] was demonstrated.

The MRI studies on estradiol-induced pituitary hyperplasia [58] and transplanted Dunning prostate tumors showed that the somatostatin analog Sandostatin significantly reduced the tumor size in rats [59].

^{1}H MRI monitored the progression of adjuvant arthritis in rat leg joints and its response to indomethacin treatment. The total hydrogen content of a defined volume was measured quantitatively in time. The differences in relaxation times allowed for characterization and quantitative separation of the fluid (long T_2) and non-fluid (relatively short T_2). The estimates of hydrogen content in both tissue components were used to assess the severity of the disease and its regression due to treatment. After 19 days of treatment, indomethacin reduced the fluid component (primarily from inflamatory edema). The method has the sensitivity required to assess the treatment efficacy in adjuvant arthritis. The authors proposed clinical MRI monitoring of therapy of arthritis in humans [60].

The *in vivo* activity of phosphate-activated glutaminase (PAG) was measured in the brain of a hyperammonaemic rat by ^{15}N NMR [61]. Brain glutamine was enriched in ^{15}N by intravenous infusion of ^{15}NH$_4^+$, further glutamine synthesis was stopped by injection of methionine-DL-sulphoxime and the infusate was changed to ^{14}NH$_4^+$. The decrease of ^{15}N-glutamine, the production of ^{15}NH$_4^+$ and its assimilation into glutamate were monitored in the brain. The *in vivo* PAG activity was lower than that *in vitro*. The result suggested that intact brain PAG activity is maintained at a low level by a suboptimal *in situ* concentration of Pi and the strong inhibitory effect of glutamate and also that PAG-catalyzed hydrolysis of glutamine is not the sole provider of glutamate used for γ-aminobutyrate (GABA) synthesis.

Silicones were assumed to be biocompatible and have been extensively used in plastic and reconstructive surgery (breast augmentation). However, these materials deteriorate and toxic effects from implants begin to appear. In order to study the ageing process of silicone in a living object, ^{1}H chemical shift-imaging has been developed [62], which allows selective mapping of silicone protons *in vivo*, while suppressing the contributions of water and fat. A rat with a surgically placed silicone implant exhibited three resonances: water protons (at 4.7 ppm), fat protons (1.5 ppm) and protons of the methyl groups of the silicone structural unit $-Si(CH_3)_2-O-$ at 0.35 ppm. The sensitivity of the experiment allows the detection of chemically unchanged silicone concentration of 5% in a voxel of 0.9 mm^3. Although the authors were not able to detect any silicone migration in the tissue around the implant during the implantation time studied, the *ex vivo* high-field multinuclear NMR studies indicated the presence of silicone in tissues and the migration of lipids into the implants. The ^{29}Si MRS studies [63] have shown that there is a detectable amount of silicon compounds in the blood of women bearing implants, thus providing evidence of the biotransformation of silicone.

Pharmacological applications of multinuclear MRS/MRI are in accordance with the new trends towards non-invasive techniques that reduce the number of animals tested. Because „identical" rats or mice are not in fact identical, the bioassays have a statistical nature and require a large number of animals. Usually, numerous animals must be sacrificed for one pharmacokinetic study where data should be collected in time, as results obtained on small populations contain uncertainties because of the individual variability of living organisms. The activity of a drug, its distribution and elimination could be more reliably determined using even a single animal but monitoring *in vivo* using MRI and MRS.

The authors are of the opinion that the techniques of magnetic resonance *in vivo* will be increasingly employed and are particularly important for the testing of new and/or potential drugs in preclinical research in the pharmaceutical industry.

13.6 MRI and MRS in Humans

13.6.1 Studies of brain.

In vivo MR has been extensively used in the study of the human brain, its functions and metabolism. MRI has a significant role to play in our understanding of how the brain works. In the last few years the most interesting application of MRI is the construction of functional maps of the brain (functional MRI, fMRI). In response to brain activation, an increase in the proportion of oxyhemoglobin with respect to the deoxyhemoglobin takes place.

The changes in regional cerebral blood flow following task activation have been studied [64], [65]. Single motor events were monitored by asking healthy volunteers to push a button once in response to displayed „5" (go) and to do nothing when „2" (no go) was presented [66]. The three-dimensional activation data were collected as sets of 12 contiguous slices every 3 s and were superimposed on reference images. Apparently, activation was observed during a „go" response in the primary motor area, but differences were also discerned in the supplementary motor area [67]. In these single motor event studies, there is a delay of several seconds following the button push before the activity is detected. This is due to the latency of the hemodynamic response, which is related to the transit time of the blood through the capillary network. However, despite the temporal limitation of the blood flow response it is possible to use functional MRI to study cognitive processes associated with single events. The ability to study various components of a cognitive and motor network was enhanced by the application of cluster algorithms [68]. The identification of groups of pixels, which shared a common time course, eliminated the need to rely on the image differencing. It enabled the observation of how the brain responded to increasing cognitive complexity in a series of motor tasks to be made [69]. Earlier studies of thebrain overestimated the selective localization of a specific function in a particular tissue. More recently, it has been recognized that different parts of the brain activate in a large number of situations and that subtle changes in the pattern of activation underlie different processes.

Functional MR was applied mainly to the studies of the auditory, sensory, motor and visual realms of the brain. Experiments on the response of the brain to intestinal sensation and to pain are in progress [70]. It is obvious that human brain mapping should include studies on the effect of drugs: stimulating, seductive, anxiolytic etc. It opens a new research field of great importance for pharmacy.

Little is known about how the human brain encodes smells and odor-associated emotions. The studies are complicated because the area of cerebral activation is adjacent to bone and air-containing interfaces and, additionally, because odorants can be also mediated by the trigeminal nerve. Five right-handed men underwent studies on a 1.5-T system (multisection, gradient-echo, echo-planar imaging with a BOLD experimental paradigm); the T_2-weighted MR images were acquired [71]. Two sets of odorants were used on two different days 1 week apart. For the first week the olfactory system-mediated odorants such as eugenol, geraniol and methyl salicylate were used. The olfactory and trigeminal nerve-mediated odorants used for the second were ylang ylang, patchouli and rosemary oil. Each subject underwent two separate, identical imaging series for each set of odorants. The two imaging series were spaced 5 min apart without changing imaging variables. Overall time in the imager was approximately 35 min per subject for each session. The olfactory system-mediated odorants produced extensive activation mainly localized in the orbitofrontal region and the cerebellum; the second exposure to the same odorants showed diminished activation relative to that in the first experiment. The paradigm with olfactory system-mediated and trigeminal nerve-mediated odorants showed wider activation of different areas of the brain (visual, precuneate, temporal, cingulate), and the second exposure produced amplification of the stimulation. Orbitofrontal activation was predominantly right-sided. The results suggest that as olfactory-mediated and trigeminal nerve-mediated odorants persist they become more irritating, whereas the pleasant odorants tend to have less of an effect with time. The sense of smell is common in the animal kingdom and is localized in a portion of brain thought to develop early in the phylogeny. The reaction to smell may be an evolutionary adaptation. The perception of danger from the unpleasant (trigeminal) odors leads to increased overall activation. The subject searches for escape, and muscular coordination and anticipation regions are activated. There is large variation in odor perception between men and women, and between young and old subject and the problem needs deeper investigation.

The studies on odor perception may be important mainly for the cosmetic industry but also for the pharmaceutical industry in developing suitable odor additives to a drug (a pleasant smell could play a role in drugs administrated to children). As one moves from healthy volunteers to specific disease states in which olfaction is affected, one encounters a further large potential for MR imaging in odor stimulation studies. There have been a number of articles that have noted a decrease in olfactory function in patients with various neurodegenerative processes [72], [73] such as Alzheimer's disease, Parkinson's disease (the loss of olfaction is unresponsive to pharmacologic intervention), schizophrenia and multiple sclerosis. However, a study of these disease states with odor-stimulated functional MR imaging requires extensive healthy control subject studies across age and sex.

Localized MR spectroscopy and MR spectroscopic imaging have been employed to demonstrate an altered biochemical composition of brain tissue in various diseases (tumors, stroke, Alzheimer's disease, AIDS, epilepsy, multiple sclerosis and brain injuries). Major metabolites detected by ^{1}H included: N-acetyl compounds, primarily N-acetylaspartate (NAA, a neuronal marker), glutamine and glutamate (GLx), phosphocreatine and its precursor creatine (Cr, bioenergetic metabolites), choline-containing compounds (Cho, comprising free choline, phosphoryl and glycerophosphoryl choline, released during membrane disruption), myo-inositol and lactate (accumulates in response to tissue damage in anaerobic metabolism). In the spectra of ^{31}P the most intense signals are from: phosphomonoesters (PMEs), phosphocreatine (PCr), α-, β- and γ-adenosine triphosphate (ATP), and inorganic phosphate (Pi).

However no clear-cut data were available concerning the possible metabolic profile asymmetry in a healthy subject. It was, therefore, necessary to establish normative data for volumes of the temporal lobes and hippocampal formations according to sex and lateral dominance. Twenty-eight healthy adults underwent MR imaging (FLASH, 2D sequence) and single voxel spectroscopy (STEAM sequence), 17 right-handers and 11 non-right-handers [74]. The single voxel (8 mL) was successively positioned within the right and left temporal lobes on the mesial side. A line width of 4–6 Hz was achieved; the water resonance was suppressed. The total time of examination was 45 min per subject. The MR-imaging morphometry and ^{1}H spectroscopy revealed an asymmetry not only in volumes but also in spectroscopic data in the temporal lobes. The volume of hippocampal formations was larger on the right in right-handers and in non-right-handers. The metabolic profiles are influenced by the lateral dominance. NAA/Cho was significantly higher in the left temporal lobe of right-handers. One can hypothesize that the asymmetry is due to the asymmetric distribution of the NAA and that neuronal density is different in right and left structures. The asymmetry is statistically significant in right-handers, with more NAA in the left side (devoted to language and verbal memory functions and dominant motor activities) than in the right nondominant side (devoted to visual memory). Left-handers form a heterogeneous population when considering the functional dominance. It is of paramount importance that these differences be taken into account when analyzing clinical data of patients with cerebral disfunction.

failure can be used as a prognostic indicator of a later outcome. In developmentally normal children age-related changes in the metabolite concentrations were observed by ^{31}P MRS [77] and ^{1}H MRS [78]. The changes were related to the maturational processes in the brain. Most neurons are formed before term birth; however, after birth there is an important increase in the number and size of neuronal dendrites and a progressive differentiation; the major portion of brain growth is myelination. The ^{31}P spectra of the selected volume in the paraventricular region (predominantly white matter) were obtained [77] and quantified by peak area measurements. The investigations were performed in 41 healthy children aged 1 month to 16 years, without sedation. With advancing age, phosphorus spectra revealed a decrease in the ratios of phosphomonoesters (PMEs) to β-adenosine triphosphate (ATP) and PMEs to phosphocreatine (PCr) and an increase in the ratios of phosphodiesters to β-ATP, PCr to β-ATP, and PCr to inorganic phosphate (Pi). The most rapid changes were observed during the first 3 years of life. ^{1}H MR spectroscopy was also performed [78] in a 1.5-T whole body imaging system in healthy children of different age. A plot of the metabolite ratios versus age showed that NAA/Cr and NAA/Ch ratios increased rapidly in the neonatal period, continued to rise during infancy (1–18 months) and began to level off at 18–20 months, whereas the Ch/Cr ratio decreased and reached a plateau at 18–20 months.

The findings of MRI and MRS can be useful in predicting outcome in neurologically impaired infants. More than 150 infants and children with brain injuries were studied by ^{1}H MRI and MRS [78]. The patients were medically stable; sedation with chloral hydrate (40–80 mg/kg) or midazolam (dose 0.1 mg/kg) was provided as needed. The MR images were used to define the optimal place for ^{1}H MRS examination. Localized water-suppressed spectra were obtained using the STEAM sequence. The spectra were acquired in a 8-cm^3 region of the brain in the occipital region (primarily gray matter) and in the parietal region (primarily white matter). The patients underwent MRS at different times relative to their injury; this affected the detection of lactate and the magnitude of metabolite ratio changes. MRS performed 36 h after an accident showed a small lactate peak and slightly reduced NAA/Cr ratio; the same study repeated 4 days later showed a large lactate peak with NAA/Cr ratio significantly reduced (the patient remained in a permanent vegetative state). The most important was the presence of lactate; this level in patients with a poor outcome was markedly higher than the level in patients with good outcome. The usefulness of MRS was clear in predicting the long-term neurological outcomes in children after hypoxic-ischemic brain injury such as cardiorespiratory arrest.

Proton MRS was used to laterize the seizure focus in temporal lobe epilepsy [79]. Sixteen patients underwent MR imaging in a 1.5-T system and MR spectroscopic imaging [using a 2D imaging sequence with point-resolved (PRESS) volume selection]. The unsuppressed water signal intensity was used as an internal standard and metabolite concentrations (in mmol/L) were calculated. Lateralization using MRI was performed with the *N*-acetylaspartate/(creatine+choline) ratio and with the absolute quantitation of NAA concentration. In the patients, the NAA concentration or the ratio of the NAA to the sum of creatine and choline were substantially decreased in the ipsilateral hippocampus. There was no difference in the concentration of choline between patients and control subjects, and a trend towards decreased creatine concentration was found. It was interesting to check whether decreased hippocampal NAA concentration is simply due to hippocampal atrophy. The patient groups showed slightly increased water signals (7.9%) probably resulting from an increase in the amount of cerebrospinal fluid. However, the water signal increase accounts for only 6.5% of the 34.2% NAA concentration decrease.

MR spectroscopic results were frequently expressed in terms of peak ratios; however, this resulted in doubt about whether one metabolite is elevated or another decreased. Absolute concentration values can eliminate such problems; this is especially important when metabolite changes are studied in disease process. The advantage of quantitating MRI data by using water signal *in vivo* is that the water signal and metabolite signals are obtained under the same conditions.

^{1}H MRS and MRI studies were frequently devoted to differentiating *in vivo* normal ageing of the human brain from age-associated memory impairment and dementia of Alzheimer's type. ^{1}H MR spectroscopic imaging demonstrated differences in brain metabolites between patients with Alzheimer disease (AD) and elderly control subjects. Significantly lower N-acetylaspartate concentration and higher inositol concentration were observed in pathological conditions [80]. AD and subcortical ischemic vascular dementia (SIVD) are associated with measures of cerebral atrophy. MR imaging studies suggested that brain volume loss in patients with AD is due to the loss of gray matter. Nine patients with AD and eight with SIVD were studied on a 2-T MRI/MRS system [81]. Immediately after imaging, a point-resolved spatially localized spectroscopic (PRESS) volume of interest was selected for 2D ^{1}H imaging. The voxels (size of approximately 2.2 mL) were selected from the midline of the brain to include as much gray matter as possible. Significantly lower NAA/choline-containing (Cho) metabolites and higher Cho/creatine-containing metabolites in posterior mesial gray matter in AD versus control subject were found. Combined measures allowed correct classification of AD and control subjects; however, none of the MR measures allowed accurate discrimination between AD and SIVD subjects. In the next study [82], in which similar ^{1}H MRI/MRS methods were used, the observation was extended to a larger population of patients. In agreement with previous results, NAA levels were statistically significantly reduced in the frontal and posterior mesial cortex of AD patients, presumably due to the neuronal loss. NAA is specifically located in neurons but is absent in glia. In the presence of gliosis, ^{1}H spectroscopic imaging may be a more specific marker of neuronal loss than atrophy observed with MRI. NAA level reductions were mildly correlated with dementia severity.

The majority of patients infected with the human immunodeficiency virus (HIV) have abnormalities in the central nervous system. It is frequently possible to provide a specific diagnosis on the basis of abnormalities seen by MR imaging; the diagnosis is difficult in the presence of focal abnormalities with mass effect, as for example the differentiation between toxoplasmosis and cerebral lymphomas. Some clinicians recommend that all patients with AIDS and brain masses should first receive antibiotics for toxoplasmosis [83]. If improvement does not occur after medication, biopsy is considered. The most accurate diagnosis could be achieved by brain biopsy, but this can lead to substantial morbidity and mortality. Any technique that allows earlier diagnosis would enable earlier commencement of appropriate therapy. This is of particular importance in the case of lymphoma because untreated mean survival is short whereas radiation therapy and steroids may improve survival. ^{1}H MRS showed significantly different biochemical profiles for AIDS-related brain lesions in 26 patients, which helped in correct diagnosis [83]. HIV-positive patients (109) were found to have focal intracranial lesions [84]; 56 of these patients also underwent MR spectroscopy. MRI/MRS was performed with a 1.5-T imager with a standard, circularly polarized head coil. ^{1}H MR spectra were acquired at an echo time of 135 ms (before administration of gadopentetate dimeglumine as contrast agent); the voxel size was from 2.8 to 8.0 mL. The spectra were compared with those typical of normal parieto-occipital white matter in the sero-negative control group. Abnormal metabolite ratios were found in all the lesion spectra. There were three broad categories of spectra: one type showed very large lipid peaks with relative suppression of all other metabolites, another showed elevation of the Cho/Cr ratio with diminished NAA, and the third category was less coherent. Unfortunately, neither visual analysis nor fitting of metabolite ratios obtained from spectra of intracranial mass lesions is sufficiently specific to differentiate between lymphoma and toxoplasmosis in HIV infection. In tumors with increased cellular synthesis the concentrations of Cho-containing compounds are increased; the neuronal damage caused by *Toxoplasma gondii* should result in spectra with a diminished NAA/Cr ratio; the lipid peak was also expected because of degradation of phospholipids and sphingomyelin. However, this hypothesis seemed too simplistic because an overlap between the lesion types exists. *T. gondii* in the brain causes a necrotic lesion with inflammatory response followed by abscess formation. Lymphoma in AIDS patients is more aggressive and central necrosis is common. Necrotic lymphomas and necrotic abscesses have similar composition, which explains why the spectra look similar. Another confusing variable is the relative maturity of the lesion; each lesion has a necrotic center and an active edge tissue. The most suitable voxel size and position are not evident. A voxel placed at the center of a large lesion may sample only necrotic tissue, whereas a voxel placed at the edge may contain a mixture of tumor and brain parenchyma. It is possible that absolute molar quantification of MRS may enhance diagnostic specificity.

^{1}H MR spectroscopy provided useful information on brain tumors [83], [84], [85], [86], [87]. MRS contributed to the biochemical characterization of different oncotypes; furthermore, metabolic changes induced in the lesion by tumor growth may be followed and should improve diagnostic and therapeutic strategies. The differentiation of neoplastic from non-neoplastic processes in the brain was possible [85] by analysis of metabolite changes in the spectra as compared with the levels in a healthy subject. In patients with cerebral tumors the Cho/NAA ratio was increased and was greater than 1 whereas for a control subject the average ratio was 0.54. The elevation of the Cho peak suggests increased cell membrane synthesis, characteristic of neoplasia. MR spectroscopy may be used to confirm that the lesion is neoplastic before biopsy or demonstrate a low likehood of neoplasia. ^{1}H MR spectra permitted differentiation of brain abscess from necrotic or cystic tumors [86]. Before MRS, T_1-weighted or T_2-weighted images were obtained to define a 2·2·2 cm volume of interest. The spectra were recorded by using a point-resolved sequence with an echo time of 270 ms. The water signal was suppressed in the standard spectra, but, for quantitative analysis, the integrals of metabolite peaks were compared to the integral of the unsuppressed water peak. In the cases of abscesses multiple resonance peaks were observed, attributed to lactate, amino acids (valine, alanine, leucine) and organic acids such as acetate and succinate. Spectra of patients with a tumor showed mainly a peak of lactate. Lactate, acetate and succinate probably originated from enhanced glycolysis and the fermentative pathway in anaerobic conditions. The authors [86] believe that acetate is a key marker of infection. The results suggest that ^{1}H MRS detected the end products of bacterial fermentation. The presence of amino acids can be explained by tumor necrotic conditions, particularly after chemotherapy.

The patients with brain astrocytomas were examined by MRS [87] on a 2-T whole body system with PRESS or STABLE sequences for volume selection and with water signal suppression. The ^{1}H spectra were acquired before and 1 day after the treatment with a ^{125}I seed. The seed was placed in the center of the tumor for 18 to 42 days; the therapy was calculated to apply a total dose of 60 or 80 Gy. Prior to therapy a slight decrease of NAA and the presence of small amounts of lactate and a strong increase in choline were noticed. After treatment with the ^{125}I seed, the spectra localized in the center of the tumor showed broad signals from necrosis due to the tissue destruction and the presence of blood in this region. The choline signal decreased after treatment, which may show that the increased membrane activity is suppressed. The strongest decrease of choline signal in the case with a higher dose might be a sign that the response to treatment of astrocytomas with the ^{125}I seed can be monitored. Furthermore, monitoring the outcome of ongoing therapy can demonstrate if the drop in choline under therapy correlates to the success of the treatment or a re-increase in choline is observed with re-growth of cancer.

Phenylketonuria is caused by a deficiency of the enzyme phenylalanine hydroxylase (which converts the amino acid to tyrosine); this disease is characterized by mental retardation in untreated patients. ^{1}H MRS was applied to the study of patients with different levels of phenyloketonuria; the changes in metabolite concentrations were observed in the spectra measured with a stimulated echo acquisition mode (STEAM) sequence with a short echo time [88]. An increase in the N-acetylaspartate/choline-containing compounds (NAA/Cho) ratio in a group of 69 patients compared with age-matched controls was found. A significant difference in choline-containing compounds (1.33 mM in patients vs 1.53 mM in controls) could be employed as a „marker" of phenylketonuria. However, the changes are quite opposite to what was expected for the demyelination process. The authors suppose that relative changes of choline signal intensity are due to the decreasing T_2 relaxation time and reflect the changes in mobility in white-matter brain tissue. The above result is interesting because it shows that the difference in the ratios of signal intensities (which is most frequently used for the description of pathologies) can be explained not only by the changes in the absolute metabolite concentration, but also by changes in the mobility reflected by the relaxation times.

13.6.2   MR Imaging of Breast

Physiological changes of breast parenchyma due to age or menstrual cycle have been observed. Mammograms are commonly recorded in premenopausal women during the first half of the menstrual cycle because beneficial effects on diagnosis are assumed due to the lower breast density. MR imaging examinations were performed on 175 patients aged 15–79 years [89]. Gadopentate dimeglumine was given at a dose of 0.16 mmol/kg body weight by means of bolus injection. Parenchymal contrast medium enhancement was significantly lower in cycle days 7–20 than in 21–6. The enhancement was higher in patients aged 35–50 than in patients younger or older. Secretory, menstrual and proliferative phases (days 21–7) all displayed histologic findings commonly associated with higher metabolic activity and consequently higher tissue perfusion. This is in agreement with the increased contrast medium enhancement. Parenchymal enhancement in relation to the menstrual cycle should be further evaluated; nevertheless, it opens an interesting research field — the influence of hormonal replacement therapy or the effects of birth control pills could be examined. Hormone-dependent metabolic changes in the normal breast were monitored by ^{31}P MR spectroscopy [90]; an interesting observation was that the level of phosphomonoesters in normal breast tissue changes significantly during the menstrual cycle.

Cancer of the breast is the most common malignancy in women, and early detection and treatment can reduce mortality. Mammography is widely applied, but MRI, as a noninvasive method, is an excellent alternative. MR imaging was performed during intravenous administration of gadopentate dimeglumine in 66 women suspected to have breast cancer [91]. The conclusion of this study was that MR imaging has greater sensitivity and is as specific as scintimammography, probably because of the better spatial resolution.

13.6.3   Musculoskeletal Tumors

The prognosis for patients with osteogenic sarcoma and other malignant muscoskeletal tumors was poor before the advent of chemotherapy and most patients died. A marked improvement in long-term survival was achieved after the introduction of neoadjuvant chemotherapy. It is important to predict therapeutic efficacy at an early stage of treatment or even before the drugs were chosen. The patients with malignant bone tumors and soft-tissue tumors underwent MRI and MRS on a 1.5-T whole body system [92]. Tumor sizes determined in 34 patients from ^{1}H MR imaging, prior to chemotherapy, varied from 29 to 2350 ml. ^{31}P spectra were acquired with surface-coil placed close to the tumor site, using fast rotating gradient (FROGS) or two-dimensional chemical shift imaging (CSI). A serial study was performed in a patient with a rhabdomyosarcoma of the hand during 1–84 days of chemotherapy and in a patient with osteosarcoma during 65 days; the spectra are illustrated in Figure 13-5:a and Figure 13-5b, respectively. Patients with rhabdomyosarcoma were treated with vincristine, actinomycin D, ifosfamide, adriamycine, and patients with osteosarcomas received preoperative chemotherapy, which included ifosfamide, doxorubicin D, methotrexate and cisplatin.

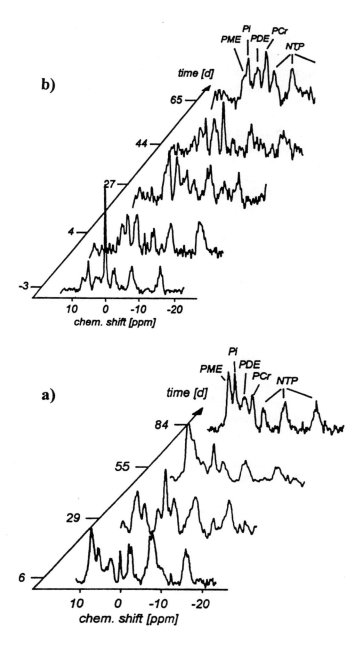

Figure 13-5: ^{31}P MR spectra obtained during chemotherapy of a) rhabdo-myosarcoma of the hand, b) osteosarcoma of the femur. Signal intensities are given in arbitrary units, spectra are plotted as a function of time; origin of the time axis, scaled in days, is the first day of chemotherapy (reproduced with permission, from [92])

Compared with normal muscle, characteristics of the tumor ^{31}P spectra were strong signals in the phosphomonoesters PME, and phosphodiesters PDE regions and low intensities from phosphocreatine PCr. None of the spectroscopic parameters could be considered as a diagnostic discriminant between different malignant tumors. The observed spectral differences seemed to be caused by differences in the tumor stage rather than by differences in the tumor type. As a trend, nonresponders to therapy showed higher PME and lowest PCr. No other metabolite ratio that may be calculated from ^{31}P spectra was found to yield significant differences between responders and nonresponders. The mean intracellular pH of the cancers was slightly alkaline (7.20, in normal muscles 7.05), pH was alkaline in responding tumors but the prognostic value of pH was poor. Anticancer drugs must be transported into cells, and their cytotoxic activity depend on extra- and intracellular pH. A low extracellular pH decreased doxorubicin activity but increased that of cisplatin. Most patients were treated with a combination of drugs and the influence of pH on the real *in vivo* activity is complex. Series of spectra obtained during chemotherapy showed that energy-rich phosphates (PCr and ATP) tended to decline in nonresponders, while the intensities of PME, Pi and PDE increased. It suggests that metabolite ratios calculated from low-energy phosphates (Pi, PME, PDE) and high-energy phosphates (PCr, NTP) may be used as sensitive predictors of eventual response to treatment. In the early stages of growth, characteristics of both aerobic respiration (high PCr) and anaerobic glycolysis (high Pi) were observed in *in vivo* ^{31}P spectra. As the tumor grew larger, levels of high-energy phosphates decreased, PCr being lost in preference to NTP [93], [94]. The changes result from the metabolic demands arising from tumor growth exceeding the available blood supply, which finally leads to hypoxia. Pretreatment differences in PCr/NTP and PCr+Pi between responders and nonresponders may reflect an increased proportion of hypoxic, metabolically inactive cells in nonresponding tumors. ^{31}P MRS is therefore likely to become a promising tool for assessing prognosis in human cancer [95].

The special MR imaging and spectroscopic methods have been developed based on relaxation time and/or on the chemical shift difference. At 1.5-T field strength the frequency difference between water and methylene protons is 215 Hz. The SENEX sequence for chemical shift selective imaging was applied to the examination of bone tumors [96] in various parts of the body. Fat-selective images showed signal-free areas for most osteogenic tumors, which could be distinguished from surrounding fat-containing soft tissues. Chemical shift-selective imaging is very sensitive to low amounts of lipids.

13.6.4   Other Studies

MRI offers unique advantages for measuring body fat distribution. The different proton relaxation properties have been used to develop a fast-scan MRI technique for adipose tissue quantification [97]. The images were acquired with four sequences and fast scan sequence was compared with the conventional T_1-weighted spin-echo one. The fast scan sequence minimized artifacts from motions (respiratory, intestinal) by acquiring images in shorter times than the respiratory cycle, and gave excellent visceral adipose tissue differentiation and tissue contrast. In comparison, the chemical shift imaging technique, which also can selectively image body fat, requires a very homogeneous magnetic field, difficult to achieve when scanning across the whole body. There is increasing evidence that obesity has important effects on health and the accumulation of abdominal fat has been linked to the development of major Western diseases. Diet and exercise are advised to manage this problem. However, the effects of diet and exercise on body fat distribution are as yet unknown, since this requires a method of adipose tissue analysis which can be repeatedly applied and which can be used to quantify internal fat compartments. MRI analysis of body fat distribution is a technique; which in the future may address adipose tissue routinely. Analysis of the effect of obesity treatment on body fat distribution may become of particular importance in the development of pharmacological therapy as well as in the testing of popular slimming agents.

Phosphorus MRS has been widely used to study energy metabolism in healthy and diseased human skeletal muscle *in vivo*. The technique can monitor non-invasively, almost in real time, the changes in phosphate-containing metabolites (the alternative standard method is needle biopsy). ^{31}P MRS is the only method for measurement of intracellular pH *in vivo* in functioning, working muscle. During exercise, adenosine triphosphate (ATP) generation comes from oxidative metabolism, from glycolysis and from phosphocreatine. Over 90% of the total creatine (phosphorylated + unphosphorylated) in the body is located in the skeletal muscle, which offers a tremendous advantage in signal to noise ratio. ^{1}H MRS was used for the quantification and imaging of the distribution of total creatine in the human leg [98]. Total creatine was represented by the *N*-methyl resonance at 3.0 ±0.1 ppm in the water-suppressed spectra. Images of water, lipid and total creatine were extracted from chemical shift image data sets; water was distributed throughout the axial section, lipid was limited to a subcutaneous surface layer and bone marrow, while total creatine was localized to muscle bundles. Assuming a concentration of water of 42.4 mol/kg wet weight, a total creatine value of 36.2 mmol/kg was determined. The differences in total creatine level are linked to differences in fiber type, diet and exercise habits. Creatine dietary supplementation can produce 20% increase in muscle creatine and a nearly 40% increase when supplementation is accompanied by exercise. The fast and noninvasive estimation of total creatine level could improve the understanding of the role of altered creatine metabolism in muscle disease; it also enables quantification of the response to creatine therapies.

The bioenergetics of human skeletal muscle was studied by ^{31}P MRS and surface electromyography [99]; the measurements were performed simultaneously, permitting accurate studies of the correlation between metabolic and electrical changes in exercising and recovering human skeletal muscle; this relationship is still poorly understood.

Sport and physical activity have become increasingly popular, but this trend is accompanied by an increased risk of muscle and muscoskeletal injuries. Strenuous exercise can result in muscle injury that may persist for 2 weeks. ^{31}P MRS can detect evidence of muscle injury completely noninvasively. One hour after exercise, a significant increase in the inorganic phosphate to phosphocreatine ratio was observed; Pi/PCr remained elevated for 3–10 days [100]. The studies of acute knee injury on a large population of patients (840) showed that MRI increased clinical diagnostic certainty and reduced the need for arthroscopy [101].

Osteoarthritis, a degenerative joint disease affects the lives of millions of people. It is characterized by loosening of the cartilage extracellular collagen matrix and substantial loss of glycosaminoglycan. Effectively imaging the concentration of glycosaminoglycan would have a major influence on the ability to assess the natural progression of the disease, the timing of therapeutical interventions and their efficacy. Glycosaminoglycan macromolecules contained numerous negatively ionized side groups. Therefore the negatively charged ion of $Gd(DTPA)^{-2}$ should be excluded from normal cartilage. A measurement of gadolinium complex concentration can serve as a surrogate for a measurement of glycosaminoglycan concentration [102]. The gadolinium compound was distributed into degenerated areas at a higher concentration than in nondegenerated. Normal human cartilage has a glycosaminoglycan concentration in the order of 6 mg/mL, whereas in osteoarthritis the concentration can decrease to nearly 0 mg/mL. The studies were performed at 1.5 T with a dedicated knee coil; a gadolinium compound with saline solution (epinephrine was added to slow resorption of the contrast) was administrated intraarticularly or intravenously. The intravenous administration has the advantage of achieving more rapid penetration of the cartilage and is more comfortable, but involves a higher dose. The authors conclude that the ability to image cartilage glycosaminoglycan concentration *in vivo* and to observe contrast-indicating differences in glycosaminoglycan content in anatomically intact cartilage is exciting! The studies involving *in vivo* imaging and *in vitro* validation after total joint replacement surgery are currently under way [102]. As a next step, the effects of therapy with anti-osteoarthritis drugs should be followed.

MRI and multinuclear MRS studies were performed on prostate [103], [104], [105], [106]. Relatively large concentrations of citrate, found by ^1H MRS, provided a potential method for the discrimination of prostatic disease. A significant correlation between water T_2 relaxation time and citrate concentration in the normal prostate was demonstrated, and this relationship is maintained in benign prostatic hyperblasia and prostatic carcinoma [103]. MR imaging and ^1H MR spectroscopic imaging were performed in patients with prostate cancer [104]. Statistically significant higher choline levels and significantly lower citrate levels were observed in the regions of cancer. The ratio of these metabolites provided a specific marker of cancer. MRI/MRS examination is valuable in defining both the presence and spatial extent of cancer. Combined MRI and MRS were performed in 25 patients who underwent cryosugery with a 1.5-T MR imager and with the use of a body coil and an expandable endorectal coil [105]. Water suppression and lipid suppression were achieved. The mean (choline + creatine)/citrate ratio for cancer in patients was 1.9, for benign prostatic hyperplasia 0.6 and for the normal peripheral zone 0.5. After cryosurgery the prostate was significantly smaller, and in most patients the distinction between normal prostatic anatomy and diseased anatomy was lost. Successful cryosurgery also had significant effects on cellular metabolism; the spectra demonstrated an absence of all prostatic metabolites. In some cases, necrotic spectra showed significant levels of lipids. Using MR spectroscopic imaging data volumes of cancer as small as 0.24 cm^3 were detected as areas of high choline and low citrate. MRS has the ability to help discriminate cancer from normal tissue even in the presence of postbiopsy hemorrhage [106]. MR spectroscopic examination could monitor the presence and spatial extent of recurrent local disease, also after hormonal therapy or changes in lifestyle and diet.

Metabolic processes in humans can be also analyzed using ^{13}C MRS spectroscopy. Because of low abundance and low magnetogyric ratio of the ^{13}C nucleus, the spectra are characterized by low signal-to-noise ratio; nevertheless, several metabolites can be detected at 1.5 T within reasonable measurement times. In order to increase the sensitivity of *in vivo* ^{13}C MRS, ^{13}C-enriched substrates have to be administrated.

^{13}C MRI was used to follow glucose metabolism in the human brain [107]; the spectra were obtained with a resolution time of 9 min using intravenous infusions of 1-^{13}C-glucose. At an isotopic enrichment level of 20%, the signals of α- and β-glucose at 92.7 and 96.6 ppm, respectively, were detected; an increasing the enrichment level to 99%, the metabolic breakdown products 1-^{13}C-glucose, glutamate, glutamine and lactate, could be recorded in the human brain.

Another interesting application of ^{13}C MRS to humans was the analysis of glycogen formation in the liver [108]. ^{13}C MRS measurements were performed on healthy volunteers with a whole-body system operating at 1.5 T in combination with ^{1}H and ^{13}C surface coils. The intravenous administration of 1-^{13}C-glucose (enrichment of 99%) was performed in combination with an hyperinsulinemic and hyperglycemic clamp. The C1 signals of α- and β-glucose could already be detected in the human liver after an infusion period of 8 min; after prolonged infusion an increase in the glycogen signal was observed. Liver glycogen formation can also be followed by using non-labeled glucose or 1-^{13}C-glucose with only 6.6 % enrichement.
This technique may provide many new insights into the physiology of carbohydrate metabolism and also into pathophysiologic changes in patients with diabetes, liver cirrhosis or hepatitis.

The idea of spatially resolved NMR was first described by Lauterbur in 1973 [109], and the first report on NMR study of human tumors *in situ* was published 10 years later [110]. ^{31}P NMR was used to follow the progress of a human tumor during chemotherapy with doxorubicin, and the authors [110] wrote: "NMR, which has been used successfully to monitor the metabolism of diseased tissue repeatedly and non-invasively, offers an exciting radiation-free technique for monitoring the effects of cancer therapy on the metabolism of tumors". Twenty-five years later, magnetic resonance imaging became the routinely used method of clinical diagnostics, and a dramatic increase in the number of diagnostic studies dealing with MRI has been observed.
The authors are of the opinion that the potential of the MRI/MRS method is still insufficiently used in pharmaceutical studies. We hope that this survey, presenting some interesting applications will encourage to further investigations.

References

1. F W Wehrli, D Shaw and J B Kneeland, Biomedical Magnetic Resonance Imaging, VCH, Weinheim, 1988
2. B Blümich and W Kuhn, Magnetic Resonance Microscopy: Methods and Applications to Material Science, Plants and Biomedicine, VCH, Weinheim, (1992)
3. P T Callaghan, Principles of Nuclear Magnetic Resonance Microscopy, Clarendon Press, Oxford 1993
4. R Kimnich, NMR Tomography, Diffusiometry, Relaxometry, Springer Verlag, Berlin, Heidelberg, New York, 1997
5. P G Morris, Nuclear Magnetic Resonance Imaging in Medicine and Biology, Clarendon Press, Oxford,1986
6. A Haase, J Frahm and D Matthaei, W Hänicke and K D Merbold, *J Magn Reson* **67**, 258-266 (1986)
7. R Kreis and Boesch, *J Magn Reson* **B113**, 103-118 (1996)
8. L N Ryner, J A Sorenson and M A Thomas, *J Magn Reson* **B107**, 126-137 (1995)
9. L A Ryner, J Sorenson and M A Thomas, *Magn Reson Imag* **13**, 853-869 (1995)
10. R L Ehman, J P Felhlee, *Radiology* **173**, 255 (1989)
11. A W Anderson, J C Gore, *Magn Reson Med* **32**, 379 (1994) G Nebgen, D Gross, V Lehmann and F Müller, *J Pharm Sci* **84**, 283-291 (1995)
12. W M M J Bovee, O T Karlsen and T H Rozijn, *Anticancer Res* **16**, 1515-1520 (1996)
13. P Macheras, P Argyrakis, *Pharm Res* **14**, 842-847 (1997)
14. G von Köller, E Köller, F Moll, *Pharm Ind* **53**, 955-958, (1991)
15. G Nebgen, D Gross, V Lehmann and F Müller, *J Pharm Sci* **84**, 283-291 (1995)
16. A R Rajabi-Siahboomi, R W Bowtell, P Mansfield, M C Davies and C D Melia, *Pharm Res* **13**, 376-380 (1996)
17. M Kojima, S Ando, K Kataoka, T Hirota, K Aoyagi, H Nakagami, *Chem Pharm Bull* **46**, 324-328 (1998)
18. V Christmann, J Rosenberg, J Seega, C -M Lehr, *Pharm Res* **14**, 1066-1072 (1997)
19. K Mäder, Y Crémmillieux, A J Domb, J F Dunn, H M Swartz, *Pharm Res* **14**, 820-826 (1997)
20. C D Melia, A Rajabi-Siahboomi, R W Bowtell, *PSTT* **1**, 32-39 (1998)
21. H Günther, NMR Spectroscopy An Introduction, J Wiley & Sons, Chichester, New York, Brisbane, Toronto, (1980)
22. C D Eccles, P T Callaghan, *J Magn Reson* **68**, 393-398 (1986)
23. C D Eccles, P T Callaghan and C F Jenner, *Biophys J* **53**, 77-81 (1988)
24. P A Bottomley, H H Rogers, T H Foster, *Proc Natl Acad Sci USA* **83**, 87-89 (1986)
25. G A Johnson, J Brown, P J Kramer, *Proc Natl Acad Sci USA* **84**, 2752-2755 (1987)
26. W Köckenberger, A Metzler, *Bruker Report* **141/95**, 21-23 (1995)
27. A Ziegler, A Metzler, W Köckenberger, M Izquierdo, E Komor, A Haase, M Décorps, M von Kienlin, *J Magn Reson, ser B* **112**, 141-150 (1996)
28. A C Kuesel, K M Briere, W C Halliday, D R Sutherland, S M Donnelly and I C P Smith, *Anticancer Res* **16**, 1485-1490 (1996)
29. G Carpinelli, C Carapella, L Palombi, F Carilli and F Podo, Anticancer Res 16 (1996) 1559-1564
30. Y Kinoshita and A Yokoto, *NMR in Biomed* **10**, 2-12 (1997)

31. M O Leach, *Anticancer Res* **16**, 1503-1514 (1996)
32. J Engelmann, J Henke, W Willker, B Kutscher, G Nössner, J Engel and D Leibfritz, *Anticancer Res* **16**, 1429-1430 (1996)
33. L L Hansen and J W Jaroszewski, *NMR in Biomed* **9**, 73-78 (1996)
34. A Moreno and C Arus, *NMR in Biomedicine* **8**, 33-45 (1996)
35. D Leibfritz, *Anticancer Res* **16**, 1317-1324 (1996)
36. R M Dixon, *Anticancer Res* **16**, 1351-1356 (1996)
37. S Tran-Dinh, A Courtois, J Wietzerbin, F Bouet and M Herve, *Biochimie* **77**, 233-239 (1995)
38. M O Pulkkinen, M M Hämäläinen, S Nyman, K Pihlaja and J Mattinen, *NMR in Biomed* **9**, 53-58 (1996)
39. S Sardon, J A Sanchez-Alvarez, S W Collier, J L Lopez-Lacomba, Cortijo M and J Ruiz-Cabello, *J Pharm Sci* **87**, 249-255 (1998)
40. R Deslauriers et al Preceedings of the 1st Kraków-Winnipeg Workshop on Biomedical Applications of MRI and MRS, (1997)
41. M Rudin, Sauter A, *Magn Reson Imag* **10**, 723-731 (1992)
42. M Hoehn-Berlage, *NMR in Biomed* **8**, 345-358 (1995)
43. J Kucharczyk, J Mintorovitch, M E Moseley, H S Asgari, R J Sevick, N Derugin and D Norman, *Radiology* **179**, 221-227 (1991)
44. J Seega and B Elger, *Magn Reson Imag* **11**, 401-409 (1993)
45. H Kobayashi, H Ide, T Kodera, Y Handa, M Kabuto, T Kubota and M Maeda, *Acta Neurochir* **60S**, 228-230 (1994)
46. M R Del Bigio, H J Yan, R Buist and J Peeling, *Stroke* **27**, 2312-2319 (1996)
47. J Peeling, Preceedings of the 1st Kraków-Winnipeg Workshop on Biomedical Applications of MRI and MRS, 1997
48. R M Dijkhuizen, M van Lookeren Campagne, T Niendorf, W Dreher, A van der Toorn, M Hoehn-Berlage, H B Verheul, D Leibfritz, K A Hossmann and K Nicolay, *NMR in Biomed* **9**, 84-92 (1996)
49. A C Silva, W Zhang, D S Williams and A P Koretsky, *Magn Reson Med* **33**, 209-214 (1995)
50. Y Xu, P Tang, W Zhang, L Firestone and P M Winter, *Anestesiology* **83**, 766 (1995)
51. R P Mason, W Rodbumrung and P P Antich, *NMR in Biomed* **9**, 125-134 (1996)
52. Y Kanazawa, K Umayahara, T Shimmura and T Yamashita, *NMR in Biomed* **10**, 35-41 (1997)
53. R J Maxwell, T A Frenkiel, D R Newell, C Bauer and J R Griffiths, *Magn Reson Med* **17**, 189-196 (1991)
54. U Noth, L J Jager, J Lutz and A Haase, *Magn Reson Imaging* **12**, 149-153 (1994)
55. L J Jager, U Noth, A Haase and J Lutz, *Adv Exp Med Biol* **361**, 129-134 (1994)
56. W J Manning, J Y Wei, E T Fossel and D Burstein, *Am J Physiol* **258**, H1181-H1186 (1990)
57. K Umemura, W Zierhut, M Rudin, D Novosel, E Robertson, B Pedersen and R P Hof, *J Cardiovasc Pharmacol* **19**, 375-381 (1991)
58. M Rudin, U Briner and W Doepfner, *Magn Reson Med* **7**, 285-289 (1988)
59. M Rudin, L Tolcvai, S Qureshi and R A Siegeln, *Prostate* **12**, 333-341(1988)

60. Borah, M D Francis, K Hovancik, J T Boyce and N M Szeverenyi, *J Rheumatol* **22**, 855-862 (1995)
61. K Kanamori and B D Ross, *Biochem J* **305**, 329-336 (1995)
62. B Pfeiderer, J L Ackerman and L Garrodo, *Magn Reson Med* **29**, 656-659 (1993)
63. L Garrido, B Pfleiderer, E Tokareva, J L Ackerman and T J Brady, SMR Meeting p 154 Nice, 1995
64. J W Belliveau, D N Kennedy, et al McKintry, *Science* **254**, 716-719 (1991)
65. S Ogawa, T Lee, A Nayak, P Glynn, *Magn Reson Med* **14**, 68-78 (1990)
66. P G Morris, Preceedings of the 1st Kraków-Winnipeg Workshop on Biomedical Applications of MRI and MRS, 1997
67. J Hykin, S Clare, R Bowtell, M Humberstone, Coxon R, B Worthington, L Blumhardt and P Morris, *MAGMA supp* **v4**, 180-184 (1996)
68. G Scarth, R Somorjai, M Alexander, B Wowk, A W A Wennerberg and M C McIntyre, *Human Brain Mapping* **S1**, 158 (1995)
69. A B A Wennerberg, B Wowk, G Scarth, J K Saunders, O Williams and M C McInyre, *Human Brain Mapping* **S1**, 308 (1995)
70. J K Saunders, Preceedings of the 1st Kraków-Winnipeg Workshop on Biomedical Applications of MRI and MRS, 1997
71. D M Yousem, S C R Williams, R O Howard, C A A Simmons, M Allin, D Suskind, E T Bullmore, M J Brammer and R L Doty, *Radiology* **204**, 833-838 (1997)
72. R L Doty, *Ann NY Acad Sci* **640**, 20-27 (1991)
    J P Kesslak, O Nalcioglu and C W Cotman, *Neurology* **41**, 51-54 (1991)
73. L C Kopala, K Good and W G Honer, *Biol Psychiatry* **38**, 57-63 (1995)
74. D Bernard, P M Walker, N Baudouin-Poisson, M Giroud, H Fayolle, R Dumas, D Martin, D Binnert and F Brunotte, *Radiology* **199**, 381-389 (1996)
75. D Gadian and A Connelly, *MAGMA* **2**, 219-223 (1994)
76. D Azzopardi, J S Wyatt, E B Cady, D T Delpy, J Baudin, A L Steward, P L Hope, P A Hamilton and E O Reynolds, *Pediatr Res* **25**, 440-444 (1989)
77. M S van der Knaap, J van der Grond, P C van Rijen, J A J Faber, J Valk, and K Willemse, *Radiology* **176**, 506-515 (1990)
78. B A Holshouser, S Ashwal, G Y Luh, S Shu, S Kahlon, K L Auld, L G Tomasi, R M Perkin and D B Hinshaw, *Radiology* **202**, 487-496 (1997)
79. G R Ende, K D Laxer, R C Knowlton, G B Matson, N Schuff, G Fein and M W Weiner, *Radiology* **202**, 809-817 (1997)
80. B L Miller, R A Moats, T Shonk, T Ernst, Woolley S and B D Ross, *Radiology* **187**, 433-437 (1993)
81. S MacKay, F Ezekiel, D J Meyerhoff, J Gerson, D Norman, G Fein, M W Weiner, *Radiology* **198**, 537-545 (1996)
82. N Schuff, D L Amend, D J Meyerhoff, J L Tanabe, D Norman, G Fein, M W Weiner, *Radiology* **207**, 91-102 (1998)
83. L Chang, B L Miller, D McBride, M Cornford, G Oropilla, S Buchtal, F Chiang, H Aronow and T Ernst, *Radiology* **197**, 525-531 (1995)
84. R J S Chinn, I D Wilkinson, M A Hall-Craggs, M N J Paley, R F Miller, B E Kendall, S P Newman, M J G Harrison, *Radiology* **197**, 649-654 (1995)
85. R Prost, V Haughton, S -J Li, *Radiology* **204**, 235-238 (1997)

86. S H Kim, K H Chang, I C Song, M H Han, H C Kim, H S Kang and M C Han, *Radiology* **204**, 239-245 (1997)
87. O Speck, T Thiel and J Hennig, *Anticancer Res* **16**, 1581-1586 (1996)
88. M Dezortova, L Hejcmanova and M Hajek, *MAGMA* **4**,181-186 (1996)
89. M Müller-Schimpfle, K Ohmenhäuser, P Stoll, K Dietz, C D Claussen, *Radiology* **203**, 145-149 (1997)
90. G S Payne, M Dowsett and M O Leach, *The Breast* **3**, 20-23 (1994)
91. T H Helbich, A Becherer, S Trattnig, T Leitha, P Kelkar, M Seifert, M Gnant, A Staudenherz, M Rudas, G Wolf, G H Mostbeck, *Radiology* **202**, 421-429 (1997)
92. H E Möller, P Vermathen, E Rummeny, K Wörtler, P Wuisman, A Rössner, B Wörmann, J Ritter and P E Peters, *NMR in Biomed* **9**, 347-358 (1996)
93. T C Ng, W T Evanochko, R N Hiramoto, V K Ghanta, M B Lilly, A J Lawson, T H Corbett, J R Durant and J D Glikson, *J Magn Reson* **49**, 271-286 (1982)
94. M Stubbs, L M Rodrigues, B A Gusterson and J R Griffiths, *Adv Enzyme Regul* **39**, 217-230 (1990)
95. R G Steen, *Cancer Res* **49**, 4075-4085 (1992)
96. F Schick, S Duda, O Lutz, C Claussen, *Anticancer Res* **16**, 1569-1574 (1996)
97. M L Barnard, J E Schwieso, E L Thomas, J D Bell, N Saeed, G Frost, S R Bloom and J V Hajnal, *NMR in Biomed* **9**, 156-164 (1996)
98. P A Bottomley, Y Lee and R G Weiss, *Radiology* **204**, 403-410 (1997)
99. P Vestergaard-Poulsen, C Thomsen, T Sinkjaer and O Heriksen, *Magn Reson Med* **31**, 93-102 (1994)
100. K McCully, Z Argov, B P Boden, R L Brown, W J Bank and B Chance, *Muscle & Nerve* **II**, 212-216 (1988)
101. E J Maurer, P A Kaplan, R G Dussault, D R Diduch, A Schuett, F C McCue, P P Hornsby, B J Hillman, *Radiology* **204**, 799-805 (1997)
102. A Bashir, M L Gray, R D Boutin and D Burstein, *Radiology* **205**, 551-558 (1997)
103. G P Liney, M Lowry, L W Turnbull, D J Manton, A J Knowles, S J Blackband and A Horsman, *NMR in Biomed* **9**, 59-64 (1996)
104. J Kurhanewicz, D Vigneron, H Hricak, P Carroll, P Narayan, S Nelson, *Radiology* **198**, 795-805 (1996)
105. J Kurhanewicz, D Vigneron, H Hricak, F Parivar, S J Nelson, K Shinohara and Carroll P, *Radiology* **200**, 489-496 (1996)
106. Y Kaji, J Kurhanewicz, H Hricak, D L Sokolov, L R Huang, S J Nelson, D B Vigneron, *Radiology* **206**, 785-790 (1998)
107. N Beckmann, I Turkalj, J Seelig and U Keller, *Biochemistry* **30**, 6362-6366 (1991)
108. N Beckmann, R Fried, I Turkalj, J Seelig, U Keller and G Stalder, *Magn Reson Med* **29**, 583-590 (1993)
109. P Lauterbur, *Nature* **242**, 190 (1973)
110. J R Griffiths, E Cady, R H T Edwards, V R McCready, D R Wilkie and E Wiltshaw, *Lancet* **25**, 1435-1436 (1983)

Chapter 14

U. Holzgrabe

14 Concluding remarks

NMR spectroscopy has a long-standing tradition in the elucidation and confirmation of the structure of synthetic products as well as compounds isolated from a variety of natural sources. The methods used for this purposes are described in a multitude of textbooks. However, this book attempted to illustrate the power and versatility of modern high-field NMR techniques in the typical pharmaceutical fields of drug analysis and drug development. Whereas for a long time the characterisation of drugs with regard to their purity, stability and isomeric composition is state of the art and increasingly used, the monitoring of the pharmacokinetic parameters and of the metabolism, by means of NMR spectroscopy, especially in combination/hyphenation with HPLC and other chromatographic or electrophoretic methods is a rather new field. As soon as 600-MHz and 750-MHz instruments become more general equipment in analytical laboratories, the application of those techniques will increase rapidly, particularly in the case of new drugs whose metabolism is unknown. In parallel, the importance of the non-invasive technique of NMR imaging will enhance in monitoring the distribution and fate of drugs or naturally occurring compounds in humans. In this context, the imaging technique will be more and more applied to plants in order to visualise the distribution of substances and time-dependently the changes of the distribution during a vegetation period. Even though NMR spectroscopy can be equal to powder diffraction spectroscopy and corresponding techniques the power of solid state measurements has not yet been fully recognised by the scientific community of pharmaceutics.

However, with an increasing number of NMR instruments equipped with a solid-state probe the technique will be increasingly used in pharmaceutics, especially because it is a fast techniques.At the latest it is possible to study the structure of huge peptide by means of the three-dimensional techniques, the NMR spectroscopy is an essential part of drug design departments in both industry and research laboratories. The interaction between drugs and proteins, e. g. enzymes, as well as drugs and membranes can be studied in solution and the findings can be often online used in molecular modelling studies. The more 600-MHz and 750-MHz instrument will become a general equipment in research laboratories the higher is the chance that the NMR spectroscopy will surpass the X-ray structure analysis of protein-drug complexes, because the elucidation of the complex structure in solution is closer to the real conditions in the human body than the structure of a crystal. In addition, the SAR by NMR approach, developed by Fesik stress the increasing importance of the NMR spectroscopy in drug development in the future.Taken together, within the last two decades the NMR spectroscopy has gone through tremendous development which opened up inconceivable possibilities of pharmaceutical applications. Whereas the drug development cannot be imagined without the NMR spectroscopy, to date it is not routinely used in quantitative analysis. Since the sensitivity of the NMR spectroscopy was considerably increased by using new techniques and high field instruments it can be expected that the application in this area will increase.

Substance Register

Acemannan®	32; 33
Acetamide	48
Acetate	126; 127
Acetophenone	49
N-Acetylaspartate	279; 280; 281; 283
Acetylcholine	39; 217; 218
N-Acetylcysteinyl	129
Actinomycin D	284
Adamantane	236; 237; 241
Adenine	16
Adenosine	16
Adrenalin	96
Adriamycine	284
Aesculin	16
Agarose	269
Alanine	121; 123; 126; 127; 270; 282
Albumin	125
Aldoximes	39
Allicin	54
Alliin	54
Aloe vera	32; 33; 34
Alprenolol	204
Amino acids	97, 125; 126; 127; 243, 268; 269; 282
6-Amino-6-deoxy-β-CD	158
4-Aminoquinaldine	35
6-Aminopenicillanic	39
Aminophenol	129
Aminoquinaldine	35
Amiodarone	209; 226
Amoxycillin	118
Ampicillin	118; 123
Amyl nitrite	17
Amyloses	252; 253
Anacin	252
Androstanolone	247
Antazoline	47
Antifolates	276
Antipyrine	129
Antitrypanosomiasis	35
Arboxymethylcellulose	263
Arginine	127; 268
Ascomycin	142; 143; 149
Ascorbic acid	49
Asparagine	127
Aspartame	248
Aspirin	252
Atenolol	99; 204
Atropine	39
Barbiturates	97
Benoxaprofen	247
Benserazide	41; 42
Benzhydrylamine	162
Benzodiazepines	38
Benzoic acid	47; 162
Benzyl chloride	125
O^6-Benzyl-guanine	158
Benzyl penicillin	158
Betamethasone	46
Bethanechol	48
(±)-1,1´-Binaphthyl-2,2´-diylhydrogen phosphate	162
Bromoadamantane	162
Brompheniramine	168
Bufferin	252
Buspirone	242; 243
Butoxycaine	16
Caffeine	49; 77; 78
Captopril	37
Carbachol	39; 48
Carbamazepine	47; 245
Carbazic acid	124
Carboxyphosphamide	124
Cardiolipin	176; 223
Carnitine	127
Carotenoids	107
4-Chloro-5-sulfamoylanthranilic acid	52
Cefaclor	248
Cefoperazone	125
Cefotaxime	39
Celluloses	252
Certoparin	27; 29
Chalcone	162
Chamazulene	16
Chlorogenic acid	162
Chloroquine	196; 197; 198; 199
Chlorpheniramine	46; 49; 160; 161; 162

Chlorpromazine	158	Diphenhydramine	52
Cholesterol	1; 175; 176; 181; 194	Dizocilpine	274
Choline	125; 126; 127; 270; 272; 279; 280; 281; 283; 288	Doxorubicin	214; 226; 284; 286; 289
		Empirin	252
Chondroitin sulfate	31	Enalapril maleate	46; 252
Chromanol	238; 239	Enoxacin	63
Ciprofloxacin	41; 63; 79	Enoxaparin	27
Cisplatin	284; 286	Enrofloxacin	63
Citrate	121; 126; 127	Ephedrine	52; 98; 99
Citric acid	84	Epicaptopril	37
Clodronate	249	Estradiol	158
Clofibrate	47	Ethinylestradiol	46
Creatine	127; 270; 272; 279; 280; 281; 286; 287; 288	Ethyl-4-biphenylyl acetate	158
		Ethylbenzene	112
Cyanoacrylate	45; 46	Etilefrine	162
4-Cyano-N,N-dimethyl aniline	125	Eu(hfc)$_3$	85; 90
Cyclodextrin	85; 91; 96; 98; 135; 252; 253	Eugenol	278
		Excedrin®	49
Cyclopenthiazide	248	Fencamfamine	162
Cyclophilin	142; 149	Fenoprofen	109; 110; 1612
Cyclophosphamide	91; 92; 94; 124; 232; 233	Flavonoids	243
		Fleroxacin	79; 80
Cyclosporin A	142; 149	Flucloxacillin	123
Dalteparin	27; 29; 30	Fluconazole	65
Dechloroethylcyclophosphamide	124	Flufenamic acid	158
Delavirdine mesylate	245	Flunarizine	220; 221; 223
5′-Deoxy-5-fluorouridine	123	Fluorouracil	123
Dequalinium chloride	35; 36; 37	α-Fluoro-β-alanine	123
Dermatan sulfate	31	Fluorocarbons	60; 61
Dexamethasone	158; 272	α-Fluoro-β-guanidinopropanoic acid	123
Dextran sulfate	31		
Diacetylhydrazine	124	Fluoroquinolones	41; 78; 79
Diatrizoate	48	Flupentixol	213; 214; 215; 216; 226
3,5-Dichlorobenzyl-N-butylamine	219	Flurazepam	38
Diazepam	158	Flurbiprofen	98; 124; 129; 162; 252
Dicyclomine	48	Fosinopril	248
Diethylcarbamazepine	119	Furusemide	52; 248
Diflunisal	252	Gadopentate dimeglumine	284
Digitoxin	158	Galangin	240
5,6-Dihydrofluorouracil	123	G-aminobutyrate	276
Dihydrolovastatin	18; 21	Garlic	54; 55
Dihydroquinidine	52	Gentamicin	158
Diltiazem hydrochloride	97	Geraniol	278
Dimethindene	162	Glibornuride	158
Dimethylamine	125; 127	Gliclazide	99; 162
1,4-Dimethyl-bicyclo [2.2.2] octane	162	Glucosamine	31
Dinitrobenzoyl-methylbenzylamine	85		

Glucose	121; 125; 126; 243; 268; 269; 271; 272; 275; 289	Iso-amyl nitrite	17
		Isoleucine	127; 268
Glucuronide	119; 121; 123; 124; 128; 129	Isoniazide	124
		Isradipine	273
Glutamate	268; 276; 279; 289	Itraconazole	63; 64; 65;158
Glutamine	268; 276; 279; 289	Ketoconazole	158
Glycerol	1; 84;125; 175; 192; 194; 195;199; 206; 208; 222	Ketocyclo-phosphamide	124
		Ketoprofen	85
Glycerophosphocholine	270	Lactate	121; 126; 127
Glycine	123; 127	Lactide	23
Glycolic acid	201; 202	Lactose	243
Glycolipids	175; 177; 179; 206	Lactulose	243
Glycosaminoglycan	287	Leucine	68; 69;127; 268; 282
Gossypol	271	Levelopamil	274
Grippostad C®	49; 50	Levodopa	41
Guaiazulene	16	Levofloxacin	79
Halothane	274	Lomefloxacin	78; 79
Heparin	22; 27; 28; 29; 31	Lovastatin	18; 20; 248; 252
Heparinoides	31	Lutein	107
Heptakis-(2,3,6-tri-O-methyl)-β-CD	168	Lysine	127; 268
		Madopar®	41; 42; 43; 44
Heptakis(2,3-di-O-acetyl)-β-CD	98; 99	Maleic acid	47; 48
		Mannitol	274
Hexafluorobenzene	264; 275	Mefenamic acid	252
Hexetil hydrochloride	158	Mefloquine	196; 197; 199
Histidine	126; 127	Meropenem	249
HMCTS	47; 48; 52	meso-Hexestrol	241
Hydralazine	124	Metaphosphate	62
Hydrocortisone	17; 158	Methicilline	158
Hydroquinone	44; 45; 46	Methimazole	47
3-Hydroxy-antipyrine	129	Methionine	127; 231; 232; 243; 244
4-Hydroxy-cyclophosphamide	124	Methotrexate	284
4′-Hydroxy-flurbiprofen	124	Methylamine	125; 127
5′-Hydroxymethylflucloxacillin	123	Methylbenzylamine	85
N-Hydroxyparacetamol	129	Methylphenidate	97
Hydroxypropylcellulose	264	Metoclopramide	48
Hydroxypropylmethylcellulose	263	Metomidate	163
HyTEMPO	149	Metoprolol	204
Ibuprofen	109; 110;120; 121; 128; 129; 252	Mexiletine	87; 88; 89; 90
		Mianserine	163
Iduronic acid	31	Miltefosine	270
Ifosfamide	284	Monohydroxy-pyridines	163
Ifosphamide	91	Mosher′s reagent	85
Iloperidone	129	Myo-inositol	127
Indomethacin	158; 162; 252; 276	Nabilione	247
Indoxyl sulphate	126	Nadolol	97
Inositol	270; 272; 279; 281	Nadroparin	27

Nalidixic	78	Phosphatidylinositol	175; 176; 177; 270; 272
(S)-1-(1-Naphthyl)-ethylamine	85	Phosphatidylserine	209
Napopxren	109; 110; 119; 163	Phosphocholine	126; 270
1-Naphtalenesulfonate	163	Phosphocreatine	270; 272; 279; 280; 286; 287
Nicardipine	158		
Nicotine	98	α-Pinene	163
Nifedipine	39; 40; 41; 158	Piperine	111; 112
Nitrilotriacetic acid	16	Piroxicam	163; 252
N-Leucine-enkephalin	158	Pluronic	22; 23
N-Methylciprofloxacin	63	Poloxamer	10; 22
N-methyl-D-aspartate	274	Polylactide	23; 24
Norephedrine	52	Polylactide/glycollide	25; 23
Norflex®	18; 19	Polymannose	32; 33
Norfloxacin	47; 63; 78; 79; 123	Polytetrafluoroethylene	264
Ofloxacin	41; 47	Porphyrin	242
Orphenadrine	17; 18; 19	Pr^{3+}fot	203
ortho-Dimethoxybenzene	85	Prednisolone	247; 252
Osteosarcomas	284	Pregnane	210
Oxoglutarate	124	Procaine	49; 51
Oxopefloxacin	123	Progesterone	158
Oxpentifylline	119	Propranolol	163; 204; 205
Oxprenolol	204	Prostaglandin	98
Paracetamol	49; 119; 121; 129; 252; 253	Pseudoephedrine	52
		Pyrophosphate	61; 62
Paraquat	121	Quercetin	241
Parnaparin	27	Quinacrine	196; 197; 198; 199
Pefloxacin	47; 63; 79; 123	Quinidine	52; 82; 83
Penicillin	118; 123; 242; 248; 250	Quinine	82; 83; 196; 197; 198; 199
Penicilloic acid	39; 118; 123	Salicylate	278
Perfluorobromooctane	60; 276	Sandostatin	276
Perfluorodecalin	60	Selegiline	98; 99
Perfluorooctane	60	Selenomethionine	243; 244
Phenacetin	129	Serine	175; 176; 177; 179; 212; 215; 221; 223
Phenetylamines	96		
Phenobarbital	246	Serylhydrazide	41
Phenomethylpenicillin	250	Sevoflurane	275
Phenylacetylglutamine	129	Silicone	1; 57; 58; 59
Phenylalanine	127	Siloxanes	59
Phenylephrine	69; 70; 71; 72; 73; 74; 75	Simethicone	57; 58
Phenytoin	48	Simvastatin	252
Phosphatidic acid	177	Somatostatin	276
Phosphatidylcholine	56; 57; 175; 176; 177; 178; 179; 180; 186; 195; 201; 203; 207; 209; 215; 221	Spermidine	127
		Spermine	127
		Sphingolipids	177
Phosphatidylethanolamine	56; 57; 176; 177; 178; 179; 180194; 221; 223	Sphingomyelin	176; 177; 178; 179
		Spirapril	276

Starches	252	2,2,2-Trifluoro-1-(9-anthrylethanol	85
Steroids	163; 209; 211	Trifluoromethylbenzoic acid	123
Sterols	177; 183; 188	Trimethylamine	127
Sulindac	252	Trimethylamine-N-oxide	125; 127
Taurine	126; 127; 270; 272	Triphenylphosphate	55
Testosterone	158	Triphosphate	62
Tetracaine	207; 208	Trofosphamide	91; 92; 93; 94
Tetracyclines	78	Trolox	234
TFAE	85; 87; 88; 97	Tryptophan	127
TFMBA	123	Tyrosine	127
Theophylline	77; 263	Urea	124
Threonine	127	Ureido sugar	234; 235; 236
Thymopentin	158	Uridine	127
Timolol maleate	97	Valine	126; 127; 268; 282
Tinzaparin	27; 29; 30	Verapamil	163; 209; 220
TMB-4	62	Vincristine	284
Tobramycin	10; 17	Vitamin A acetate	104; 105; 106; 107; 113; 114
Tocopherol	38; 243; 244		
Tolazoline	47	Vitamins B1, B2	47
Toliprolol	204	Xanthine	77
1,1,1-Trichlor-2-methyl-2-propanol	49; 50; 51	Yb(tfc)$_3$	85
		Yt(tfc)$_3$	98
Trifluoperazine	209; 212; 213	Zeaxanthin	107; 108

Biotechnology

Second, Completely Revised Edition

Precise, Comprehensive, Concise!

Well structured, cutting edge topicality with a broad scope helps make this series a comprehensive biotech reference work. If you are looking for fast and affordable means to increase your effectiveness in tackling problems in molecular biology, chemical engineering, and industrial chemistry or just want to beef up your reference library, **Biotechnology – Second, Completely Revised Edition** puts the field at your fingertips.

edited by
H.-J. Rehm, University of Münster, Germany
G. Reed, Formerly Universal Foods Corporation, Durham, NC, USA

in cooperation with
A. Pühler, University of Bielefeld, Germany
P.J.W. Stadler, Artemis Pharmaceuticals, Köln, Germany

Special Set Price:
DM 425.–/€ 217.30/sFr 378.–
per volume (if entire set is purchased)
Set-ISBN 3-527-28310-2

Single Volume Price:
DM 545.–/€ 278.65/sFr 485.–

Putting the field at your fingertips!

Fundamentals

Volume 1: Biological Fundamentals
1993. XIII, 641 pages, 289 figs, 75 tabs. ISBN 3-527-28311-0

Volume 2: Genetic Fundamentals and Genetic Engineering
1992. XIV, 880 pages, 271 figs, 101 tabs. ISBN 3-527-28312-9

Volume 3: Bioprocessing
1993. XV, 816 pages, 270 figs, 140 tabs. ISBN 3-527-28313-7

Volume 4: Measuring, Modelling, and Control
1991. XIII, 658 pages, 334 figs, 65 tabs. ISBN 3-527-28314-5

Products

Volume 5A: Recombinant Proteins, Monoclonal Antibodies, and Therapeutic Genes
1998. XVIII, 562 pages, 102 figs, 65 tabs.
ISBN 3-527-28315-3

Volume 5B: Genomics and Bioinformatics
2000. Approx. 600 pages.
ISBN 3-527-28328-5

Volume 6: Products of Primary Metabolism
1996. XIV, 739 pages, 231 figs, 208 tabs. ISBN 3-527-28316-1

Volume 7: Products of Secondary Metabolism
1997. XV, 728 pages, 267 figs, 47 tabs. ISBN 3-527-28317-X

Volume 8A: Biotransformations I
1998. XIV, 607 pages, 440 figs, 80 tabs.
ISBN 3-527-28318-8

Volume 8B: Biotransformations II
1999. Approx. 600 pages, 200 figs, 50 tabs.
ISBN 3-527-28324-2

Special Topics

Volume 9: Enzymes, Biomass, Food and Feed
1995. XVI, 804 pages, 189 figs, 167 tabs. ISBN 3-527-28319-6

Volume 10: Special Processes
2000. Approx. 650 pages.
ISBN 3-527-28320-X

Volume 11A: Environmental Processes I – Waste and Wastewater Treatment
1999. Approx. 600 pages, 300 figs, 50 tabs.
ISBN 3-527-28321-8

Volume 11B: Environmental Processes II – Soil Decontamination
1999. Approx. 600 pages.
ISBN 3-527-28323-4

Volume 11C: Environmental Processes III – Waste and Waste Gas Treatment, Drinking Water Preparation
2000. Approx. 650 pages.
ISBN 3-527-28323-4

Volume 12: Legal, Economic and Ethical Dimensions
1995. XIII, 695 pages, 53 figs, 113 tabs. ISBN 3-527-28322-6

A Cumulative Index will be published separately.
All volumes are hardcover.

Visit our Biotech Website:
http://www.wiley-vch.de/home/biotech

WILEY-VCH

WILEY-VCH • P.O. Box 10 11 61 • 69451 Weinheim, Germany •Fax: +49 (0) 62 01-60 61 84